CRC Press
Taylor & Francis Group
6000 Broken Sound Parkway NW, Suite 300
Boca Raton, FL 33487-2742

First issued in paperback 2019

ISBN-13: 978-1-4822-3244-8 (hbk)
ISBN-13: 978-0-367-37719-9 (pbk)

Visit the Taylor & Francis Web site at
http://www.taylorandfrancis.com

and the CRC Press Web site at
http://www.crcpress.com

For my mother, who showed me
Donald in Mathmagic Land
and a great many other wonderful things.

Contents

List of Figures

List of Tables

Preface

Suppose a new deck of cards is shuffled. Then as the number of shuffles grows, the distribution of the deck becomes closer and closer to uniform over the set of permutations of the cards. On the other hand, no matter how long you shuffle the deck, it will never be perfectly uniform over the set of permutations.

This technique—of making small changes to a state in order to move it closer to some target distribution—is known today as Markov chain Monte Carlo, or MCMC. Each step brings the chain closer to its target, but it never quite reaches it. With roots in The Manhattan Project, by the early 1990s this idea had become a cornerstone of simulation. For statistical physicists it was the only way to get approximately correct samples in a reasonable amount of time, especially for high dimensional examples such as the Ising model. For frequentist statisticians, it was the only way to calculate p-values for complex statistical models. For Bayesian statisticians, it allowed the use of a dizzying array of non-conjugate priors and models.

For almost half a century MCMC had been used, and still only a finite number of steps could be taken. No one realized that it might be possible to somehow jump to an infinite number of steps in a finite number of time. That changed when James Propp and David Wilson introduced coupling from the past (CFTP) in 1996. Their elegant idea allowed users for the first time to sample exactly from the stationary distribution of a Markov chain. That paper, "Exact sampling with coupled Markov chains and applications to statistical mechanics," gave the first exact sample from the Ising model at the critical temperature on an over 16 million dimensional problem.

The list of applications started to grow. Spatial statistics models, domino tilings, linear extensions, random spanning trees... the first few years after publication saw enormous growth in the field. Along the way *exact sampling* changed to *perfect simulation*. The term was introduced by Wilfred Kendall in 1998, and immediately caught on as a way to differentiate the CFTP protocol from other Monte Carlo algorithms that returned samples exactly from the distribution, but without using the peculiar ideas central to CFTP.

David Wilson created a website entitled, "Perfectly Random Sampling with Markov Chains" at http://dimacs.rutgers.edu/ dbwilson/exact/ to inform the nascent community about developments. As of July 2015 the site still exists; a snapshot of the early days of the field.

As time went on, variants of the CFTP method were developed. Read-once coupling from the past allowed samples to be taken using less memory, Fill's method and FMMR connected CFTP to the much older (and less widely applicable) method of acceptance/rejection.

Then around 2000 something interesting happened. Protocols that were very different from CFTP, such as the Randomness Recycler, were developed. CFTP was now not the only way to draw from the stationary distribution of a Markov chain! New life was breathed into acceptance/rejection through the use of retrospective sampling and sequential acceptance/rejection, allowing for the first time perfect simulation from some diffusions and perfect matchings. Partially recursive acceptance/rejection even allowed for simulation from the classic problem of the Ising model.

CFTP was no longer unique, but fit within a larger framework of perfect simulation ideas. The connections between CFTP and acceptance/rejection became clearer, as each employs a recursive structure that could conceivably go on forever, but with probability 1 does not.

Today the set of perfect simulation algorithms continues to grow, albeit more slowly than in the early days. In researching and writing this text, I was astounded at the wide range of problems to which perfect simulation ideas had been applied. Coupling from the past was not just a protocol; it was the first to show that high dimensional simulation from interacting distributions was even possible, opening up areas of simulation that have yet to be fully explored.

Acknowledgments: My study of perfect simulation began with a special topics course taught by Persi Diaconis when I was a graduate student at Cornell. Persi later became my postdoc advisor as well, and his unflagging enthusiasm for new ideas was an inspiration to me.

I would also like to thank my Ph.D. advisor David Shmoys who took a chance in letting me change research directions in the middle of my graduate work to dive into this newly forming field.

Jim Fill introduced me to several intriguing problems in the area, and later became my first perfect simulation collaborator. Together we added the Randomness Recycler to the world of perfect simulation.

Thanks also to my many colleagues and collaborators who brought me up to speed on Bayesian statistics and patiently sat through the many talks I gave on perfect simulation while I was in the midst of figuring problems out. A great academic environment stimulates the mind in unexpected ways, and I have been privileged to belong to several.

Mark L. Huber
23 July, 2015

Chapter 1

Introduction

To improve is to change; to be perfect is to change often.
Winston Churchill

Perfect simulation algorithms draw random variates from a distribution using a random number of steps. The term perfect simulation can be used either as an adjective for a specific algorithm with these properties, or for a general protocol for creating algorithms with these properties. Perfect simulation algorithms have been developed for Markov random fields, permutation problems, spatial point processes, stochastic differential equations, and many other applications. Until the advent of perfect simulation, Markov chain Monte Carlo was often used to draw approximate samples from these distributions.

1.1 Two examples

Example 1.1. *Suppose that I have a fair six-sided die that I can roll as often as I would like, and I wish to draw a random variate that is uniform over* $\{1,2,3,4,5\}$.

The following algorithm does the job. First, roll the die. If the number that comes up is in $\{1,\ldots,5\}$, keep it as the answer, otherwise, roll the die again. Continue until a result in $\{1,\ldots,5\}$ is obtained.

This is a simple example of a *perfect simulation* algorithm, a method that draws a random variate exactly from a target distribution, but which uses a random number of draws from a second distribution in order to do so.

The example also illustrates the idea of recursion that is often found in perfect simulation algorithms. Another way to state the algorithm is as follows: roll the die once. If the roll is in $\{1,\ldots,5\}$, keep the answer, otherwise, recursively call the algorithm to generate the random variate. At each level of recursion, there is a positive chance ($5/6$ in the example) that no further recursive calls are needed. Therefore the algorithm terminates with probability one.

This example is an instance of the acceptance/rejection (AR) protocol for creating perfect simulation algorithms. A protocol is not an algorithm in itself, but a methodology for designing algorithms for specific problems. Usually this methodology works for a variety of examples.

AR was the first widely used protocol for perfect simulation, and pretty much the only one for about six decades. The next major protocol to be developed was coupling from the past (see Chapter 3). CFTP was quickly followed by FMMR (Chapter 5), the randomness recycler (Chapter 8), retrospective sampling (Chapter 10), and partially recursive acceptance/rejection (see Chapter 9). In addition there are multiple variants of each of these protocols.

Example 1.2. *Suppose I have a coin with probability p of heads that I can flip as often as I would like. The goal is to generate a random variable* $X \in \{1,2,3,\ldots\}$ *such that* $\mathbb{P}(X=i) = p(1-p)^{i-1}$. *Say that X is a geometric random variable with mean* $1/p$.

One way to generate such a random variable is to repeatedly flip the coin until a head is obtained. The number of flips needed to get a head (including the last flip) will be a geometric random variable. As with the earlier example, this can be expressed recursively. Flip the coin. If it is heads, return 1. Otherwise, return 1 plus the value of a call to the algorithm.

Unlike general protocols, the method used to simulate this random variable is very problem-dependent. Another example of this problem-dependent type of perfect simulation algorithm is the cycle popping method for generating random spanning trees of a graph (Chapter 8).

1.2 What is a perfect simulation algorithm?

Both of the examples of the last section implicitly used the notion of a stopping time that is the number of recursive calls to the algorithm that occurs.

Definition 1.1. *Say that T is a* stopping time *with respect to a sequence* $X_1, X_2, \ldots$ *if given the values* $\{X_1, \ldots, X_n\}$, *it is possible to determine if* $T \leq n$.

Throughout, the initialism *iid* will mean a sequence $X_1, X_2, \ldots$ is *independent and identically distributed.* so that each X_i has the same distribution, and for any n, the variables $X_1, \ldots, X_n$ are independent. For a finite set A, let $\mathsf{Unif}(A)$ denote the uniform distribution over A, where each element is equally likely to be chosen.

With this notation, the problem of Example 1.1 is to generate $Y \sim \mathsf{Unif}(\{1,2,\ldots,5\})$ given iid draws $X_1, X_2, \ldots$ that are $\mathsf{Unif}(\{1,2,,\ldots,6\})$. The algorithm works by using the stopping time $T = \inf\{t : X_t \in \{1,2,\ldots,5\}\}$. The output of the algorithm is just X_T. This is the idea of a perfect simulation method: it is an algorithm that, after running for a random amount of time, outputs an answer that comes exactly from the desired distribution.

The notion of an algorithm is encapsulated in a *computable function.* Several definitions of what it means to be computable have been proposed. Ideas such as λ-calculus (see [48] for an introduction) and Turing machines [123] are often used. Informally, say that a function is computable if the input to the function can be arbitrarily large, the function value is required to be output after a finite number of steps, and each step only affects a finite number of memory locations (although the total amount of memory is infinite).

Definition 1.2. *Let* $\mathscr{I}$ *be an index set such that for each* $I \in \mathscr{I}$, *there is a distribution* π_I *over a state space* Ω_I. *Given a sequence of random variables* $X_1, X_2, \ldots$, *where*

$X_I \in S_I$, *a* randomized algorithm *is a family of stopping times* $\{T_I\}_{I \in \mathscr{I}}$ *and computable functions* $\{f_{I,t}\}_{i \in \mathscr{I}, t \in \{1,2,\ldots\}}$. *The output of the algorithm is* $f_{I,T_I}(X_1, \ldots, X_{T_I})$.

Definition 1.3. *A* perfect simulation *algorithm is a randomized algorithm whose output is a random variable that comes exactly from a target distribution.*

Perfect simulation algorithms are a subclass of exact simulation algorithms, algorithms that draw precisely from a target distribution. For this reason, the term *exact simulation* is often used to describe one of these algorithms. Although allowed in a formal sense, algorithms where the stopping time T_I is deterministic so that the algorithm runs using a fixed number of random choices are generally considered exact simulation but not perfect simulation algorithms.

Say that X is uniform over $[0,1]$ (write $X \sim \mathsf{Unif}([0,1])$) if $\mathbb{P}(X \leq a) = a$ for all $a \in [0,1]$. Digital computers are assumed to be deterministic (leaving aside engineering defects) and to only have a finite amount of memory. This means that no computer can generate actual $X_1, X_2, \ldots$ that are truly iid $\mathsf{Unif}([0,1])$. However, there exist well-known pseudorandom sequences (see for instance [89]) that are actually deterministic, but behave in ways very similar to true random sequences. Therefore, the $X_1, X_2, \ldots$ sequence is usually assumed to be iid uniform random variables over $[0,1]$.

In Example 1.2, the sequence of random variables used by the algorithm are $X_1, X_2, \ldots$ that are iid $\mathsf{Unif}([0,1])$. The stopping time is $T = \inf\{t : X_t \leq p\}$, and $f_T(X_1, \ldots, X_T) = T$. The computable function is $f_t(x_1, \ldots, x_t) = t$, so $f_T = T$.

1.3 The Fundamental Theorem of perfect simulation

The two most important protocols for creating perfect simulation algorithms are acceptance/rejection (AR) and coupling from the past (CFTP). They both employ recursion in a simple way to generate random variates. The key result that gives correctness for these algorithms comes from recursion.

Recall that an algorithm is recursive if it calls itself while it is running. A classic example is the algorithm for generating $n!$ from input n that is a positive integer.

`Factorial` *Input:* n *Output:* $y = n!$
1) If $n = 1$ then $y \leftarrow 1$
2) Else $y \leftarrow n \cdot \texttt{Factorial}(n-1)$

Suppose that in order to generate a sample from the distribution of X, an iid sequence $U_1, U_2, \ldots$ is given. Then if $X \sim g(U_1)$, that immediately gives an exact simulation procedure.

However, suppose that something weaker happens. Suppose that b maps the state space of the U_i to $\{0,1\}$. Further, suppose that there are functions g and f such that $X \sim b(U_i)g(U_i) + (1 - b(U_i))f(X, U_i)$. Then the following recursive procedure generates from X. First generate $U \sim U_i$. If $b(U) = 1$, then set $X = g(U)$ and quit. Otherwise, recursively call the same algorithm to generate $X' \sim X$, and set X equal to $f(X', U)$.

This seems somewhat like cheating, as we are recursively calling our algorithm to get our final answer! Even worse, unlike the `Factorial` example, the number of recursions used by the algorithm does not have an upper bound. Still, the Fundamental Theorem of perfect simulation shows that using recursion in this sneaky way does have output that has the desired distribution.

Theorem 1.1 (Fundamental Theorem of perfect simulation). *Suppose for $U_1, U_2, \ldots$ independent and identically distributed, there are computable functions b, g, and f such that b has range $\{0,1\}$ and $\mathbb{P}(b(U)=1)>0$. Then for a random variable X that satisfies*

$$X \sim b(U)g(U) + (1-b(U))f(X,U), \tag{1.1}$$

let $T = \inf\{t : b(U_t) = 1\}$. Then

$$Y = f(\cdots f(f(g(U_T),U_T),U_{T-1}),\ldots,U_1) \tag{1.2}$$

has the same distribution as X and $\mathbb{E}[T] = 1/\mathbb{P}(b(U)=1)$.

Equivalently, this theorem states that if (1.1) holds and $\mathbb{P}(b(U)=1)>0$, then the output of the following pseudocode comes from the distribution of X.

`Perfect_simulation`	*Output: Y*
1)	Draw random variable U
2)	If $b(U)=1$
3)	$Y \leftarrow g(U)$
4)	Else
5)	$X \leftarrow$ `Perfect_simulation`
6)	$Y \leftarrow f(X,U)$

Proof. Let $X_0, X_1, \ldots$ be independent draws, each distributed as X, and $U_1, U_2, \ldots$ be iid $\textsf{Unif}([0,1])$ that are independent of the X_i. Then for X_t, set $X_{t,t} = X_t$ and recursively set $X_{t,i} = b(U_{i+1})g(U_{i+1}) + (1-b(U_{i+1}))f(X_{t,i+1},U_{i+1})$ for $i \in \{0,\ldots,t-1\}$. Then from (1.1), $X_{t,0} \sim X$.

So $X_{0,0}, X_{1,0}, X_{2,0}, \ldots$ are identically distributed as X, but are not necessarily independent. Now consider the variable Y as generated by `Perfect_simulation` where the value of U used as the tth call to the function is U_t. Then the key observation is that $X_{t,0} = Y$ if $t \geq T$. Since the U_i are independent, $\mathbb{P}(T>t) = (1-\mathbb{P}(b(U)=1))^t$, which goes to zero as t goes to infinity by the assumption that $\mathbb{P}(b(U)=1)>0$.

Therefore, for any t and any set C,

$$\begin{aligned}
\mathbb{P}(Y \in C) &= \mathbb{P}(Y \in C, t \geq T) + \mathbb{P}(Y \in C, t < T) \\
&= \mathbb{P}(X_{t,0} \in C, t \geq T) + \mathbb{P}(Y \in C, t < T) \\
&= \mathbb{P}(X_{t,0} \in C) - \mathbb{P}(X_{t,0} \in C, t < T) + \mathbb{P}(Y \in C, t < T) \\
&= \mathbb{P}(X \in C) - \mathbb{P}(X_{t,0} \in C, t < T) + \mathbb{P}(Y \in C, t < T).
\end{aligned}$$

These last two terms are each bounded in magnitude by $\mathbb{P}(t<T) = (1-\mathbb{P}(b(U)))^t$. Since the equation holds for all t, $\mathbb{P}(Y \in C) = \mathbb{P}(X \in C)$ and $Y \sim X$.

The fact that $\mathbb{E}[T] = 1/\mathbb{P}(b(U)=1)$ just follows from an elementary analysis of geometric random variables. □

Going back to Example 1.1 from earlier, let U be uniform over $\{1,2,3,4,5,6\}$, $b(u)$ be 1 when $u \in \{1,2,3,4,5\}$ and 0 otherwise, $g(u) = u$, and $f(x,u) = x$. Then for X uniform over $\{1,2,3,4,5,6\}$, let $W = b(U)g(U) + (1 - b(U))f(X,U)$. Let $i \in \{1,2,3,4,5\}$, then

$$\mathbb{P}(W = i) = \mathbb{P}(U = i) + \mathbb{P}(U > 5)\mathbb{P}(X = i) = (1/6) + (1/6)(1/5) = 6/30 = 1/5.$$

so $W \sim X$.

For Example 1.2, let U be 1 with probability p and 0 otherwise. Then $b(u) = u$, $g(u) = 1$, $f(x,u) = x + 1$. It is straightforward to check that for X geometric with parameter p, the relationship $U + (1 - U)(X + 1) \sim X$ holds, so the algorithm output has the correct distribution.

Note that in Example 1.1 the output X does not depend on $T = \inf\{t : b(U_t) = 1\}$. In the second example, $X = T$, so X and T are highly dependent!

Definition 1.4. *In Theorem 1.1, if X and T are independent, say that the algorithm is* interruptible. *Otherwise, the algorithm is* noninterruptible.

Also, in both these examples, the function $f(X,U)$ did not actually depend on U. So the value of U could be used to check if $b(U) = 1$ in line 2 and then thrown away. Thus the randomness in U only needed to be read once. If $f(X,U)$ does depend on U, the randomness needs to be used twice, making the algorithm read twice.

Definition 1.5. *A perfect simulation that uses $X \sim b(U)g(U) + (1 - b(U))f(X,U)$ is* read once *if $f(X,u) = f(X,u')$ for all u, u'. Otherwise it is* read twice.

In general, an interruptible algorithm is preferable to a noninterruptible one, and a read-once algorithm is preferable to a read-twice method.

1.4 A little bit of measure theory

This section collects some basic ideas from measure theory, which are useful in making probability results precise. Like all probability texts, we begin with a space Ω called either the *sample space* or *outcome space*, which (true to the name) will contain all the possible outcomes from an experiment.

Given a particular experimental outcome $\omega \in \Omega$, and a subset $A \subseteq \Omega$, a natural question to ask is what is the chance that $\omega \in A$? The set of A for which this question makes sense and can be answered with a number is called the set of measurable sets. Mathematically, this collection of subsets of Ω is known as a σ-algebra. If a set A can be measured, then it is important that the complement of the set A^C also can be measured. Moreover, it is helpful that if a sequence $A_1, A_2, \ldots$ of sets can be measured, then so can their union. This is made precise in the following definition.

Definition 1.6. *A collection $\mathcal{F}$ of subsets of Ω is a* σ-algebra *if*

1. *The set $\mathcal{F}$ is nonempty.*
2. *If $A \in \mathcal{F}$, then $A^C \in \mathcal{F}$.*
3. *If $A_1, A_2, \ldots \in \mathcal{F}$, then $\cup A_i \in \mathcal{F}$.*

Call a set $C \in \mathcal{F}$ a measurable set. Call $(\Omega, \mathcal{F})$ a measurable space *or* measure space.

Throughout this book, ν will be used to denote a measure. (While the Greek letter μ is often used in mathematics for measure, in this text μ is reserved for constants such as the mean of a normal or the magnetization of the Ising model.)

Definition 1.7. *For a measure space* $(\Omega, \mathscr{F})$ *a function* $\nu : \mathscr{F} \to [0,\infty) \cup \{\infty\}$ *is a* measure *if*

1. *[The measure of nothing is nothing]* $\nu(\emptyset) = 0$
2. *[Countable additivity] For a sequence* $A_1, A_2, \ldots \in \mathscr{F}$ *such that for all* $i \neq j$, $A_i \cap A_j = \emptyset$,

$$\nu(\cup A_i) = \sum \nu(A_i).$$

Definition 1.8. *If* $\nu(\Omega) < \infty$, *then* ν *is a* finite measure. *If* $\nu(\Omega) = 1$, *then* ν *is a* probability measure *or* distribution. *For a finite measure* ν, *call* $\nu(\Omega)$ *the* normalizing constant.

Usually, the symbols τ or π will be used to denote a probability measure. A simple way to turn a finite measure into a probability measure is to divide by its normalizing constant.

Lemma 1.1. *If* ν *is a finite measure for* $(\Omega, \mathscr{F})$, *then for all* $C \in \mathscr{F}$, $\tau(C) = \nu(C)/\nu(\Omega)$ *is a probability measure.*

Probability measures are also often written using $\mathbb{P}$. So $\mathbb{P}(C) = \tau(C)$ indicates the probability that the outcome falls in C. For unnormalized (but finite measures) τ and π, if $\tau(C)/\tau(\Omega) = \pi(C)/\pi(\Omega)$ for all measurable sets C, write $\tau \sim \pi$.

Definition 1.9. *Consider a probability space* $(\Omega_1, \mathscr{F}_1, \nu)$ *and a measurable space* $(\Omega_2, \mathscr{F}_2)$. *An* Ω_2-valued random variable *is a function* X *from* Ω_1 *to* Ω_2 *such that for all* $A \in \mathscr{F}_2$, $\{\omega \in \Omega_1 : X(\omega) \in C\} \in \mathscr{F}_1$.

The set $\{\omega \in \Omega_1 : X(\omega) \in C\}$ is usually abbreviated more simply as $\{X \in C\}$. The definition of a random variable ensures that $\nu(\{X \in C\})$ exists, and this number is also written $\mathbb{P}(X \in C)$. The measure $\nu_X(C) = \mathbb{P}(X \in C)$ is called the *distribution* of X, or the *law* of X. When $X \in \mathbb{R}$, then $F_X(a) = \mathbb{P}(X \leq a)$ is the *cumulative distribution function*, or cdf of X. If $\nu_X(C) = \nu'$ for some measure ν', write $X \sim \nu'$.

1.4.1 Integrals

Two of the most common measures are Lebesgue measure and counting measure. The counting measure of a finite set A is just the number of elements in set A. When ν is Lebesgue measure, then (roughly speaking) in one dimension ν corresponds to length, so $\nu([a,b]) = b - a$. In two dimensions, ν corresponds to area, in three dimensions to volume, and so on.

Integration with respect to Lebesgue measure over a closed bounded interval reduces to Riemann integration when the function is continuous over the region of integration, so

$$\int_{x \in [a,b]} f(x)\nu(dx) = \int_a^b f(x)\,dx. \tag{1.3}$$

In counting measure, for a finite set A, $\nu(A)$ denotes the number of elements in the set A. Integrals with respect to counting measure just become a summation over the set.

$$\int_{x \in C} w(x)\nu(dx) = \sum_{x \in C} w(x). \tag{1.4}$$

1.4.2 Densities

Another notion that comes in handy is the idea of a density.

Definition 1.10. *Say that a probability distribution τ has a* density $f(x)$ *with respect to a measure ν if for all measurable sets C,*

$$\tau(C) = \int_{x \in C} f(x)\nu(dx).$$

Suppose that $Z = \int_\Omega f(x)\nu(dx)$, where $Z > 0$ and is finite, but not necessarily 1. Then say that f is an unnormalized density *for the probability distribution τ defined as*

$$\tau(C) = Z^{-1} \int_{x \in C} f(x)\nu(dx).$$

Call Z the normalizing constant.

Usually densities are given with respect to Lebesgue measure or counting measure. A density with respect to counting measure is also called a *probability mass function*. When the density is with respect to Lebesgue measure over $\mathbb{R}^1$, then this is the usual notion of density for continuous distributions over the real line.

Definition 1.11. *Suppose $X \sim \tau$. If τ has a density f_X, say that f_X is the density of X.*

Integrals with respect to measures then give probabilities and expectations.

$$\int_{x \in A} f_X(x)\nu(dx) = \mathbb{P}(X \in A),$$
$$\int_{x \in \Omega} g(x) f_X(x)\nu(dx) = \mathbb{E}[g(X)].$$

Definition 1.12. *The indicator function $\mathbb{1}$(expression) evaluates to 1 when the expression inside is true, and 0 otherwise. For example, the function $\mathbb{1}(x \geq 3)$ is 1 when $x \geq 3$, and 0 when $x < 3$.*

Using the indicator function, the probability of an event can be written as an expectation:

$$\mathbb{P}(X \in A) = \mathbb{E}[\mathbb{1}(X \in A)]. \tag{1.5}$$

In most of the applications of Monte Carlo methods, only an unnormalized density of a random variable X is known, and the goal of the Monte Carlo approach is either to approximate the normalizing constant, or find $\mathbb{E}[g(X)]$ for some function g without knowing the normalizing constant of the density of X.

Table 1.1 *Initialisms.*

AR	Acceptance/rejection
cdf	Cumulative distribution function
CNF	Conjunctive normal form
CFTP	Coupling from the past
DNF	Disjunctive normal form
FMMR	Fill, Machida, Murdoch, and Rosenthal
iid	Independent, identically distributed
MA1	Move ahead 1
MAVM	Multiple auxiliary variable method
MCMC	Markov chain Monte Carlo
mgf	Moment-generating function
MLE	Maximum likelihood estimator
MRF	Markov random field
MTF	Move to front
pdf	Probability density function
PRAR	Partially recursive acceptance/rejection
ras	Randomized approximation scheme
RAR	Recursive acceptance/rejection
SAR	Sequential acceptance/rejection
SAVM	Single auxiliary variable method
SDE	Stochastic differential equation

1.5 Notation

First, some basic notation to be used.

- Throughout this work, π will denote a target distribution, and the goal of a perfect simulation algorithm is to draw a random variate from π.
- For a finite set A, $\#A$ or $\#(A)$ will denote the number of elements in A.
- $G = (V,E)$ will always refer to an undirected graph where V is the set of nodes and E is a collection of sets of two nodes. The degree of node i, written $\deg(i)$, is the number of nodes j such that $\{i, j\}$ is an edge. The maximum degree of any node in the graph will be denoted Δ. For example, in the square lattice $\Delta = 4$.
- When A and B are sets, $A \times B = \{(a,b) : a \in A \text{ and } b \in B\}$ denotes the Cartesian product of the sets.

Probability has built up a surprising number of initialisms. This text continues the trend by using initialisms for most of the major perfect simulation protocols, as well as for many models. A summary of the initialisms used in this text is given in Table 1.1. Some of these abbreviations are quite common, others pertain only to perfect simulation methods.

Table 1.2 *Discrete distributions.*

Name	Notation	Density $f(i)$
Bernoulli	$\mathsf{Bern}(p)$	$p\mathbb{1}(i=1)+(1-p)\mathbb{1}(i=0)$
Geometric	$\mathsf{Geo}(p)$	$p(1-p)^{i-1}\mathbb{1}(i\in\{1,2,\ldots\}$
Poisson	$\mathsf{Pois}(\mu)$	$\exp(-\mu)\mu^i(i!)^{-1}\mathbb{1}(i\in\{0,1,2,\ldots\}$
Uniform	$\mathsf{Unif}(A)$	$(\#A)^{-1}\mathbb{1}(i\in A)$

Table 1.3 *Continuous distributions.*

Name	Notation	Density $f(x)$
Uniform	$\mathsf{Unif}([a,b])$	$(b-a)^{-1}\mathbb{1}(x\in(b-a))$
Exponential	$\mathsf{Exp}(\lambda)$	$\lambda\exp(-\lambda x)\mathbb{1}(x>0)$
Gamma	$\mathsf{Gamma}(r,\lambda)$	$\lambda^r x^{r-1}\exp(-\lambda x)\Gamma(r)^{-1}\mathbb{1}(x>0)$
Normal (Gaussian)	$\mathsf{N}(\mu,\sigma)$	$(2\pi\sigma^2)^{-1/2}\exp(-(x-\mu)^2/[2\sigma^2])$
Inverse Gaussian	$\mathsf{InvGau}(\mu,\lambda)$	$(2\pi x^3/\lambda)^{-1/2}\exp(-\lambda(x-\mu)^2/(2\mu^2x))$

1.5.1 Distributions

Several common distributions will be used throughout this work. Discrete distributions have a density with respect to counting measure while continuous distributions have a density with respect to Lebesgue measure. See Table 1.2 and Table 1.3 for a summary of univariate distributions. For higher dimensional spaces, say that X is uniform over Ω (write $X\sim\mathsf{Unif}(\Omega)$) if for ν counting measure or Lebesgue measure, $\nu(\Omega)$ is finite and X has density $f_X(x)=\mathbb{1}(x\in\Omega)/\nu(\Omega)$.

1.5.2 Conditional distributions

Often, the distribution of a random variable depends upon one or more other random variables. For instance, if (X,Y) is chosen uniformly from the triangular region with vertices $(0,0)$, $(1,1)$, and $(1,0)$, then given the value of Y, X is uniform over the interval $[Y,1]$. This information is denoted $[X|Y]\sim\mathsf{Unif}([Y,1])$. Similarly, $[Y|X]\sim\mathsf{Unif}([0,X])$.

Formally, conditional expectations of the form $\mathbb{E}[X|Y]$ are defined using the Radon-Nikodym derivative between measures (see for instance [31]) and then conditional probabilities are defined using $\mathbb{P}(X\in C|Y)=\mathbb{E}[\mathbb{1}(X\in C)|Y]$. The key properties of conditional expectation that will be used here are that $\mathbb{E}[X|Y]=f(Y)$ for some function Y. Also, $\mathbb{E}[\mathbb{E}[X|Y]]=\mathbb{E}[X]$ for random variables X and Y with finite mean.

1.6 A few applications

Perfect simulation methods have been successfully applied to a variety of problems coming from many different areas. The list of applications includes:

- Markov random fields (specifically auto models). These are labellings of a graph where the density is a product of functions of the labels on each node, and a function of the labels on each edge. Examples of this type of problem include the Ising model, the Potts model, sink-free orientations of a graph, and the autonormal model.
- Permutations. A permutation can be thought of as an ordering of n objects. The set of permutations is easy to sample from, but once the goal is to sample from a density over permutations, the problem can become very difficult.
- Stochastic differential equations. These are used to model continuous functions that fluctuate randomly. They are used often in finance and as models of physical systems.
- Spatial point processes. These are used as models for data that is in the form of points in a space. Examples include the Strauss process, marked Hawkes processes, and Matérn processes.
- Bayesian posteriors. In Bayesian statistics, a prior is a probability distribution on parameters of interest. Conditioned on data, the prior is then updated using Bayes' rule to give an unnormalized density that is called the posterior. The normalizing constant for the posterior is often very difficult to compute exactly. Hence the ability to sample from the posterior without calculating the normalizing constant is essential to Bayesian inference. While it is not known how to perfectly sample from every Bayesian posterior, many do have perfect simulation algorithms.
- Combinatorial objects. Given a graph G, consider the number of proper colorings of the graph or the number of sink-free orientations. Sets of combinatorial objects such as these tend to grow in size exponentially fast in the size of the underlying graph. For many of these problems, the ability to draw samples from the set (and restrictions on the set) leads to approximation algorithms for counting the number of objects in the set. These types of problems can often be linked to problems coming from physics or statistics.
- Markov processes. Markov chains are powerful modeling tools that describe how a stochastic process is updated. Under simple conditions, the distribution of the state of the process will converge to what is known as a stationary distribution. Sampling from this stationary distribution directly can often be difficult. Often the stationary distribution cannot be described exactly. Examples of these kinds of problems dealt with here include the antivoter model, self-organizing lists, and perpetuities.

Each of these applications will be described more fully as we are ready to present algorithms for them. One thing all of these applications have in common is that they can be presented as an unnormalized density for which it is far too expensive computationally to calculate the normalizing constant directly.

The first two on this list, Markov random fields and permutation problems, serve as excellent examples for understanding how approximate sampling works, and so a detailed description of these models is presented here first.

1.6.1 Markov random fields

Consider a graph $G = (V,E)$.

Definition 1.13. *Say that a subset of nodes S separates i and j if every path in the graph from i to j passes through a node in S.*

A Markov random field is a probability distribution on the labellings of a graph with the property that if i and j are separated by S, then conditioned on the values of the nodes in S, the labels of i and j are independent.

Definition 1.14. *Given a graph $G = (V,E)$ with a set of node labellings Ω, say that the distribution of X over Ω is a* Markov random field *(MRF) if for all nodes i and j, and all sets S separating i and j, $[X(i)|X(j),X(S)] \sim [X(i)|X(S)]$. A state $x \in \Omega$ is called a* configuration.

To characterize distributions that are Markov random fields, it helps to have the notion of neighbors and cliques.

Definition 1.15. *The* neighbors *of a node i are all nodes j such that $\{i,j\} \in E$.*

Definition 1.16. *A* clique *is a subset of nodes such that every two vertices in the subset are connected by an edge.*

The Hammersley Clifford Theorem states that for Markov random fields with positive density, the density can be factorized over cliques.

Theorem 1.2 (Hammersley Clifford Theorem). *Given a finite graph $G = (V,E)$, the distribution π is an MRF if it has density f_X and there exist functions ϕ_C for all cliques C such that f_X can be written as*

$$f_X(x) = \frac{1}{Z} \prod_{C \in cliques(G)} \phi_C(x). \tag{1.6}$$

Here Z is the normalizing constant known as the partition function.

Following Besag [8], consider the following important class of MRFs.

Definition 1.17. *Say that an MRF is an* auto model *if there exist functions f and g so that the density of $X \sim \pi$ can be written*

$$f_X(x) = \frac{1}{Z} \left[\prod_{i \in V} f_i(X(i)) \right] \left[\prod_{\{i,j\} \in E} g_{\{i,j\}}(X(i),X(j)) \right], \tag{1.7}$$

In other words, auto models are MRFs where only the size one and two cliques affect the distribution. These types of models are used for image restoration [9] and more generally anywhere you have local spatial dependence in the model [8]. Arguably the most famous auto model is the Ising model.

Definition 1.18. *The* Ising model *is an auto model with parameters* μ *and* β, $\Omega = \{-1,1\}^V$, *and functions* $f(c) = \exp(\mu c)$, $g(c_1,c_2) = \exp(\beta \mathbb{1}(c_1 = c_2))$. *The parameter* μ *is known as the* magnetization, *and* β *as the* inverse temperature. *When* $\beta > 0$ *call the model* ferromagnetic, *when* $\mu = 0$ *say the model has* no magnetic field.

Most of the terminology comes from the initial use of the Ising model as a tool for understanding the magnetic behavior of materials. Note that when $\mu > 0$ the distribution favors 1's at a node, and when $\beta > 0$ higher probability is given to configurations where adjacent nodes have the same color.

The definition can be easily extended in several ways. Different values of μ_i for each node and β_e for each edge could be used. The algorithms given in later chapters easily handle this type of generalization but for clarity of presentation are only given here for the simpler case where the μ and β values are constant over the graph.

Another way to generalize is to allow more than two different labels of the nodes, giving what is commonly known as the Potts model. This can be used to model alloys where there are k different types of metals that can occupy a site, or discrete grayscale images where each pixel is assigned a number from 0 to 255 [9].

Definition 1.19. *The* Potts model *is an auto model with parameter* β, *where* $\Omega = \{1,2,\ldots,k\}^V$, $f(c) = 1$, $g(c_1,c_2) = \exp(\mathbb{1}(c_1 = c_2)\beta)$.

Another model is known as the hard-core gas model, because it envisioned gas molecules as behaving like billiard balls with a hard-core so that two adjacent sites in a lattice could not both be occupied.

Definition 1.20. *The* hard-core gas model *is an auto model over* $\{0,1\}^V$ *with parameter* $\lambda > 0$ *where* $f(c) = \lambda^c$ *and* $g(c_1,c_2) = 1 - c_1c_2$. *When* $x(v) = 1$, *say that node* v *is* occupied, *otherwise it is* unoccupied.

Note that if $x(i) = x(j) = 1$ for adjacent nodes, then $1 - 1 \cdot 1 = 0$ and the chance of having the configuration immediately drops to 0. Because the occupation of a site lowers the chance (in the previous case, to 0) of having occupation at a neighboring site, this is an example of a repulsive process. A generalization known as the Strauss model [119] allows for adjacent labels of 1, but assigns a penalty factor to the density each time this occurs.

Definition 1.21. *The* Strauss model *is an MRF over* $\{0,1\}^V$ *with parameters* $\lambda > 0, \gamma \in [0,1]$ *where* $f(c) = \lambda^c$ *and* $g(c_1,c_2) = 1 + (\gamma - 1)c_1c_2$. *When* $x(v) = 1$, *say that node* v *is* occupied, *otherwise it is* unoccupied.

Another way to put this: each time there is an edge $\{i,j\}$ with $x(i) = x(j) = 1$, the density acquires an extra factor of γ.

1.6.2 Permutations

A permutation can be viewed as an ordering of a set of objects. In order to keep the notation consistent it will be useful to write a permutation using a vector.

Definition 1.22. *A* permutation *on* n *positions is a vector in* $\{1,2,\ldots,n\}^n$ *where each component is labeled a different value, that is, a vector* x *where* $i \neq j \Rightarrow x(i) \neq x(j)$. *The set of permutations on* n *objects is also known as the* symmetric group, *and can be denoted by* S_n.

Typically σ will be used to denote a permutation. Different distributions on permutations arise across mathematical disciplines. A permutation is also known as an *assignment* because it can be viewed as assigning each of n workers to n different jobs. When different workers have different aptitudes for the jobs, the result is a weighted permutation, which in turn gives a distribution on permutations. Another source of distributions on permutations is nonparametric testing.

A classic distribution problem on permutations is related to finding the permanent of a nonnegative matrix.

Definition 1.23. *Given nonnegative numbers $w(i,j)$, let*

$$w(\sigma) = \prod_{i=1}^{n} w(i, \sigma(i)). \tag{1.8}$$

As long as there exists at least one σ with $w(i,\sigma(i)) > 0$ for all i, this gives an unnormalized density on the set of permutations. The normalizing constant for this density is called the permanent *of the matrix $(w(i,j))$. (If no such σ exists then the permanent is 0.)*

Another distribution of interest that arises in the context of nonparametric testing ([106]) is the uniform distribution over permutations where certain items are required to be put in a lower position than other items.

Definition 1.24. *Given a set P, a* partial order *on P is a binary relation $\preceq$ such that for all $a,b,c \in P$:*

1. *(Reflexive) $a \preceq a$.*
2. *(Antisymmetric) If $a \preceq b$ and $b \preceq a$, then $a = b$.*
3. *(Transitive) If $a \preceq b$ and $b \preceq c$, then $a \preceq c$*

A partially ordered set is also known as a poset.

Then a linear extension of a partial order can be viewed as a permutation that respects a partial order.

Definition 1.25. *A* linear extension *of a partial order on $1,\ldots,n$ is a permutation where if i and j are such that $\sigma(i) \prec \sigma(j)$, then $i < j$.*

For example, in the partial order on 4 elements where $2 \prec 4$, $(1,2,4,3)$ is a linear extension, while $(1,4,2,3)$ is not because items 2 and 4 are out of order.

1.7 Markov chains and approximate simulation

Although there is a great need for the ability to sample from high dimensional distributions, until the development of perfect simulation methods there only existed approximate methods, many of which utilized Markov chains.

The many protocols for approximate sampling using Markov chains are collectively known as Markov chain Monte Carlo. A Markov chain is a stochastic process that is memoryless, so the next state of the chain only depends on the current state, and not on any of the past history.

Definition 1.26. *A stochastic process* $\{X_t\}$ *is a* Markov chain *if for all* t,

$$[X_t|X_0, X_1, \ldots, X_{t-1}] \sim [X_t|X_{t-1}]. \tag{1.9}$$

For each state x in the state space Ω, the transition kernel of the Markov chain gives the probabilities that X_{t+1} falls into a measurable set C given that $X_t = x$. Formally, Markov kernels have two properties.

Definition 1.27. *Consider a measurable space* $(\Omega, \mathscr{F})$. *A* Markov kernel $K : \Omega \times \mathscr{F} \to [0,1]$ *satisfies:*

- *For all* $A \in \mathscr{F}$ *and* $a \in [0,1]$, *the set of* x *such that* $K(x,A) \leq a$ *is measurable.*
- *For all* $x \in \Omega$, *the map* $A \mapsto K(x,A)$ *is a probability measure.*

If for a particular Markov chain $\{X_t\}$, *it holds that* $\mathbb{P}(X_{t+1} \in A|X_t = x) = K(x,A)$ *for all* $x \in \Omega$ *and* $A \in \mathscr{F}$, *say that* K *is the kernel of the Markov chain.*

Note that it is possible to use a different Markov kernel K_t for each step of the chain. Such a Markov chain is called *time heterogeneous.* A Markov chain where the kernel is the same throughout is *time homogeneous.*

Common examples of Markov chains include shuffling a deck of cards, or mixing up a Rubik's cube by randomly twisting sides. As long as the next state only depends on the current state, and not on any states further back in time, it is a Markov chain.

In both these examples, the goal is to achieve a state that is close to uniformly distributed over the set of possible states. Say that a state of a Rubik's cube is *reachable* if there exists a finite sequence of moves starting from the initial state where each side of the cube is a solid color to the desired state. When a Rubik's cube starts in a state that is uniformly distributed over reachable states, and a random move is made, the state is still uniform over the reachable states. A distribution where taking a move in the Markov chain leaves the distribution of the state unchanged is called stationary.

Definition 1.28. *A distribution* π *for a Markov chain is* stationary *if*

$$X_t \sim \pi \to X_{t+1} \sim \pi.$$

Many protocols exist to build a Markov chain whose stationary distribution matches a particular target distribution π. Examples of such methods include Metropolis-Hastings [91, 47] and Gibbs sampling.

Under mild conditions, a Markov chain will have a limiting distribution that equals the stationary distribution. There are a variety of conditions that guarantee this; here the notion of a Harris chain is used since it applies to both discrete and continuous state space Markov chains.

Definition 1.29. *A Markov chain* $\{X_t\}$ *over* Ω *is a* Harris chain *if there exist measurable* $A, B \subseteq \Omega$ *and* $\varepsilon > 0$ *for* $x \in A$, $y \in B$, *and a probability measure* ρ *with* $\rho(B) = 1$ *where*

1. *For* $T_A = \inf\{t \geq 0 : X_t \in A\}$, $(\forall z \in \Omega)(\mathbb{P}(T_A < \infty|X_0 = z) > 0)$.
2. *If* $x \in A$ *and* $C \subseteq B$, *then* $\mathbb{P}(X_1 \in C|X_0 = x) \geq \varepsilon\rho(C)$.

In other words, a chain is a Harris chain if there is a subset of states A that the chain returns to with positive probability after a finite number of steps from any starting state. Moreover, if the state is somewhere in A, then with probability ε the next state does not depend on exactly which element of A the state is at.

The notions of recurrence and aperiodicity are needed to make the stationary distribution limiting as well. A Harris chain always has positive probability of returning to A. When the chance of returning is 1, then the chain is recurrent.

Definition 1.30. *Let* $R = \inf\{n > 0 : X_n \in A\}$. *A Harris chain is* recurrent *if for all* $x \in A$, $\mathbb{P}(R < \infty | X_0 = x) = 1$. *A Harris chain that is not recurrent is transient.*

Given a recurrent Harris chain, suppose the state space can be partitioned into $P_1, P_2, \ldots, P_k$ so that for a state in P_i, the chance of moving to P_{i+1} is 1. For a state in P_k, the next state is back in P_1. Then if k is the smallest number for which there is such a partition, call k the period of the chain. Intuitively, when k is 1, the chain is aperiodic. The following definition of aperiodicity is equivalent to this notion.

Definition 1.31. *A recurrent Harris chain is* aperiodic *if for all* $x \in \Omega$, *there exists* n *such that for all* $n' \geq n$,

$$\mathbb{P}(X_{n'} \in A | X_0 = x) > 0. \tag{1.10}$$

Theorem 1.3 (Ergodic Theorem for Harris chains). *Let* X_n *be an aperiodic recurrent Harris chain with stationary distribution* π. *If* $\mathbb{P}(R < \infty | X_0 = x) = 1$ *for all* x, *then as* $t \to \infty$, *for all measurable sets* C *and for all* x:

$$|\mathbb{P}(X_t \in C | X_0 = x) - \pi(C)| \to 0. \tag{1.11}$$

This theorem is the heart of Markov chain Monte Carlo: for a given target distribution, build a Harris chain whose stationary distribution equals the target distribution and is recurrent and aperiodic. Then the limiting distribution of the Harris chain will be the target distribution. So by running the chain an infinite number of steps, a draw from the target distribution is made.

Since very few practitioners actually have an infinite amount of time to spare, instead a finite number of steps are taken, and the user hopes that it is sufficient to make the state close to the limiting distribution. However, while it is often known that the distribution of the state of the chain converges to the stationary distribution, it is rare to be able to calculate how quickly that convergence occurs.

One of the few methods that exist for doing so is known as coupling.

Definition 1.32. *Suppose* $\{X_t\} \sim \nu_X$ *and* $\{Y_t\} \sim \nu_Y$. *Then a* coupling *of* $\{X_t\}$ *and* $\{Y_t\}$ *is a bivariate process* $\{(X'_t, Y'_t)\}$ *such that* $\{X'_t\} \sim \nu_X$ *and* $\{Y'_t\} \sim \nu_Y$.

The idea here is that X_t and Y_t both have a specified distribution, perhaps both are Markov chains with a specified transition kernel, for instance. Then a coupling is a way of advancing both X_t and Y_t forward into the future at the same time while each, when viewed individually, is following the transition probabilities of the appropriate Markov chain. Say that the two Markov chains are *coupled*.

The word couple comes from the Latin for "to fasten together," and this usage is still the important part of using coupled processes. It is important to recognize when the two processes cross paths; that is, when they are equal to one another.

Definition 1.33. *If $X_t = Y_t$, then say that the processes have* coupled *or* coalesced.

Coupling is important for assessing convergence because of the following result.

Theorem 1.4 (The Coupling Lemma [29, 1]). *Let $Y_0 \sim \pi$ and $X_0 = x_0$ both evolve according to a coupled Markov chain. Then for all measurable C,*

$$|\mathbb{P}(X_t \in C | X_0 = x) - \pi(C)| \leq \mathbb{P}(X_t \neq Y_t).$$

In other words, the distance between the distribution of X_t and the stationary distribution is bounded above by the probability that the process X_t has not coupled at time t with a stationary process Y_t.

Even if it is possible to use the coupling lemma to determine how close the law of the current state is to the target distribution, at best this can only give approximately correct samples. Still, it is a start!

1.8 Designing Markov chains

In order to use Markov chain Monte Carlo, it is necessary to build a Harris chain for which π is a stationary distribution. It is usually better to require more than just stationarity, and to build a chain that is reversible. To describe reversibility, it is helpful to have a differential notation for measures. When π has a density $f(x)$, then let $\pi(dx) = f(x)\ dx$. This differential notation means that for all measurable sets A, $\pi(A) = \int_{x \in A} \pi(dx) = \int_{x \in A} f(x)\ dx$. Using this differential notation, reversibility works as follows.

Definition 1.34. *A distribution π is* reversible *with respect to a Markov chain if $\pi(dx)\mathbb{P}(X_{t+1} \in dy | X_t = x) = \pi(dy)\mathbb{P}(X_{t+1} \in dx | X_t = y)$.*

Lemma 1.2. *If π is reversible, then it is also stationary.*

Proof. Let Ω be the state space of the Markov chain, and π be reversible for the chain. Then let $X_t \sim \pi$, and C be any measurable set. Then

$$\begin{aligned}
\mathbb{P}(X_{t+1} \in C) &= \mathbb{E}[\mathbb{1}(X_{t+1} \in C)] \\
&= \mathbb{E}[\mathbb{E}[\mathbb{1}(X_{t+1} \in C) | X_t]] \\
&= \int_{x \in \Omega} \mathbb{E}[\mathbb{1}(X_{t+1} \in C) | X_t = x] \pi(dx) \\
&= \int_{x \in \Omega} \mathbb{P}(X_{t+1} \in C | X_t = x) \pi(dx) \\
&= \int_{x \in \Omega} \int_{y \in C} \mathbb{P}(X_{t+1} \in dy | X_t = x) \pi(dx) \\
&= \int_{y \in C} \int_{x \in \Omega} \mathbb{P}(X_{t+1} \in dx | X_t = y) \pi(dy) \\
&= \int_{y \in C} \mathbb{P}(X_{t+1} \in \Omega | X_t = y) \pi(dy) \\
&= \int_{y \in C} \pi(dy) \\
&= \pi(C).
\end{aligned}$$

Hence π is stationary. The swapping of the order of the iterated integrals in the proof is allowed by Tonelli's theorem since the integrand is nonnegative; see for instance [31]. □

The main reversible Markov chain types are Gibbs samplers, Metropolis-Hastings, and shift-move chains. Auxiliary variable chains can be used together with these ideas to expand their range of applicability. Each of these methods is now discussed in turn.

1.8.1 Gibbs sampling

There are many varieties of Gibbs sampler. The simplest operates over a state space of the form C^V. Call $v \in V$ a *dimension* of the problem. Given a current state $X_t = x$, the Gibbs sampler chooses a dimension v uniformly from V, then lets $L(x,v)$ be all the states that match the label of x everywhere except possibly at dimension v. So $L(v) = \{y : (\forall w \in V \setminus \{v\})(y(w) = x(w))\}$. Given $X_t = x$, the next state X_{t+1} is chosen from π conditioned to lie in $L(x,v)$.

As an example, consider the Ising model when $\mu = 0$. In this case, a dimension of the problem consists of a single node v. So select a node v uniformly at random from V, and consider the two states that equal x at every node $w \neq v$. The value at node v can be 1 or -1. Call these two states $x_{v\to 1}$ and $x_{v\to -1}$, respectively.

Then make the next state either $x_{v\to 1}$ or $x_{v\to -1}$, where the probability of choosing $x_{v\to 1}$ is proportional to $\pi(\{x_{v\to 1}\})$. So

$$\mathbb{P}(X_{t+1} = x_{v\to 1}) = \frac{\pi(\{x_{v\to 1}\})}{\pi(\{x_{v\to 1}\}) + \pi(\{x_{v\to -1}\})}. \tag{1.12}$$

Recall that for the Ising model with $\mu = 0$,

$$\pi(\{x\}) = \frac{1}{Z} \prod_{\{i,j\}\in E} \exp(\beta \mathbb{1}(x(i) = x(j)). \tag{1.13}$$

In (1.12), the normalizing constant Z cancels out. So does the $\exp(\beta \mathbb{1}(x(i) = x(j)))$ factors unless either i or j equals v. So the fraction reduces to

$$\begin{aligned}\mathbb{P}(X_{t+1} = x_{v\to 1}) &= \frac{\prod_{j:\{v,j\}\in E} \mathbb{1}(x(j) = 1)}{\prod_{j:\{v,j\}\in E} \mathbb{1}(x(j) = 1) + \prod_{j:\{v,j\}\in E} \mathbb{1}(x(j) = -1)} \\ &= \frac{\exp(\beta n_1)}{\exp(\beta n_1) + \exp(\beta n_{-1})}\end{aligned}$$

where n_c is the number of neighbors of v that have label c.

For instance, suppose the node v is adjacent to three nodes labeled 1, and one node labeled 0. Then the chance that v should be labeled 1 is $\exp(3\beta)/[\exp(\beta) + \exp(3\beta)]$.

This gives the following pseudocode for taking one step in the Markov chain.

`Ising_Gibbs` *Input:* old state x *Output:* new state x

Draw $v \leftarrow \mathsf{Unif}(V)$, $U \leftarrow \mathsf{Unif}([0,1])$
Let n_1 be the number of neighbors of v labeled 1 in x
Let n_{-1} be the number of neighbors of v labeled -1 in x
If $U < \exp(n_1\beta)/[\exp(n_1\beta)+\exp(n_{-1}\beta)]$
    $x(v) \leftarrow 1$
Else
    $x(v) \leftarrow -1$

Note here that the node v is being chosen uniformly at random from among the nodes of the graph. This is known as using *random scan*. Another popular choice with Gibbs is to run through all the nodes one at a time in either a specified or a random order. This is known as *deterministic scan*.

More generally, suppose that the current state is a vector $(X(1),X(2),\ldots,X(n))$. Then choose some subset of states I according to some distribution. Then the next state chooses values for $X(i)$ for all $i \in I$ from π conditioned on the values of $X(i')$ for all $i' \notin I$. For instance, in the Ising model, an edge $\{i,j\}$ could be chosen uniformly from the set of edges, and the values of $X(i)$ and $X(j)$ could be chosen from π conditioned on all of the remaining states.

In random scan, the set I is chosen uniformly at random. It is easily to verify that as long as the choice of I is independent of the current state, the resulting chain is stationary with respect to π. It is usually not difficult to show that it is possible to reach any state from any starting state in the finite number of moves.

1.8.2 Metropolis-Hastings

With Metropolis-Hastings, for each state x of the chain, there is a density q_x. Choose a proposal state y according to q_x. Then the next state of the chain will be either x or y. The probability the next state is y is chosen in such a way that reversibility holds with respect to the target distribution. If the next state is y, say that the chain accepts the move to y. Otherwise, the chain rejects the move, and stays at x.

For this to work, it is necessary that if $q_x(y) > 0$, then $q_y(x) > 0$ as well. Then the move to state y occurs with probability

$$\min\left\{1, \frac{f_\pi(y)q_y(x)}{f_\pi(x)q_x(y)}\right\}. \tag{1.14}$$

Note that f_π can be either a normalized or unnormalized density for the target distribution.

To see this in action, consider again the Ising model with $\mu = 0$. Then suppose at each step a node v is selected uniformly at random, and then a candidate color for $y(v)$ is chosen uniformly from $\{-1,1\}$. Then $q_y(x) = q_x(y) = 1/2$ so those factors in (1.14) cancel out. In fact, the original Metropolis et al. method [91] just considered chains where $q_y(x) = q_x(y)$. Only later did Hastings [47] consider $q_x(y) \neq q_y(x)$.

Again we need to know n_1 (the number of neighbors of v labeled 1) and n_{-1} (the number of neighbors labeled -1) to calculate $f_\pi(y)/f_\pi(x)$, but again the remainder of the factors in f_π simply cancel out. The resulting algorithm looks like this.

Ising_MH	*Input:* old state x	*Output:* new state x
1)	Draw $v \leftarrow \text{Unif}(V)$, $U \leftarrow \text{Unif}([0,1])$, $c \leftarrow \text{Unif}(\{-1,1\})$	
2)	Let n_1 be the number of neighbors of v labeled 1 in x	
3)	Let n_{-1} be the number of neighbors of v labeled -1 in x	
4)	If $U < \exp(n_c\beta)/\exp(n_{x(v)}\beta)$	
5)	$\quad x(v) \leftarrow c$	

Now consider trying to build a Markov chain whose stationary distribution has a density with respect to the uniform distribution over permutations. It is impossible to change only one component of a permutation, so at least two values of the permutation vector must be altered.

Given a permutation σ, let

$$\sigma_{i\leftrightarrow j}(k) = \begin{cases} \sigma(j) & \text{when } k = i \\ \sigma(i) & \text{when } k = j \\ \sigma(k) & \text{when } k \notin \{i,j\} \end{cases}$$

denote the move that transposes (swaps) the items at positions i and j.

Given current state x, a common way to construct a chain on permutations is to choose i and j uniformly at random, and let the proposed state be $y = x_{i\leftrightarrow j}$. Then $q_x(y) = q_y(x)$, and the chance of accepting the move is the minimum of 1 and $f(y)/f(x)$. Since only two elements are being considered, this is usually easy to calculate.

As an example, consider the uniform distribution over linear extensions of a poset. With respect to the uniform distribution on permutations, the density of a permutation where two elements are out of order is 0, while the density of any valid linear extension is 1.

A transposition is called adjacent if $j = i+1$. By choosing only adjacent transpositions, it is easy to check if the partial order is still satisfied. For $i \in \{1,2,\ldots,n-1\}$ and $j = i+1$, let $\sigma(i) = a$ and $\sigma(i+1) = b$. Then do not transpose if $a \prec b$; otherwise, the transposition always occurs.

1.8.3 Auxiliary random variables

In most of the applications of MCMC, the target density of X has a multiplicative structure. For these densities, it is possible to add a vector Y of auxiliary (extra) random variables such that (X,Y) has a simpler joint distribution.

As an example, consider once again the Ising model with $\mu = 0$. Then for each edge $\{i,j\}$, create an auxiliary random variable $Y(\{i,j\})$ whose distribution conditioned on X is uniform over $[0,1]$ if $X(i) \neq X(j)$, and uniform over $[0,\exp(\beta)]$ if $X(i) = X(j)$.

The density of a uniform over $[a,b]$ is $f(s) = (b-a)^{-1}\mathbb{1}(s \in [a,b])$, so this choice for Y makes the joint density of (X,Y) uniform over $X \in \{0,1\}^V$ and $Y \in \{[0,\infty) : (\forall\{i,j\})(y(\{i,j\}) \leq \min(\exp(\beta),1))\}$.

The Markov chain then is simple: given the state x, choose a new value of y conditioned on x, and then choose a new x conditioned on y. (This can be viewed as a deterministic scan Gibbs sampler applied to the new joint distribution.)

`Auxiliary_chain` *Input:* old state (x,y) *Output:* new state (x,y)

1) Draw y conditioned on x
2) Draw x conditioned on y

From the construction, drawing y conditioned on x is straightforward. Drawing x conditioned on y can be more complex. Consider again the Ising model with $\beta > 0$. Then $\exp(\beta) > 1$. When $y(\{i,j\}) \in [0,1]$, then it could be true that $x(i) = x(j)$ or $x(i) \neq x(j)$. But when $y(\{i,j\}) > 1$, it must be true that $x(i) = x(j)$. The edges with $y(\{i,j\}) > 1$ break the graph into clusters of connected components. Each component must have the same x value across the component, but otherwise is uniform.

Therefore, break the graph into connected components using the edges with $y(\{i,j\}) > 1$, then for each connected component, choose a label uniformly from $\{-1,1\}$ and apply that label to every node in the component.

This example was one of the first examples of an auxiliary variable chain, and is due to Swendsen and Wang [121]. For this reason, auxiliary variable chains are sometimes known as Swendsen-Wang chains.

`Swendsen_Wang_chain` *Input:* old state (x,y) *Output:* new state (x,y)

1) For each edge $\{i,j\} \in E$
2) Draw $y(\{i,j\}) \leftarrow \text{Unif}([0,\exp(\beta \cdot \mathbb{1}(x(i) = x(j)))])$
3) Let $A \leftarrow \{\{i,j\} : y(\{i,j\}) > 1\}$
4) For each connected component C in the graph (V,A)
5) Draw $B \leftarrow \text{Unif}(\{-1,1\})$
6) For each $v \in B$, set $x(v) \leftarrow B$

1.8.4 Slice sampler

The slice sampler (see [95] for more details) is a special case of auxiliary variables that uses just a single component Y with conditional distribution $[Y|X = x] \sim \text{Unif}([0, f_X(X)])$. Then $[X|Y = y] \sim \text{Unif}(\Omega \cap \{x : f_X(x) \geq y\})$.

This last set $\Omega \cap \{x : f_X(x) \geq y\}$ is a slice through the density of f_X at height y, hence the name slice sampler.

1.8.5 Shift chains

Although Gibbs, Metropolis-Hastings, and auxiliary random variables are the most widely used types of Markov chains, they are not the only types of chains.

An important type of chain that is useful for repulsive processes is the shift move. The canonical example is permutations: they are repulsive in the sense that two components of the permutation cannot both be assigned the same value. The solution to creating a chain that can move under repulsion is simple: choose two components uniformly and then swap the labels in those components. In permutations, this is called a transposition.

This is a special case of the far more general random walk on a group Markov chain, but this particular move will suffice for the examples used here.

Such shifts of one value from one dimension to another can be useful for all repulsive processes, not just permutations. Consider the hard-core gas model (Definition 1.20) where each node is either occupied ($x(v) = 1$) or unoccupied ($x(v) = 0$) but no two adjacent nodes can both be occupied. Each occupied node contributes a factor of λ to the density. Then constructing a Gibbs step is straightforward.

`HCGM_Gibbs` *Input:* old state x *Output:* new state x	
1)	Draw $v \leftarrow \mathsf{Unif}(V)$, $U \leftarrow \mathsf{Unif}([0,1])$
2)	Let n_1 be the number of neighbors of v labeled 1 in x
3)	If $U < \lambda/(\lambda+1)$ and $n_1 = 0$
4)	$\quad x(v) \leftarrow 1$

As soon as even a single neighbor of v is 1, the chain cannot alter the value of v. A shift move allows that if a single neighbor of v has label 1, then with some chance, remove the neighbor and label v with a 1. That is, take the 1 at the neighbor, and shift it over to node v. The resulting update takes parameter λ and probability of shifting p_{shift} and works as follows. [32]

`HCGM_shift` *Input:* old state x *Output:* new state x	
1)	Draw $v \leftarrow \mathsf{Unif}(V)$, $U \leftarrow \mathsf{Unif}([0,1])$, $S \leftarrow \mathsf{Bern}(p_{\text{shift}})$
2)	Let n_1 be the number of neighbors of v labeled 1 in x
3)	If $U \leq \lambda/(\lambda+1)$ and $n_1 = 0$
4)	$\quad x(v) \leftarrow 1$
5)	Else if $U \leq \lambda/(\lambda+1)$ and $n_1 = 1$ and $S = 1$
6)	$\quad$ Let w be the unique neighbor of v with $x(w) = 1$
7)	$\quad x(v) \leftarrow 1, x(w) \leftarrow 0$

When two or more neighbors are occupied, shifting does not happen and the node remains unoccupied.

1.9 Uniform random variables

Note that in Sections 1.8.1 and 1.8.2, all of the random choices were made in the very first line. Once these random choices had been made, the rest of the update proceeded deterministically.

As noted earlier, the random choices made by a computer are of course never truly random. Instead, pseudorandom numbers are generated that approximate an iid sequence of random $\{0,1\}$ bits. Consider the `Ising_Gibbs` step from Section 1.8.1. It first drew $v \leftarrow \mathsf{Unif}(V)$. This requires $\lceil \log_2(V) \rceil$ random bits. Next it says to draw $U \leftarrow \mathsf{Unif}([0,1])$. Of course no real-world digital computer can do this, as U consists of an infinite number of bits! See Figure 1.1.

The good news is that the full value of U was never used. In fact, the only use of

$$U = 0.1111111000000001111000110\ldots$$

Figure 1.1 *Written in binary, an actual uniform over* $[0,1]$ *contains an infinite sequence of bits, each of which is uniform over* $\{0,1\}$. *Any computer-generated randomness only contains a finite number of such bits.*

U was in line 4, where U was compared to a number. On average, only two bits will be needed to make such a comparison.

To see why, consider a concrete example. Suppose it is necessary to determine if $U \leq 2/3$. In binary, $2/3$ is written $0.101010101010\ldots$. Suppose the first bit of U is 0. Then $U = 0.0\ldots$ and no matter how the rest of the digits are filled out, $U \leq 2/3$. So only if the first bit is 1 will more bits of U be needed. Suppose the first and second bits are 1. Then $U = 0.11\ldots$ and now no matter how the rest of the bits are filled out, $U > 2/3$.

After k bits are chosen, the only way to need the $k+1$st bit is if the first k bits match the number to be compared against exactly. Each bit has a $1/2$ chance of failing to match the bit of the number, and so a geometric number of bits with mean $1/(1/2) = 2$ are needed before being able to determine if $U < a$.

So it is possible to treat the uniform U as $\sum_{i=1}^{\infty} B_i/2^i$, where B_i is the ith bit in the binary expansion of U. Then $B_1, B_2, \ldots$ are an iid sequence of Bernoulli random variables with mean $1/2$.

Of course, since there are an infinite number of bits, it is possible to take this single uniform, and break it up into two uniforms. The first uniform uses the odd bits in U, and the second uses the even. So $U_1 = \sum_{i=1}^{\infty} B_{2i-1}/2^i$ and $U_2 = \sum_{i=1}^{\infty} B_{2i}/2^i$. Since U_1 and U_2 are using independent Bernoulli random variables, they are independent of each other.

This can be taken further. Along the idea of Cantor diagonalization, this sequence of bits can be turned into an iid sequence of uniforms.

Here is how this works. Let B_1 be the first binary digit of U_1. Then B_2 is the second binary digit of U_1 and B_3 is the first binary digit of U_2. Then B_4 is the third binary digit of U_1, B_5 is the second binary digit of U_2, and B_6 is the first binary digit of U_3. Continuing in this fashion, the $\{B_i\}$ sequence of random bits allows construction of $U_1, U_2, U_3, \ldots$, each of which is independent and has $\textsf{Unif}([0,1])$ distribution.

Therefore a single $U \sim \textsf{Unif}([0,1])$ can be viewed as an infinite sequence of iid $\textsf{Unif}([0,1])$ random variables! The point is not to do this in practice, but by realizing this equivalence, all the randomness used by any randomized algorithm can be summarized by a single uniform, which simplifies the notation considerably.

In other words, any randomized algorithm can be written as a deterministic function $\phi(x, U)$, where x is the deterministic input to the problem, and $U \sim \textsf{Unif}([0,1])$ represents all the randomness used by the algorithm.

When a randomized algorithm that takes one step in a Markov chain is written in this fashion, call the algorithm an update function. Update functions are the building blocks of many perfect simulation algorithms (see Chapter 3).

1.10 Computational complexity

An important facet of evaluating algorithms is determining their running time as a function of input size. For example, suppose to generate a particular permutation on n elements, $n^3 + n^2$ random uniform numbers need to be generated. As n gets large, the n^3 term dominates the number of uniforms, and the n^2 term is insignificant.

If $2n^3$ terms were needed, then the constant has changed, but the way the running time scales with the problem input remains the same. Therefore, it is common to use an asymptotic notation that does not depend either on the constants in front of the largest term, or on the insignificant terms.

This asymptotic notation consists of sets of functions that behave the same way in a large limit. There are three types of sets. The first type, called Big-O, gives an upper limit on the rate of growth of the function. The second type, called Big-Omega, gives a lower limit on the rate of growth of the function. The third type, called Big-Theta, gives both upper and lower limits on the rate of growth.

For instance, say that n^3, $2n^3$, and $n^3 + n^2$ are all in the set of functions denoted $O(n^3)$. Using proper set notation, this would be written $n^3 + n^2 \in O(n^3)$. However, it is customary in this setting to abuse notation and use an equality sign $=$ rather than a set inclusion sign $\in$. Formally, these notations can be defined for our purposes as follows. (Note that while more general definitions are possible, the definition here is best suited to analyzing running times of algorithms.)

Definition 1.35. *Given functions $f(x)$ and $g(x)$ with common domain that is a subset of $[0, \infty)$, say that*

$$f(x) = O(g(x)) \Leftrightarrow (\exists n)(\exists c)(\forall x > n)(f(x) \leq c \cdot g(x))$$
$$f(x) = \Omega(g(x)) \Leftrightarrow (\exists n)(\exists c)(\forall x > n)(c \cdot g(x) \leq f(x))$$
$$f(x) = \Theta(g(x)) \Leftrightarrow (\exists n)(\exists c_1)(\exists c_2)(\forall x > n)(c_1 \cdot g(x) \leq f(x) \leq c_2 \cdot g(x)).$$

Of course the smaller the running time of an algorithm, the better! Running time functions in the sets $O(n)$, $\Omega(n)$, and $\Theta(n)$ are referred to as *linear time*. Similarly, running time functions in the sets $O(1)$, $\Omega(1)$, and $\Theta(1)$ are called *constant time*.

Chapter 2

Acceptance/Rejection

'Tis a lesson you should heed:
Try, try, try again.
If at first you don't succeed,
Try, try, try again.

Thomas H. Palmer

The acceptance/rejection (AR) method was the first major protocol for perfect simulation. The basic method gives a way of sampling from a random variable given some condition on that variable. When used with an auxiliary variable, this method can be extended to sampling from any bounded density. Other bounds on probabilities such as Markov's inequality and Chernoff bounds can actually be viewed as AR algorithms, and these bounds assist in simulating from rare events.

2.1 The method

In Example 1.1 of the previous chapter, the goal was to draw uniformly from $\{1,2,\ldots,5\}$ given a fair six-sided die. The solution was to roll the die until the result fell into $\{1,\ldots,5\}$. Intuitively this gives the desired distribution; the following theorem proves this fact and generalizes the result.

Theorem 2.1. *Suppose that ν is a finite measure over B, and A is a measurable subset of B with $\nu(A) > 0$. Let $X_1, X_2, \ldots$ be iid draws from ν over B, and $T = \inf\{t : X_t \in B\}$. Then*

$$X_T \sim [X_1 | X_1 \in A]. \tag{2.1}$$

That is, if you repeatedly draw from the measure ν over B, and return the first value that lies in A, the resulting measure is from ν conditioned to lie in A.

Proof. This follows from the Fundamental Theorem of perfect simulation (Theorem 1.1), using $U_1, U_2, \ldots$ iid draws from ν over B, $b(U) = \mathbb{1}(U \in A)$, $g(U) = U$, and $f(X,U) = X$. □

Since $f(X,U)$ does not depend on U, this is an interruptible algorithm.

The above theorem gives a mathematical description of the algorithm; see Figure 2.1 for a graphical illustration. In pseudocode it is quite simple.

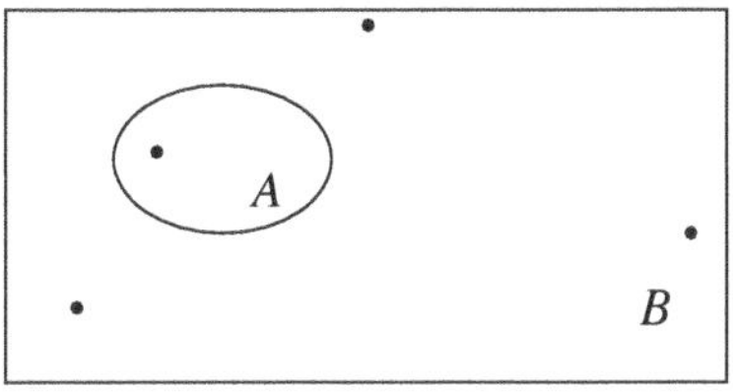

Figure 2.1 *Acceptance/rejection draws from the larger set B until the result falls in the smaller set A. The final point comes from the distribution over B conditioned to lie in A.*

AR	
1)	Repeat
2)	Draw X from ν over B
3)	Until $X \in A$

The number of steps taken by this algorithm (on average) depends on how large A is compared to B.

Lemma 2.1. *The probability that $X \in A$ is $\nu(A)/\nu(B)$. The expected number of draws used by the algorithm is $\nu(B)/\nu(A)$.*

Proof. From the last theorem, $\mathbb{P}(X \in A) = \nu(A)/\nu(B)$. This is just how probability measures induced by finite measures are defined. Given this probability of success, the number of draws of X is a geometrically distributed random variable with parameter $\nu(A)/\nu(B)$. The mean of a geometric is just the inverse of the parameter. □

2.1.1 Example: drawing from the unit circle

Consider the problem of drawing uniformly from the unit circle. Because the distribution of the angle and radius in the unit circle can be found analytically, it is possible to give a direct method for doing so. Let U_1 and U_2 be iid uniforms over $[0,1]$. Then let $\theta = 2\pi U_1$ and $r = \sqrt{U_2}$. Then $(r\sin(\theta), r\cos(\theta))$ will be uniform over the unit circle. AR gives a means for generating from the circle without the need to compute functions other than additions and multiplications.

Begin with a bounding region for the unit circle that is easy to sample from. The square $[-1,1] \times [-1,1]$ does the trick. Using this bounding region makes the algorithm as follows.

AR_unit_circle	*Output:* $(X,Y) \sim \text{Unif(unit circle)}$
1)	Repeat
2)	Draw $U_1 \leftarrow \text{Unif}([0,1])$, $U_2 \leftarrow \text{Unif}([0,1])$
3)	$X \leftarrow 2U_1 - 1$, $Y \leftarrow 2U_2 - 1$
4)	Until $X^2 + Y^2 \leq 1$

This method does not need to evaluate the trigonometric functions sine and cosine that the exact method did, so it can be useful on processors of limited capability.

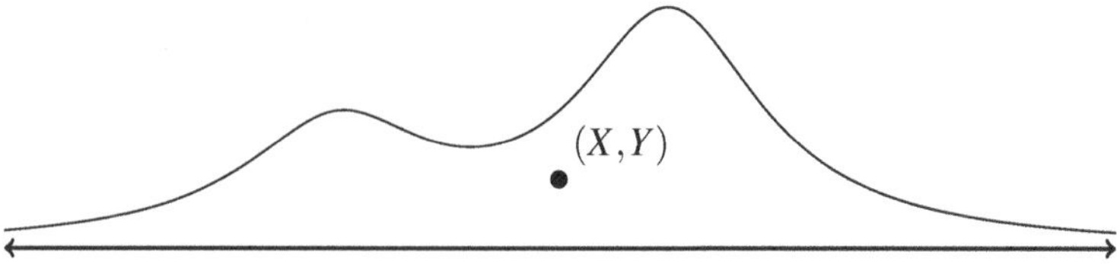

Figure 2.2 *To draw X with density f, draw* (X,Y) *uniformly from the region under the density, then throw away the Y value. This procedure works whether or not the density is normalized.*

The area of $[-1,1]\times[-1,1]$ is 4, while the unit circle has area π. Therefore the expected number of uniforms drawn is $(2)(4/\pi)\approx 2\cdot 1.273$ compared to the just two uniforms that the exact method uses.

One more note: for (X,Y) uniform over the unit circle, X/Y has a standard Cauchy distribution. Therefore this gives a perfect simulation method for Cauchy random variables that does not utilize trigonometric functions.

2.2 Dealing with densities

The basic AR method presented does not handle densities; however, densities are easily added to the mix using auxiliary random variables. Consider the following theorem from measure theory, known also as the fundamental theorem of Monte Carlo simulation (see [115], p. 47).

Theorem 2.2 (Fundamental theorem of Monte Carlo simulation)**.** *Suppose that* X *has (possibly unnormalized) density* f_X *over measure* ν *on* Ω*. If* $[Y|X]\sim$ *Unif*$([0,f_X])$*, then* (X,Y) *is a draw from the product measure* $\nu\times$ *Unif over the set* $\{(x,y):x\in\Omega,0\le y\le f_x\}$.

Why is this fundamental to Monte Carlo methods? Because this theorem in essence says that all densities are illusions, that every problem in probability can be reduced to sampling over a uniform distribution on an augmented space. Since pseudorandom sequences are iid uniform, this is an essential tool in building Monte Carlo methods. Moreover, the density of X can be unnormalized, which is essential to the applications of Monte Carlo.

A simple example of this theorem is in one dimension, where ν is Lebesgue measure. This theorem says that to draw a continuous random variable X with density f, choose a point (X,Y) uniformly from the area between the x-axis and the density, and then throw away the Y value. Here it is easy to see that for any $a\in\mathbb{R}$, $\mathbb{P}(X\le a)=\int_{-\infty}^{a}f(x)\,dx$ as desired. (See Figure 2.2.)

To see how this operates in practice, let μ be a measure over Ω, and suppose for density g it is possible to draw from the product density $\mu\times$ Unif over $\Omega_g=\{(x,y):x\in\Omega,0\le y\le g\}$.

Once this draw is made, and the Y component is thrown away, then the remaining X component has density g. But suppose instead that the goal is to draw X with density f instead.

Well, if (X,Y) has measure $\mu\times$ Unif over Ω_g, then $(X,2Y)$ has measure $\mu\times$ Unif over $\Omega_{2g} = \{(x,y) : x \in \Omega, 0 \leq y \leq 2g\}$. Of course, there is nothing special about the use of 2 there: for any constant $c > 0$, (X, cY) has measure $\mu \times \text{Unif}$ over Ω_{cg}.

Suppose $c \geq \sup_{x \in \Omega} f(x)/g(x)$. Then

$$\Omega_f \subseteq \Omega_{cg}.$$

Therefore, AR can be used to draw from Ω_f: simply draw X from density g, draw Y uniformly from $[0, c \cdot g(X)]$, and if the resulting Y lands in $[0, f(X)]$ (so that $(X,Y) \in \Omega_f$) then accept (X,Y) as a draw from Ω_f, which means X has density f.

Of course, to test if a uniform from $[0, c \cdot g(X)]$ falls in $[0, f(X)]$, it is not necessary to actually generate the uniform. Simply draw a Bernoulli random variable with mean $f(X)/[c \cdot g(X)]$. If successful, consider X a draw from f. Note that this method applies even when f and g are unnormalized densities. It is given in the following pseudocode.

`AR_densities`
Input: f and g densities (possibly unnormalized) over ν
Output: X with density f over ν

Let c be any number greater than $f(x)/g(x)$ for all $x \in \Omega$
Repeat
  Draw X from density g over ν
  Draw C as $\text{Bern}(f(X)/[cg(X)])$
Until $C = 1$

Since the probability of accepting varies inversely with c, the best possible choice of c is $\sup_{x \in \Omega} f(x)/g(x)$. It is not always possible to calculate this value exactly. In this case, an upper bound of the supremum can be used; it just makes for a slower algorithm.

In `AR_densities`, let $Z_f = \int_{x \in \Omega} f(x)\nu(dx)$ and $Z_g = \int_{x \in \Omega} f(x)\nu(dx)$ be the normalizing constants for the possible unnormalized densities f and g.

Lemma 2.2. *The probability that* `AR_densities` *accepts a draw X is $Z_f/[cZ_g]$. The expected number of draws of X used by the algorithm is cZ_g/Z_f.*

Proof.

$$\begin{aligned}
\mathbb{P}(C = 1) &= \mathbb{E}[\mathbb{E}[\mathbb{1}(C = 1)|X]] \\
&= \int_{x \in \Omega} \mathbb{P}(C = 1|X = x)\mathbb{P}(X \in dx) \\
&= \int_{x \in \Omega} f(x)/[cg(x)](g(x)/Z_g)\nu(dx) \\
&= [cZ_g]^{-1} \int_{x \in \Omega} f(x)\nu(dx) \\
&= Z_f/[cZ_g].
\end{aligned}$$

□

2.2.1 *Example: from Cauchy to normal*

To see how this works in practice, suppose that a user has the ability to generate from the standard Cauchy distribution with density $g(x) = [\pi(1+x^2)]^{-1}$, and wishes to generate a standard normal Z with density $f(x) = (2\pi)^{-1/2}\exp(-x^2/2)$.

First, the normalization is unnecessary, so let $g_1(x) = (1+x^2)^{-1}$ and $f_1(x) = \exp(-x^2/2)$. Now $f_1(x)/g_1(x) \leq 2$ for all x, so a valid algorithm is as follows.

```
Normal_from_Cauchy
```

Repeat
  Draw X as a standard Cauchy
  Draw C as $\textsf{Bern}((1/2)(1+X^2)\exp(-X^2/2))$
Until $C = 1$

The number of X draws used by the algorithm has a geometric distribution with parameter equal to the chance of accepting, which is

$$\mathbb{P}(C=1) = \int_{x\in\mathbb{R}} \mathbb{P}(X \in dx, C=1)$$
$$= \int_{x\in\mathbb{R}} g(x)\frac{1}{2}(1+x^2)\exp(-x^2/2)\,dx = (2\pi)^{-1/2} = 0.3989\ldots$$

The average number of X draws used by the algorithm equal to $\sqrt{2\pi} = 2.506\ldots$.

A faster algorithm can be created by using a better bound on $(1+x^2)\exp(-x^2/2)$. In fact, $\sup_x (1+x^2)\exp(-x^2/2) = 2/\sqrt{e}$. This makes the algorithm faster by a factor of square root of e, which means the algorithm requires $\sqrt{2\pi/e} = 1.520\ldots$ draws from X on average.

As another example, suppose $U \sim \textsf{Unif}([0,1])$, and $X = \lfloor 1/U \rfloor$, where $\lfloor x \rfloor$ is the floor function that is the greatest integer at most x.

Then $\mathbb{P}(X=i) = \mathbb{P}(i \leq 1/U < i+1) = \mathbb{P}(1/i \geq U > 1/(i+1)) = 1/(i^2+i)$ for $i \in \{1,2,\ldots\}$. Suppose the goal is to generate W so that $\mathbb{P}(W=i) = 1/i^2$ (this is a Zipf distribution).

Then X has density $f_X(i) = 1/(i^2+i)$ and the goal is to produce W with density $f_W(i) = 1/i^2$ for counting measure over $\{1,2,3,\ldots\}$. To use AR for this task, note

$$\sup_{i\in\{1,2,\ldots\}} \frac{1/i^2}{1/(i^2+i)} = 2.$$

Hence the AR approach is as follows.

```
Zipf_from_Uniform
```

1) Repeat
2)   Draw $U \leftarrow \textsf{Unif}([0,1])$
3) $X \leftarrow \lfloor 1/U \rfloor$
4)   Draw C as $\textsf{Bern}((1/2)(1+1/X))$
5) Until $C = 1$

The chance that $C=1$ is $\sum_{i=1}^{\infty} \mathbb{P}(C=1, X=i) = \sum_{i=1}^{\infty} 1/(2i) = \pi^2/12$. Hence the expected number of draws of X used is $12/\pi^2 < 1.216$.

2.3 Union of sets

Another way to employ acceptance/rejection is to account for multiple ways of reaching a sample. For instance, consider the problem of sampling from measure ν over $S_1 \cup \cdots \cup S_n$, where set S_i has finite measure $\nu(S_i)$.

Consider the following algorithm: draw I from $\{1,2,\ldots,n\}$, where $\mathbb{P}(I = i) \propto \nu(S_i)$. Then conditioned on I, draw X from S_I using measure ν.

Of course, X does not come from ν over $S_1 \cup \cdots \cup S_n$ because there are multiple ways that X could have been selected. For instance, if $X \in S_1$ and $X \in S_2$, but $X \notin S_3,\ldots,S_n$, then I could have been either 1 or 2 in selecting X.

Lemma 2.3. *For the procedure above, X is a draw from ν over $S_1 \cup \cdots \cup S_n$ with density $f(x) = \#\{i : x \in S_i\}/\nu(S_1 \cup \cdots \cup S_n)$.*

Proof. Let $x \in S_1 \cup \cdots \cup S_n$, and $C = \nu(S_1 \cup \cdots S_n)$. Then

$$\begin{aligned}\mathbb{P}(X \in dx) &= \sum_{i=1}^{n} \mathbb{P}(X \in dx, I = i) = \sum_{i=1}^{n} \mathbb{P}(I = i)\mathbb{P}(X \in dx | I = i) \\ &= \sum_{i=1}^{n} C^{-1}\nu(S_i)\mathbb{1}(x \in S_i)\nu(dx)\nu(S_i)^{-1} = C^{-1}\sum_{i=1}^{n} \mathbb{1}(x \in S_i) \\ &= C^{-1}\#\{i : x \in S_i\}.\end{aligned}$$

□

The next step is to use AR to obtain samples from the target density, which is just proportional to 1. Again it is not necessary to learn the value of $\nu(S_1 \cup \cdots \cup S_n)$ before using AR, as knowledge of the normalizing constant of the density is not necessary.

`AR_union_sets`

1) Repeat
2) Draw I so that $\mathbb{P}(I = i) \propto \nu(S_i)$
3) Draw X from ν over S_I
4) Draw C as $\mathsf{Bern}(1/\#\{i : X \in S_i\})$
5) Until $C = 1$

For instance, suppose that the goal is to sample uniformly from the union of the unit circles centered at $(0,0)$, $(1/2,0)$, and $(1/2,1/2)$. Then draw $I \sim \mathsf{Unif}(\{1,2,3\})$. If $I = 1$, draw X from the unit circle centered at $(0,0)$. For $I = 2$ or 3, draw X from the unit circle centered at $(1/2,0)$ and $(1/2,1/2)$ respectively. Accept X as a draw from the union with probability equal to the inverse of the number of unit circles that X lies in (see Figure 2.3.)

2.3.1 *Example: AR for disjunctive normal form*

An example of the union of sets method in action comes from Boolean logic, which deals with formulas that are either true or false. Begin with some definitions.

Definition 2.1. *A* propositional variable *is a variable that can be either true or false.*

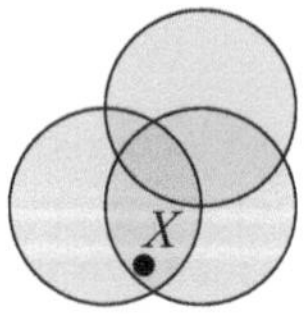

Figure 2.3 *Drawing uniformly from the union of three circles of equal area. First draw I uniformly from* $\{1,2,3\}$, *then the point X uniformly from circle I. Finally, accept X with probability 1 over the number of circles that X falls into. In this example, circle 2 in the lower right was chosen, then the point X that was uniform over this circle falls into two of the circles, so it would be accepted as a draw with probability 1/2.*

Definition 2.2. *The* negation *of a propositional variable is false if the variable is true, and true if the variable is false. The negation of* x *will be denoted here as* $\neg x$. *A formula that is of the form* x *or* $\neg x$ *is called a* literal.

Definition 2.3. *The logical operator* or *is a binary operation on literals, written* $x \vee y$, *and defined as true if either* x *or* y *(or both) are true, and false when both* x *and* y *are false.*

Definition 2.4. *The logical operator* and *is a binary operation on literals, written* $x \wedge y$, *and defined as true if both* x *and* y *are true, and false otherwise.*

Definition 2.5. *A formula is in* disjunctive normal form (DNF) *if it is written as clauses of literals connected by and operators, and each clause is connected by or operators.*

For example, $(x_1 \wedge x_3) \vee (x_2 \wedge x_4)$ is in DNF form, but $(x_1 \vee x_3) \wedge (x_2 \vee x_4)$ is not.

Definition 2.6. *A* satisfying assignment *for a formula is a configuration where each propositional variable is assigned either true or false in such a way that the formula is true.*

For a formula in DNF, at least one clause must be true. Consider the problem of drawing uniformly from the set of satisfying assignments. The basic acceptance/rejection technique would randomly assign to each variable either true or false, and then accept the result if the formula is satisfied.

Unfortunately, the chance that a single clause is satisfied goes down exponentially in the length of the clause. To be precise, for a clause with k literals, there is exactly a $1/2^k$ chance that the clause evaluated to true. For a single clause that contains all n variables, the chance of evaluating to true is $1/2^n$.

Let S_i denote the set of assignments that satisfy clause i. If there are m clauses in the formula, the goal is to sample uniformly from $S_1 \cup \cdots \cup S_m$.

The size of each S_i can be calculated as follows. Suppose clause i contains $\ell(i)$ literals. In a DNF formula all the literals in the clause must be set to true, so there is only one way to set each of the $\ell(i)$ variables to make the literals true.

The other variables can be set in any fashion desired. The result is that $\#S_i = 2^{n-\ell(i)}$. Now simply apply the `AR_union_sets` procedure.

Since there are at most m clauses, the chance of accepting at each step is at least

$1/m$. Therefore this method draws at most (on average) m DNF formulas before accepting, making this a polynomial time algorithm.

2.4 Randomized approximation for #P complete problems

The problem of whether a satisfying assignment to a problem in DNF exists is an example of the problem class called NP. Intuitively speaking, a yes-no problem is in NP if whenever the answer is yes, there is a way to prove the answer that can be checked in polynomial time.

Definition 2.7. *A* decision problem *is a question in a formal system (such as a Turing machine) such that given the values of input parameters, the answer is either yes or no.*

Definition 2.8. *A decision problem is in the class* nondeterministic polynomial (NP) *if whenever the answer to the problem is yes, there always exists a certificate that can be used to prove the answer true or false in a number of steps that is polynomial in the input to the problem.*

Definition 2.9. *A decision problem is in the class* polynomial (P) *if there is a way to determine if the answer is yes or no in a polynomial number of steps in the input.*

The DNF problem is: given a DNF formula, does there exist an assignment for which the formula evaluates to true? This problem is in NP, since when the answer is yes, it is possible to give an assignment that can be checked in polynomial time.

This problem is also trivially in P, since every such formula has a satisfying assignment if and only if there exists a clause that does not contain both a variable and its negation. This condition is easily checked in polynomial time.

One of the most interesting questions in mathematics is: does $P = NP$? That is, can every problem whose answer can be checked in polynomial time have a solution algorithm that runs in polynomial time?

Problems in Number-P are slightly different. These problems are related to NP, but are not yes-no; instead they count the number of solutions to a problem.

Definition 2.10. *A problem is in the class* Number-P *(abbreviated #P) if it consists of counting the elements of a discrete set, where it is possible to test membership in the set in time polynomial in the input to the problem.*

An example of a problem in #P is counting the number of satisfying assignments to a DNF problem. Each individual satisfying assignment can be checked in polynomial time.

In order to understand these complexity classes further, the notion of reducing problems is needed.

Definition 2.11. *Say that decision problem A is* polynomial time reducible *to decision problem B if given a problem input I_A to problem A, it is possible to build in polynomial time an input I_B of size polynomial in I_A where the answer to problem B with input I_B is the same as the answer to problem A with input I_A.*

In other words, A reduces to B if knowing how to solve B gives a method for solving A.

Definition 2.12. *A problem is* NP complete *if all problems in NP are polynomial time reducible to it. A problem is* #P complete *if all problems in #P are polynomial time reducible to it.*

In an NP complete problem, the goal is to determine if there is at least one solution. In #P complete problems, the goal is to count exactly the number of solutions. Therefore #P complete problems are in one sense more difficult that NP complete problems.

So why bring all this up now? Because counting the number of satisfying assignments to a DNF formula is a #P complete problem.

This means that it is unlikely that an efficient algorithm for counting the number of such solutions exactly will be found. It turns out that having an AR algorithm means that there is an efficient method for estimating the number, though!

The number of draws used by an AR algorithm is geometric with parameter equal to the probability of success. So in basic AR where draws from ν over B are made until the result falls in A, the chance of success is $\nu(A)/\nu(B)$. Say that T draws are used by the algorithm. If $\nu(B)$ is known, then $1/T$ is an estimate of $\nu(A)/\nu(B)$, so $\nu(B)/T$ is an estimate of $\nu(A)$.

Definition 2.13. *Let $\mathscr{P}$ be a problem with input in $\mathscr{I}$ and output in $\mathbb{R}$. Call a computable function $f : \mathscr{I} \times [0,1]$ an* (ε, δ)-randomized approximation scheme (ras) *if for $U \sim$ Unif$([0,1])$*

$$(\forall I \in \mathscr{I}) \left(\mathbb{P}\left(\left| \frac{f(I,U)}{\mathscr{P}(I)} - 1 \right| > \varepsilon \right) \leq \delta \right).$$

For an (ε, δ)-ras, the chance that the relative error is greater than ε is at most δ.

Consider the problem of estimating the probability of heads for a coin that can be flipped as often as needed. This is equivalent to estimating the mean of Bernoulli random variables $X_1, X_2, \ldots$ which are identically distributed with mean p. Dagum, Karp, Luby, and Ross gave an (ε, δ)-ras for this problem that uses $O(p^{-1}\varepsilon^{-2}\ln(\delta^{-1}))$ samples [26]. Rather than describe this algorithm here, a more advanced version is given in Section 2.5.

This is exactly what is needed when AR is being used to estimate the solution to a problem in #P. The value of p is $\nu(A)/\nu(B)$, therefore, AR gives an (ε, δ)-ras that uses $O(\nu(B)\nu(A)^{-1}\varepsilon^{-2}\ln(\delta^{-1}))$ samples.

Fact 2.1. *An (ε, δ)-ras for counting a DNF satisfying assignments, where the formula contains n variables and m clauses, exists that uses $O((n+\ln(m))m\varepsilon^{-2}\ln(\delta^{-1})$ Bernoulli random variables on average.*

Proof. The underlying measure ν here is just counting measure on the 2^n possible true-false values for the variables. The target measure is $f(x) = \mathbb{1}(x \text{ is satisfying})$, and so Z_f is the number of satisfying assignments. Now $g(x) = \#\{\text{satisfied clauses}\}$, and so $Z_g \leq mZ_f$. Finally, $c = 1$. Hence $Z_f/[cZ_g]geq1/m$.

Each draw requires choosing $I \sim$ Unif$(\{1, \ldots, m\})$ (which requires $\ln(m)$ Bernoullis) and then an assignment of up to n variables (which requires n Bernoullis). □

2.5 Gamma Bernoulli approximation scheme

As seen in the last section, using AR to approximate integrals or sums reduces to the question of estimating the probability p of heads on a coin that can be flipped as many times as necessary. The Gamma Bernoulli approximation scheme (GBAS) [56] is a simple way of building an estimate $\hat{p}$ for p with the unique property that the relative error in the estimate does not depend in any way upon the value of p.

First, the estimate. In pseudocode, it looks like this.

```
GBAS   Input: k
1)  S ← 0, R ← 0.
2)  Repeat
3)      X ← Bern(p), A ← Exp(1)
4)      S ← S + X, R ← R + A
5)  Until S = k
6)  p̂ ← (k − 1)/R
```

To understand what is happening in this algorithm, it helps to introduce the notion of a Poisson process.

Definition 2.14. *Suppose that* $A_1, A_2, \ldots$ *are iid* $\mathsf{Exp}(\lambda)$. *Then the points* $A_1, A_1 + A_2, A_1 + A_2 + A_3, \ldots$ *form a* Poisson point process of rate λ *on* $[0, \infty)$.

Poisson point processes are often used as models of customer arrivals. For instance, suppose that customers are arriving at a rate of 6 per hour. Since this is a Poisson point process, this does not tell us that exactly 6 arrive every hour, instead it only tells us that on average 6 arrive in the first hour. In the first half-hour, on average $6(1/2) = 3$ customers arrive. In the first quarter-hour, on average $6(1/4) = 1.5$ customers arrive, and so on.

It could be that in the first hour, no customers actually arrive or it could be that ten arrive. Any nonnegative integer number of arrivals is possible during this time period, but of course some are more likely than others.

Now suppose that for every customer that arrives, only one in six actually buys something when they are in the store. Then intuitively, if overall customers are arriving at a rate of 6 per hour, and each has a $1/6$ chance of making a purchase, then the rate at which customers that buy something arrive is only one per hour. The following lemma shows that this intuition is correct. It says that if each point in the Poisson point processes is marked with a Bernoulli random variable (a coin flip), then the time from 0 until the first point that gets a 1 is itself an exponential random variable, albeit one with a rate proportional to the old rate times the probability of a 1.

Fact 2.2. *Let* $A_1, A_2, \ldots$ *be iid* $\mathsf{Exp}(\lambda)$, *and* $G \sim \mathsf{Geo}(p)$. *Then* $A_1 + \cdots + A_G \sim \mathsf{Exp}(\lambda p)$.

The easiest way to show this result is with moment-generating functions.

Definition 2.15. *The* moment-generating function *of a random variable X is*

$$\text{mgf}_X(t) = \mathbb{E}[\exp(tX)]. \tag{2.2}$$

Moment-generating functions are unique when they are defined over a nontrivial interval of t values.

Lemma 2.4 (Moment-generating function uniqueness). *Suppose* $\text{mgf}_X(t) = \text{mgf}_Y(t)$ *for all* $t \in [a,b]$ *where* $a < 0 < b$. *Then* $X \sim Y$.

The important idea needed here is that the sum of a random number of copies of a random variable has moment-generating function equal to a composition of moment-generating functions.

Lemma 2.5. *If* $Y \in \{0,1,2,\ldots\}$ *and* $X_1, X_2, \ldots$ *have finite moment-generating functions for* $t \in [a,b]$, *then* $X_1 + \cdots + X_Y$ *has moment-generating function*

$$\text{mgf}_{X_1+\cdots+X_Y}(t) = \text{mgf}_Y(\text{mgf}_{X_i}(t)). \tag{2.3}$$

Now to the proof of our fact.

Proof. A geometric random variable G with mean $1/p$ has $\text{mgf}_G(t) = pe^t/(1-(1-p)e^t)$, at least when $t < -\ln(1-p)$. The mgf of an exponential random variable with rate λ is $\text{mgf}_A(t) = \lambda(\lambda - t)^{-1}$ for $t < \lambda$. So the mgf of $A_1 + \cdots + A_G$ is

$$\text{mgf}_{A_1+\cdots+A_t}(t) = \frac{p\lambda(\lambda-t)^{-1}}{1-(1-p)\lambda(\lambda-t)^{-1}} = \frac{p\lambda}{p\lambda - t}.$$

But this is just the mgf of an exponential of rate $p\lambda$, which completes the proof. □

This procedure of marking points in the Poisson point process with independent coin flips is called *thinning*. This concept is discussed in greater detail in Chapter 7.

Now consider the GBAS algorithm from earlier. It does this thinning procedure starting from a rate of 1, and thinning using the p-coin flips. Therefore the new rate of the thinned process is p. This is done k times, giving k iid $\mathsf{Exp}(p)$ random variables that are then added together.

When you add independent exponentials together, you get a gamma distribution with shape parameter equal to the number of exponentials added and a rate parameter equal to the rate of the exponentials. This establishes the following fact.

Fact 2.3. *In GBAS,* $R \sim \mathsf{Gamma}(k,p)$.

When a gamma distributed random variable is scaled by a constant, the rate changes by the inverse of the constant. So $R/(k-1) \sim \mathsf{Gamma}(k, p(k-1))$. The actual output of GBAS is $(k-1)/R$, which is the multiplicative inverse of a gamma distributed random variable. This gives a distribution called the inverse gamma.

Fact 2.4. *The output* $\hat{p}$ *of GBAS is inverse gamma with shape parameter* k *and scale parameter* $p(k-1)$. *The estimate is unbiased:* $\mathbb{E}[\hat{p}] = p$. *Also,* $(\hat{p}/p) \sim \mathsf{InvGamma}(k, 1/(k-1))$, *so the relative error* $(\hat{p}/p) - 1$ *is a shifted inverse gamma function.*

Proof. The mean of $X \sim \mathsf{InvGamma}(\alpha, \beta)$ is $\beta/(\alpha-1)$, so $\mathbb{E}[\hat{p}] = p(k-1)/(k-1) = p$. Dividing $\hat{p}$ by p is the same as scaling it by $1/p$, so the scale parameter is multiplied by the same amount. □

The key point of the fact is that $\hat{p}/p-1$ has a known distribution: the value of p does not affect it in any way! The $\mathsf{InvGamma}(k,k-1)$ distribution is concentrated around 1, and is more tightly concentrated as k increases. So GBAS can be used to obtain an (ε,δ)–ras as follows.

GBAS_ras *Input:* ε,δ

1) Let $k=\min\{k: X\sim \mathsf{InvGamma}(k,k-1)\rightarrow \mathbb{P}(|X-1|>\varepsilon)<\delta\}$
2) $S\leftarrow 0,\ R\leftarrow 0$
3) Repeat
4) $\quad X\leftarrow \mathsf{Bern}(p),\ A\leftarrow \mathsf{Exp}(1)$
5) $\quad S\leftarrow S+X,\ R\leftarrow R+A$
6) Until $S=k$
7) $\hat{p}\leftarrow (k-1)/R$

2.5.1 *Application: exact p-values*

As an application of this method, consider the problem of estimating exact p-values. Recall that a statistical model for data is a probability distribution such that the data is believed to be a draw from the distribution. For instance, suppose an observer believes that the number of taxicabs in a town is 200, and that the number of a randomly spotted taxi is equally likely to be any integer from 1 to 200. Suppose that the observer sees one taxi, and it is numbered 14.

The observer wishes a principled method for determining if their statistical model should be rejected based upon the data. One such method for doing so is called the p-value approach. Let S be any statistic, that is, a function of the data. Then if D is the data, and the hypothesis to be tested (often called the null hypothesis) is that $D\sim\pi$, then the p-value is just the probability that $S(X)$ is at least as extreme as $S(D)$ where $X\sim\pi$.

For instance, in the taxicab example, it would be unusual to see a low-numbered taxi if the number of cabs in the town is very large. Therefore, let $S(x)=x$, and consider $S(X)$ to be as extreme than $S(D)$ if $S(X)\leq S(D)$.

Then $S(D)=14$, and so the p-value for the hypothesis that there are 200 cabs and that the likelihood of seeing each cab number is the same is $\mathbb{P}(S(X)\leq 14)=14/200=7\%$.

In general, if $X_1,X_2,\ldots\sim\pi$ using a perfect simulation algorithm, then the p-value is exactly the chance that $S(X)$ is at least as extreme as S applied to the data, and so constitutes the exact type of experiment where GBAS can be applied.

2.5.1.1 *Testing for Hardy-Weinberg equilibrium*

An *allele* is a variant of a gene in an organism. For instance, one allele might cause fur to become brown, while a different allele of a gene causes fur to be black in color.

If a gene is in Hardy-Weinberg equilibrium, the fitness (propensity to reproduce) of the organism is independent of which allele the organism carries. Such alleles will not be selected either for or against in the population.

A *diploid* organism carries two copies of each gene, each of which could come from different alleles. The order of the alleles does not matter. Hence if the two alleles for a gene are denoted A and a, the genotype of the organism is either AA, Aa, or aa. In Hardy-Weinberg equilibrium, the frequency of allele AA will be the square of the frequency of allele A in the population. The frequency of Aa will be twice the frequency of A times the frequency of a, and that of aa will be the square of the frequency of a in the population.

Consider the problem of testing whether a population is in Hardy-Weinberg equilibrium. For m alleles $A_1, \ldots, A_m$, the data consists of the number of sampled individuals with genotype A_iA_j for $1 \leq i < j \leq m$.

Given data table D, Guo and Thompson [43] suggested using the statistics $S(X) = \mathbb{P}(X = D)$, where X is a random draw from tables of genetic data from populations in Hardy-Weinberg equilibrium. When there are n individuals, this made the statistic

$$S(D) = \frac{n! \prod_{i=1}^{m} \sum_{j=i+1}^{m} D_{ij}!}{(2n)! \prod_{j>i} D_{ij}!} 2^{\sum_{j>i} D_{ij}}. \tag{2.4}$$

The exact p-value is then $P(S(Y) \leq S(D))$, where Y is a perfect draw from tables under the Hardy-Weinberg equilibrium model. The Guo and Thompson method for directly sampling from such tables was later improved in [59].

2.5.1.2 *Testing for differential gene expression*

A related task is to test for how gene expression changes under environmental conditions. Each gene encodes proteins that are produced by the cell: the amount of protein produced can be affected by the environment of the cell. Consider an experiment (such as in [122]) where the expression rate of a gene is measured conditioned on different environmental factors.

The data is a matrix where K_{ij} is the number of experiments under condition j that resulted in expression of gene i. If the gene is equally likely to be expressed under all conditions, then (as in the Hardy-Weinberg example), it is easy to generate data tables in linear time.

The test statistic is once again the probability of the table being generated, and the p-value is the probability that a randomly drawn table has test statistics at most equal to the test statistic applied to the data. Once again, GBAS can be used to estimate this p-value with exact confidence intervals.

2.6 Approximate densities for perfect simulation

The key step in AR for sampling from density f using density g is when for $X \sim g$, the $C \sim \text{Bern}(f(X)/[cg(X)]$ coin is flipped. However, it is not necessary to know $f(X)/[cg(X)]$ exactly in order to flip the C coin.

To draw $C \sim \text{Bern}(p)$, draw $U \sim \text{Unif}([0,1])$, and let $C = \mathbb{1}(U \leq p)$. Now suppose that for all n, $a_n \leq p \leq b_n$. Then it is not necessary to know p exactly in order to determine if $U \leq p$.

To be precise, suppose that $a_1 \leq a_2 \leq a_3 \leq \cdots$ and $b_1 \geq b_2 \geq b_3 \geq \cdots$ have $\lim a_i = \lim b_i = p$. Then draw $U \sim \text{Unif}([0,1])$. If $U \leq \lim a_i = p$, then $C = 1$, if

$U > \lim b_i = p$, then $C = 0$. Fortunately, it is not necessary to compute the entire infinite sequence of $\{a_i\}$ and $\{b_i\}$ in order to determine where U falls with respect to the p. If $U \leq a_n$ for some n, then certainly $U \leq p$. On the other hand, if $U > b_n$, then definitely $U > p$, and the algorithm can stop.

`Bounded_mean_Bernoulli`	*Input:* p, $\{a_n\}$, $\{b_n\}$	*Output:* $C \sim \text{Bern}(p)$

1) $n \leftarrow 1, U \leftarrow \text{Unif}([0,1])$
2) While $U \in (a_n, b_n]$ do
3) $\quad n \leftarrow n+1$
4) $C \leftarrow \mathbb{1}(U \leq a_n)$

While the algorithm only needs to generate one uniform random variable U, it might take quite a few floating point comparisons to find C.

Lemma 2.6. *The expected value of n is $1 + \sum_{i=1}^{\infty}(b_i - a_i)$.*

Proof. The value of $n = 1$ initially. For fixed i, the probability that $i > n$ is just $\mathbb{P}(U \in (a_i, b_i]) = b_i - a_i$. The result then follows from the tail sum formula for expectation of nonnegative integer valued random variables. □

Therefore, in order for this approach to be efficient, it is necessary to have bounds $\{a_i\}$ and $\{b_i\}$ that approach each other rapidly.

2.6.1 *Example: First passage time of Brownian motion*

Standard Brownian motion $\{B_t\}_{t \in \mathbb{R}}$ is a stochastic process that starts at 0 (so $X_0 = 0$), is continuous with probability 1, has independent increments (so for any $a < b < c < d$, $B_d - B_c$ and $B_b - B_a$ are independent), and normal increments (so for any $a < b$, $B_b - B_a \sim \text{N}(0, b-a)$). See Chapter 10 for a formal definition and a more detailed description of the properties of Brownian motion.

Because the map $t \mapsto B_t$ is continuous with probability 1, it makes sense to define stopping times such as $T_1 = \inf\{t : B_t \in \{-1, 1\}\}$. Stopping times of this form are known as a *first passage time*. Simulation of T_1 arises in mathematical finance applications when pricing barrier-options.

Lemma 2.7. *The density of T_1 is given by the following alternating infinite series*

$$g_{T_1}(t) = \frac{1}{\sqrt{2\pi t^3}} e^{-1/(2t)} + \sum_{j=1}^{\infty} \frac{(-1)^j}{\sqrt{2\pi t^3}} c_j$$

$$c_j = (2j+1)\exp\left(-\frac{(2j+1)^2}{2t}\right) - (2j-1)\exp\left(-\frac{(2j-1)^2}{2t}\right).$$

Burq and Jones [19] showed that

$$g_{T_1}(t)/[c g_{\lambda,b}(t)] \leq 1 \tag{2.5}$$

for a gamma density $g_{\lambda,b}(x) = \lambda^b x^{b-1} e^{-\lambda x}/\Gamma(b)$ for $c = 1.25$, $b = 1.09$, $\lambda = 1.24$.

For $t \in [0,1]$, let $a_1(t) = (2\pi t^3)^{-1/2}\exp(-1/(2t))$. For $i \geq 1$, let $b_i(t) = a_i(t) +$

$(2\pi t^3)^{-1/2}(2i-1)\exp(-(2i-1)^2/(2t))$ and $a_{i+1}(t) = b_i(t) - (2\pi t^3)^{-1/2}(2i+1)\exp(-(2j+1)^2/(2t))$.

Since $\max_{x\geq 0} x\exp(-x^2/(2t))$ occurs at $x = \sqrt{t}$ and the function is strictly decreasing thereafter, for $2i-1 > \sqrt{t}$ the sequences a_i and b_i become monotone and oscillate around the target value. Moreover, the difference $b_i - a_i$ goes down faster than exponentially, therefore only a constant number of steps (on average) are necessary to generate one draw from the distribution using `Bounded_mean_Bernoulli`.

2.7 Simulation using Markov and Chernoff inequalities

In full generality, any upper bound on the size or number of a set of objects is equivalent to some AR-style algorithm.

As an example of this, Bucklew [18], showed how Markov's inequality and Chernoff inequalities [23] could actually be viewed as AR algorithms.

2.7.1 *Markov's inequality as an AR algorithm*

Begin with Markov's inequality.

Lemma 2.8 (Markov's Inequality). *Consider a random variable X that is nonnegative with probability 1 and has finite expectation $\mathbb{E}[X]$. Then for all $a > 0$,*

$$\mathbb{P}(X \geq a) \leq \mathbb{E}[X]/a. \tag{2.6}$$

From a simulation perspective, Markov's inequality allows us to sample from $[X|X \geq a]$. In other words, if X has density $f_X(x)$, then the goal is to sample from the unnormalized density $f_X(x)\mathbb{1}(x > a)$. To use AR, use the unnormalized density $xf_X(x)$. Then $f_X(x)\mathbb{1}(x \geq a)/[xf_X(x)]$ is 0 when $x < a$, and $1/x$ when $x \geq a$, and so is never greater than $1/a$. This gives rise to the following algorithm.

`AR_Markovs_Inequality`	*Input:* density f, a	*Output:* $X \sim f$ given $X > a$

1) Repeat
2) Draw X from unnormalized density $xf(x)$
3) If $X \geq a$, draw $C \sim \text{Bern}(a\mathbb{1}(X \geq a)/x)$
4) Until $C = 1$

Note the normalizing constant for $xf(x)$ is just $\int xf(x)\nu(dx) = \mathbb{E}[X]$. So the chance of accepting is just

$$\int_{x\geq 0} \frac{xf_X(x)}{\mathbb{E}[X]} \frac{a\mathbb{1}(X > a)}{x} \nu(dx) = \int_{x>a} \frac{af_X(x)}{\mathbb{E}[X]} \nu(dx) = \frac{a\mathbb{P}(X > a)}{\mathbb{E}[X]}. \tag{2.7}$$

Since this chance of accepting must be at most 1, this fraction is at most 1, which proves Markov's inequality.

Using the ideas of Section 2.4, this can also then be used to build an (ε, δ)-ras $\mathbb{P}(X > a)$ in $O([\mathbb{E}[X]/\mathbb{P}(X > a)]\varepsilon^{-2}\ln(\delta^{-1}))$ samples.

2.7.2 *Chernoff bounds as an AR algorithm*

Chernoff bounds [23] give upper limits on the probability that a sum of random variables is greater (or smaller) than a set value. Let $S_n = X_1 + \cdots + X_n$, where $X_1, \ldots, X_n$ are iid; then the analog sampling goal to generate a draw from S_n conditioned on $S_n \geq a$ or $S_n \leq a$.

Markov's inequality can also be applied to the sum of random variables, but the bound only goes down as the inverse of n. Chernoff bounds, on the other hand, tend to decrease exponentially in n, thereby giving much closer matches to the actual behavior of the tails of S_n and a much faster way of generating samples from the tails.

Chernoff bounds can be viewed as an application of Markov's inequality to the moment-generating function of X.

Lemma 2.9 (Chernoff bounds). *Consider a random variable X where $\mathbb{E}[e^{tX}]$ is finite for $t \in [a,b]$. Then for $a \in \mathbb{R}$,*

$$\mathbb{P}(X \geq a) \leq \text{mgf}_X(t)/\exp(ta) \textit{ for all } t \in [0,b], \tag{2.8}$$

$$\mathbb{P}(X \leq a) \leq \text{mgf}_X(t)/\exp(ta) \textit{ for all } t \in [a,0]. \tag{2.9}$$

To get the best possible bound on the right-hand side, minimize over the allowed values of t.

Proof. First $\mathbb{P}(X \geq a) = \mathbb{P}(tX \geq a)$ for any positive t. Next $\mathbb{P}(tX \geq ta) = \mathbb{P}(\exp(tX) \geq \exp(ta))$. Finally, apply Markov's inequality to get $\mathbb{P}(\exp(tX \geq \exp(ta))) \leq \mathbb{E}[\exp(tX)]/\exp(ta)$. The result for $\mathbb{P}(X \leq a)$ is shown similarly. □

Now consider what happens when Chernoff bounds are used on a sum of independent random variables. First a fact about moment-generating functions.

Lemma 2.10. *For $S_n = X_1 + \cdots + X_n$, where the $\{X_i\}$ are iid with finite moment-generating functions, $\mathbb{E}[\exp(tS_n)] = [\mathbb{E}[\exp(tX_i)]]^n$.*

Applying this to Chernoff bounds immediately yields the following.

Lemma 2.11. *For $S_n = X_1 + \cdots + X_n$, where the $\{X_i\}$ are iid with $\text{mgf}_{X_i}(t)$ is finite for $t \in [a,b]$,*

$$\mathbb{P}(S_n \geq \alpha n) \leq \left(\frac{\text{mgf}_{X_i}(t)}{\exp(t\alpha)}\right)^n \textit{ for all } t \in [0,b], \tag{2.10}$$

$$\mathbb{P}(S_n \leq \alpha n) \leq \left(\frac{\text{mgf}_{X_i}(t)}{\exp(t\alpha)}\right)^n \textit{ for all } t \in [a,0]. \tag{2.11}$$

Say that $X_i \sim f_X$. The goal is to use the moment-generating function idea that led to the Chernoff bound to obtain a better AR-type algorithm. To accomplish this goal, it must be possible to draw from the *twisted* density $g_t(x) \propto e^{tx} f_X(x)$. When t is large, this density pushes the probability in density g_t out to larger values. When t is small and negative, this pushes the probability in g_t to more negative values.

Let $t > 0$. For $x \geq a$, then $g_t(x) = e^{tx} f_X(x) \geq e^{ta} f_X(x)$. Therefore a draw from g_t

can be accepted as a draw from f_X with probability e^{ta}/e^{tx}. If the probability that x is very much larger than a is small, then this acceptance probability will be very close to 1. A similar method applies when $t < 0$.

`AR_Chernoffs_bound` *Input:* density f_X, a, t
Output: $S_n = X_1 + \cdots + X_n$ given $S_n \geq a$ (when $t > 0$) or $S_n \leq a$ (when $t < 0$)

---

1) If $t > 0$, then $A \leftarrow [a, \infty)$, else $A \leftarrow (-\infty, a]$
2) Repeat
3) Draw $X_1', \ldots, X_n'$ iid from unnormalized density $e^{tx} f(x)$
4) $S_n \leftarrow X_1' + \cdots + X_n'$
5) Draw $C \leftarrow \mathsf{Bern}(\exp(t(a - S_n))\mathbb{1}(S_n \in A))$
6) Until $C = 1$

Fact 2.5. *Suppose* $\text{mgf}_{X_i}(t)\exp(-ta/n) < 1$. *Then* `AR_Chernoff_bound` *generates* S_n *from* $[S_n | S_n > a]$ *when* $t > 0$ *or from* $[S_n | S_n < a]$ *when* $t < 0$.

Proof. Consider any vector $(x_1, \ldots, x_n)$ so that $\sum x_i = s$. Then consider $X_1, \ldots, X_n$ iid with density $f_X(x)$, and $X_1', \ldots, X_n'$ iid with density $\exp(tx) f_X(x)$. Then the density of $S_n' = X_1' + \cdots + X_n'$ will just be the density of $S_n = X_1 + \cdots + X_n$ with an extra factor of $\exp(tx_1)\exp(tx_2)\cdots\exp(tx_n) = \exp(ts)$.

Hence AR has generation density $g(s) = \exp(ts) f_{S_n}(s)$ and the target density is $f(s) = f_{S_n}(s)\mathbb{1}(s \in A)$. Therefore $f(s)/g(s)$ is either 0 (when $s \notin A$), or $\exp(-ts)$ when $s \in A$.

When $A = [a, \infty)$, then $t > 0$ and $\exp(-ts) \leq \exp(-ta)$. Similarly, when $A = (-\infty, a]$, then $t < 0$ and $\exp(-ts) \leq \exp(-ta)$. Hence $c = \exp(-ta)$ works, and $f(s)/[cg(s)] = \exp(ta)\exp(-ts)\mathbb{1}(s \in A)$, as given in the algorithm. □

2.7.3 *Example: generating from the tails of a binomial random variable*

Suppose X is a binomial random variable that is the number of successes on n independent Bernoulli trials, each of which is a success with probability p. Write $X \sim \mathsf{Bin}(n, p)$. Then $\mathbb{E}[X] = np$. For $a > np$, consider drawing samples from $[X | X \geq a]$.

The simplest way to draw a binomial is to draw $B_1, \ldots, B_n$ iid $\mathsf{Bern}(p)$, and then let $X = B_1 + \cdots + B_n$. Therefore this method works very well with the Chernoff bound approach.

The density of a Bernoulli is p at 1 and $1 - p$ at 0. Therefore, the mgf of $B \sim \mathsf{Bern}(p)$ is $mgf_B(t) = e^t p + (1 - p)$. Hence

$$\mathbb{P}(X \geq \alpha n) \leq \left(\frac{e^t p + (1 - p)}{\exp(t\alpha)} \right)^n. \tag{2.12}$$

The value of t that minimizes the right-hand side (giving the best bound) occurs when $e^t = \alpha(1 - p)/[p(1 - \alpha)]$.

Hence the density of the B_i' should be $e^t p = (1 - p)\alpha/(1 - \alpha)$ at 1 and $1 - p$ at 0. Normalizing, this gives $\mathbb{P}(B_i' = 1) = \alpha$ and $\mathbb{P}(B_i' = 0) = 1 - \alpha$.

For drawing $[X | X \geq a]$, $a = \alpha n$ so $\alpha = a/n$, which gives the following algorithm.

```
AR_Chernoffs_bound_binomial
```
Input: n,p,a
Output: $[X|X \geq a]$, where $X \sim \mathsf{Bin}(n,p)$

1) $\gamma \leftarrow (a/n)(1-p)/[p(1-(a/n))]$
2) Repeat
3) Draw $B'_1,\ldots,B'_n$ iid $\mathsf{Bern}(a/n)$
4) $X \leftarrow B'_1 + \cdots + B'_n$
5) Draw $C \leftarrow \mathsf{Bern}(\gamma^{a-X}\mathbb{1}(X \geq a))$
6) Until $C = 1$

For example, when $n = 10$ and $p = 0.1$, consider drawing $[X|X \geq 5]$. The basic AR method of drawing X until $X \geq 5$ requires on average over 6807.2 times through the repeat loop to generate one sample. The `AR_Chernoffs_bound_binomial` routine requires on average at most 3.7 times through the repeat loop to generate one conditional sample from X.

2.8 Where AR fails

So why does our journey through perfect simulation not end here with the AR method? Because for many problems of interest, the chance of acceptance goes down exponentially with the size of the problem input.

Consider again the problem of generating a linear extension of a permutation by acceptance/rejection. Suppose there are an even number n elements, and in the partial order $1 \preceq 2, 3 \preceq 4, \ldots, n-1 \preceq n$.

A permutation can be drawn uniformly at random by inserting item 1 into one location, then choosing 2 to be either before or after 1, then 3 to be in any of three positions, and so on. Then the chance that the partial order is respected by this random permutation is only $(1/2)^{n/2}$.

Or consider the Ising model with no magnetic field from earlier. The unnormalized density of a configuration has a minimum of 1 and a maximum of $\exp(\beta \# E)$. Therefore an acceptance/rejection-style algorithm draws a configuration uniformly from $\{-1,1\}^V$, and then accepts with probability $\exp(-\beta H(x))/\exp(\beta \# E)$, where $H(x) = -\sum_{\{i,j\} \in E} \mathbb{1}(x(i) = x(j))$. If the graph consists of an even number of nodes where the degree of each node is exactly 1, then each edge has $\mathbb{1}(x(i) = x(j)) = 1$ with probability $1/2$ independently of the others. For $\beta > 0$ the chance that $H(x) > (3/4)\# E$ declines exponentially in $\# E$, which in turn means with high probability there is only a $\exp(-\beta \# E/4)$ chance of accepting the result.

It turns out that a more advanced form of AR for the Ising model exists, which does only require a polynomial number of draws on average for a nontrivial set of β. This method is discussed in Chapter 9.

Chapter 3

Coupling from the Past

Life can only be understood backwards; but it must be lived forwards

Søren Kierkegaard

Coupling from the past (CFTP) allowed the creation of perfect simulation algorithms for problems where only Markov chain methods existed before. The easiest way to use this approach is to take advantage of monotonicity in the update function for a Markov chain. However, even for nonmonotonic problems, the use of bounding chains can combine with CFTP to obtain perfect samples. Basic CFTP does have drawbacks, though, the two biggest being noninterruptibility and the fact that the random choices made by the algorithm must be used twice.

3.1 What is a coupling?

Coupling from the past, hereafter referred to as CFTP, is a method that allows ideas from Markov chain theory to be used to create perfect simulation algorithms. When introduced by Propp and Wilson [112] in the mid 1990s, it greatly extended the applicability of perfect simulation, and made efficient simulation of problems such as the Ising model feasible for the first time.

The idea is deceptively simple, and begins with the notion of an update function.

Definition 3.1. *Say that* $\phi : \Omega \times [0,1] \to \Omega$ *is an* update function *for a Markov chain* $\{X_t\}$ *if for* $U \sim \textsf{Unif}([0,1])$, $[X_{t+1}|X_t] \sim \phi(X_t, U)$.

The function ϕ is deterministic—all of the randomness is contained in the value of U. Any Markov chain that can be simulated on a computer is an example of an update function, so creating an update function representation of a Markov chain is done every time a Markov chain is actually simulated. Update functions for a given transition kernel are not unique. In fact, a given Markov chain usually has an infinite number of different update functions that result in the same transition kernel.

Given an iid sequence $U_0, U_1, U_2, \ldots$, the entire process can be simulated by letting $X_1 = \phi(x_0, U_0)$, and for $i > 1$, $X_i = \phi(X_{i-1}, U_{i-1})$. Let U denote the entire sequence, then let $\phi_t(x_0, U) = \phi(\phi(\phi(\cdots(\phi(x_0, U_0), U_1), \ldots, U_{t-1}))$ be the state after t steps in the chain starting from x_0.

Suppose x_0 and y_0 are any two states in Ω. Let $X_t = \phi_t(x_0, U)$ and $Y_t = \phi_t(y_0, U)$

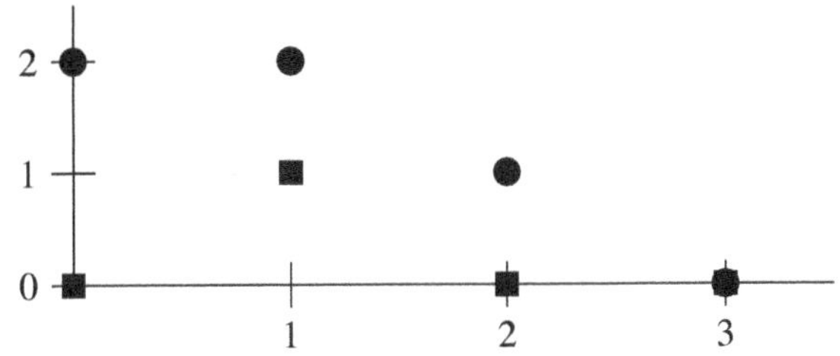

Figure 3.1 *A completely coupled set of processes. X (represented by squares) and Y (represented by circles) are simple symmetric random walks with partially reflecting boundaries on* $\Omega = \{0,1,2\}$. *They have coupled at time* $t = 3$.

(using the same values of U). Then $\{X_t\}$ and $\{Y_t\}$ are both instances of the Markov chain, just with (possibly) different starting points x_0 and y_0.

A Markov chain can be created for every possible starting state $x \in \Omega$ using the U values. A family of Markov chains that share a common kernel is known as a *coupling*.

Note that with this coupling, if $X_t = Y_t = x$ for some $t \geq 0$, then $X_{t'} = Y_{t'}$ for all $t' \geq t$. Just as two train cars that are coupled together stay together, two processes at the same state remain together no matter how much time passes.

An update function and the use of common U_t values gives a coupling that applies to the family of Markov chains started at every possible state in the set Ω, so this is also sometimes called a *complete coupling* [51].

Definition 3.2. *Let* $\mathscr{S}$ *be a collection of stochastic processes defined over a common index set* $\mathscr{I}$ *and common outcome space* Ω. *If there is an index* $i \in \mathscr{I}$ *and state* $x \in \Omega$ *such that for all* $S \in \mathscr{S}$, $S_i = x$, *then say that the stochastic processes have* coupled *or* coalesced.

Example 3.1. *Consider the simple symmetric random walk with partially reflecting boundaries on* $\Omega = \{0,1,2\}$. *This Markov chain takes a step to the right (or left) with probability 1/2, unless the move would take the state outside of* Ω. *An update function for this chain is*

$$\phi(x,U) = x + \mathbb{1}(x < 2, U > 1/2) - \mathbb{1}(x > 0, U \leq 1/2). \tag{3.1}$$

Figure 3.1 is a picture of two Markov chains, $\{X_t\}$ and $\{Y_t\}$, where $X_0 = 0$ and $Y_0 = 2$. The first random choice is $U_1 = 0.64\ldots$, so both chains attempt to increase by 1. The next random choice is $U_2 = 0.234\ldots$, so both chains attempt to decrease by 1. The third random choice is $U_3 = 0.100\ldots$, so again both chains attempt to decrease by one. At this point, both $X_3 = Y_3 = 0$. So the two chains have coupled at time 3.

In fact, we could go further, and consider a chain started in state 1 at time 0. The first move takes this chain up to state 2, then the second move brings it down to 1, and then the last move takes it to state 3. No matter where the chain started, after three moves the state of the chain will be state 0.

3.2 From the past

Propp and Wilson started with this notion of coupling through use of an update function, and showed how this could be used to obtain perfect samples from the target distribution.

To see how this works, consider a stationary state X. Then for a fixed t, let $Y = \phi_t(X,U)$. Then Y is also stationary, because ϕ_t is the composition of t stationary updates.

The output of CFTP will always be Y. Let $W = (U_1, U_2, \ldots, U_t)$ be uniform over $[0,1]^t$. Consider a measurable set $A \subset [0,1]^t$. Then either $W \in A$ or $W \notin A$. So

$$Y = \phi_t(X,W)\mathbb{1}(W \in A) + \phi_t(X,W)\mathbb{1}(W \notin A). \tag{3.2}$$

Suppose we could find a set A, such that when $W \in A$, $\phi(X,W)$ did not depend on X. That is to say, whenever $W \in A$, it holds that $\phi(x,W) = \phi(x',W)$ for all $x, x' \in \Omega$. Such a Markov chain has completely forgotten its starting state through its random moves. If this happens, then there is no need to know the value of X in order to find the value of Y!

Recall the example of the simple symmetric random walk with partially reflecting boundaries on $\{0,1,2\}$ illustrated in Figure 3.1. In that example

$$\phi_3(0,W) = \phi_3(1,W) = \phi_3(2,W) = 0. \tag{3.3}$$

So $\phi_3(\{0,1,2\},W) = \{0\}$.

So what is a valid choice of A here? Suppose that the first three steps were (up, down, down). This sequence of moves corresponds to the set of values $A_1 = (1/2,1] \times [0,1/2] \times [0,1/2]$. If $W \in A_1$, then $\phi_3(\{0,1,2\},W) = \{0\}$.

Another set of moves that moves all starting states together is (down, down, down). Here $A_2 = [0,1/2] \times [0,1/2] \times [0,1/2]$. Of course, if A_1 and A_2 are both valid, so is $A_3 = A_1 \cup A_2$. So $\phi_3(\{0,1,2\},W) = \{0\}$ for all $W \in A_3$. The point is that the set A does not have to be exactly the set of moves that coalesce down to a single point. However, the bigger A is, the more likely it will be that $W \in A$.

Back to Equation (3.2). In the first term, the value of X does not matter, so $\phi_t(X,W)\mathbb{1}(W \notin A) = \phi_t(x_0,W)\mathbb{1}(W \in A)$ where x_0 is an arbitrary element of the state space. That is,

$$Y = \phi_t(x_0,W)\mathbb{1}(W \in A) + \phi_t(X,W)\mathbb{1}(W \notin A). \tag{3.4}$$

So when $W \in A$, there is no need to figure out what X is, in order to compute Y. Okay, that is all well and good, but what happens when $W \notin A$?

Then it is necessary to evaluate $\phi(X,W)$ to obtain Y. The central idea behind CFTP is to find $X \sim \pi$ by recursively calling CFTP (that is the "from the past" part of the algorithm), and then evaluate $Y = \phi(X,W)$ as before.

To see how this works in practice, it helps to alter our notation somewhat by adding a time index. Let $Y_0 = Y$, $W_0 = W$, and $Y_{-1} = X$. With this notation,

$$Y_0 = \phi_t(x_0,W_0)\mathbb{1}(W_0 \in A) + \phi_t(Y_{-1},W_0)\mathbb{1}(W_0 \notin A). \tag{3.5}$$

So if $W_0 \in A$, we are done. Otherwise, it is necessary to draw Y_{-1}. But this is done in exactly the same way as drawing Y_0:

$$Y_{-1} = \phi_t(x_0, W_{-1})\mathbb{1}(W_{-1} \in A) + \phi_t(Y_{-2}, W_{-1})\mathbb{1}(W_{-1} \notin A), \tag{3.6}$$

where $W_{-1} \sim \text{Unif}([0,1]^t)$ and is independent of W_0.

In general,

$$Y_{-i} = \phi_t(x_0, W_{-i})\mathbb{1}(W_{-i} \in A) + \phi_t(Y_{-i-1}, W_{-i})\mathbb{1}(W_{-i} \notin A), \tag{3.7}$$

This can be taken back as far as necessary. The end result is an infinite sequence representation of a stationary state.

Theorem 3.1 (Coupling from the past). *Suppose that ϕ is an update function for a Markov chain over Ω such that for $U = (U_0, \ldots, U_{t-1}) \sim \text{Unif}([0,1]^t)$ the following holds.*

- *For $Y \sim \pi$, $\phi(Y, U_0) \sim \pi$.*
- *There exists a set $A \subseteq [0,1]^t$ such that $\mathbb{P}(U \in A) > 0$ and $\phi_t(\Omega, A) = \{x\}$ for some state $x \in \Omega$.*

Then for all $x_0 \in \Omega$,

$$\begin{aligned} Y_0 = {} & \phi_t(x_0, U_0)\mathbb{1}(U_0 \in A) + \phi_t(\phi_t(x_0, U_{-1}), U_0)\mathbb{1}(U_{-1} \in A) + \\ & \phi_t(\phi_t(\phi_t(x_0, U_{-2}), U_{-1}), U_0)\mathbb{1}(U_{-2} \in A) + \cdots \end{aligned}$$

has $Y_0 \sim \pi$.

Proof. Fix $x_0 \in \Omega$. Then the result follows from the Fundamental Theorem of perfect simulation (Theorem 1.1) using $g(U) = \phi_t(x_0, U)$, $b(U) = \mathbb{1}(U \in A)$, and $f(X, U) = \phi_t(X, U)$. □

The key insight of Propp and Wilson was that it is not necessary to evaluate every term in the sequence to find Y_0. We only need $U_0, \ldots, U_{-T}$, where $U_{-T} \in A$. As long as $\mathbb{P}(U \in A) > 0$, T will be a geometric random variable with mean $1/\mathbb{P}(U \in A)$. The following code accomplishes this task.

`Coupling_from_the_past` *Output:* $Y \sim \pi$

1) Draw $U \leftarrow \text{Unif}([0,1]^t)$
2) If $U \in A$
3) Return $\phi_t(x_0, U)$ (where x_0 is an arbitrary element of Ω)
4) Else
5) Draw $X \leftarrow$ `Coupling_from_the_past`
6) Return $\phi_t(X, U)$

This is a recursive algorithm; the algorithm is allowed to call itself in line 5. When the algorithm recursively calls itself, it does not pass the values of U as a parameter. The new call to `Coupling_from_the_past` generates in line 1 its own value of U that is indpendent of the choice of U in the calling function. This is important to do, as otherwise the Fundamental Theorem of Perfect Simulation would not apply.

Table 3.1 *For simple symmetric random walk on* $\{0,1,2\}$, *this lists the possible outcomes and their probabilities for taking two steps in the chain given* $U \notin A$.

state X	Moves		
	up-up	up-down	down-up
0	2	0	1
1	2	1	1
2	2	1	2

Example 3.2. *Use CFTP to draw uniformly from* $\{0,1,2\}$ *by using the Markov chain with update function*

$$\phi(x,U) = x + \mathbb{1}(x<2, U>1/2) - \mathbb{1}(x>0, U \leq 1/2),$$

$t=2$, *and* $A = [0,1/2] \times [0,1/2]$.

As noted earlier, $\phi_2(\{0,1,2\},U) = \{0\}$ whenever $U \in A$. So to use CFTP, first draw U. If $U \in A$, output $\{0\}$ and quit. Otherwise recursively call the algorithm in order to find $X \sim \pi$. Then output $\phi(X,U)$. Note that because this second step was only executed when $U \notin A$, the output of $\phi(X,U)$ in this second step is not stationary, but instead is the distribution of $\phi(X,U)$ conditioned on $U \notin A$.

The theory guarantees that the output of this procedure has the correct distribution, but it can also be verified directly in this example.

What is the probability that the output of the algorithm is 0? The event $U \in A$ occurs with probability $(1/2)(1/2) = (1/4)$, so there is a 1/4 chance that the state is 0 after the first step, and a 3/4 chance that the algorithm continues to the next step. During this next step, the state X is a draw from the stationary distribution, and so has a one-third chance of being either 0, 1, or 2. Conditioned on $U \notin A$, the uniforms could either result in the chain moving up-up, up-down, or down-up. Therefore there are nine possibilities to consider, the outcomes of which are collected in Table 3.1. For instance, if the state starts at 0, then the chain moves down-up, then moving down leaves the state at 0 and then moving up brings the chain to 1. So the final state is 1.

Each of these outcomes occur with probability $1/9$. There is a $1/3$ chance for each starting state and (conditioned on $U \notin A$) a $1/3$ chance for each pair of moves. Therefore, the chance of ending in state 0 is

$$\frac{1}{4} + \frac{3}{4} \cdot \frac{1}{9} = \frac{4}{12} = \frac{1}{3}. \tag{3.8}$$

The chance the algorithm ends in state 1 or 2 is

$$0 + \frac{3}{4} \cdot \frac{4}{9} = \frac{1}{3}. \tag{3.9}$$

So the output does have the correct distribution! Now consider the running time, as measured by the number of evaluations of the update function ϕ that is needed.

Lemma 3.1. *The number of calls to ϕ needed by* `Coupling_from_the_past` *is at most* $t[2\mathbb{P}(U \in A)^{-1} - 1]$.

Proof. In the call to the algorithm, if $U \notin A$, then the recursive call in line 5 calls `Coupling_from_the_past` again. Each of these recursive calls is independent of the time before, so the number of calls is a geometric random variable with parameter $\mathbb{P}(U \in A)$. That makes the mean number of calls $\mathbb{P}(U \in A)^{-1}$.

Line 3 is a call to ϕ_t, which requires t calls to ϕ. If line 6 is executed, then a second call to ϕ_t is needed. Therefore, the expected number of calls to ϕ in any level of the algorithm is $t[1 + \mathbb{P}(U \notin A)] = t[2 - \mathbb{P}(U \in A)]$. Therefore the overall number of expected calls is $t[2\mathbb{P}(U \in A)^{-1} - 1]$. □

To keep the running time small, it is important that t be large enough that $\mathbb{P}(U \in A)$ is reasonably sized, but once that is achieved t should be kept as small as possible. It is not always easy to determine what such a t should be. Fortunately, there is a way around this problem.

3.2.1 *Varying the block size*

In building CFTP, a single step update ϕ was composed with itself t times. Call this set of t steps a *block*.

Suppose there is a threshold value t' such that $\mathbb{P}(U \in A) = 0$ if $t < t'$. In practice, it is usually very difficult to compute t' explicitly; therefore, it is common to use blocks of increasing size to make sure that eventually t is greater than the threshold.

In general, there is no reason why the same number of steps needs to be used for each block. The statement and proof of Theorem 1.1 easily adapts to the more general situation where the functions b, g, and f change with each level of recursion.

Theorem 3.2 (Fundamental Theorem of perfect simulation (2nd form)). *Suppose for $U_1, U_2, \ldots$ independent and identically distributed, there are sequences of computable functions $\{b_t\}$, $\{g_t\}$, and $\{f_t\}$ such that each b_t has range $\{0,1\}$ and $\prod_{t=1}^{\infty} \mathbb{P}(b_t(U) = 0) = 0$. Let X be a random variable such that for all t:*

$$X \sim b_t(U)g_t(U) + (1 - b_t(U))f_t(X,U). \tag{3.10}$$

Let $T = \inf\{t : b_t(U_t) = 1\}$. Then

$$Y = f_0(\cdots f_{T-2}(f_{T-1}(g_T(U_T),U_T),U_{T-1}),\ldots,U_1) \tag{3.11}$$

has the same distribution as X and $\mathbb{E}[T] = 1/\mathbb{P}(b(U) = 1)$.

Proof. The proof is nearly identical to that of Theorem 1.1. □

Using this for general version of the fundamental theorem for CFTP gives the following result.

Theorem 3.3. *Let $t(0) < t(1) < t(2) < \cdots$ be positive integers. Let $W_0, W_1, \ldots$ be independent with $W_i \sim$ Unif$([0,1]^{t(i)})$, and $A_{t(i)}$ a measurable subset of $[0,1]^{t(i)}$ such that $\phi_{t(i)}(\Omega, W_i) = \{y\}$. Suppose $\prod_{i=1}^{\infty} \mathbb{P}(W_i \in A) = 0$. Then for any $x_0 \in \Omega$,*

$$Y_0 = \phi_{t(0)}(x_0, W_0)\mathbb{1}(W_0 \in A) + \phi_{t(1)}(\phi_t(x_0, W_{-1}), W_0)\mathbb{1}(W_{-1} \in A) + \phi_{t(2)}(\phi_t(\phi_t(x_0, W_{-2}), W_{-1}), W_0)\mathbb{1}(W_{-2} \in A) + \cdots$$

has $Y_0 \sim \pi$.

A common choice is to make $t(i) = 2t(i-1)$, so that the number of steps taken doubles at each step. If the initial $t = t(0)$ is 1, then doubling each time reaches t' after only a logarithmic number of recursions. Moreover, the total work done in all of the blocks will be about $1 + 2 + 4 + \cdots + t' = 2t' - 1$.

```
Doubling_Coupling_from_the_past    Input: t    Output: Y

Draw U ← Unif([0,1]^t)
If U ∈ A_t
    Return φ_t(·,U)
Else
    Draw X ← Doubling_Coupling_from_the_past(2t)
    Return φ_t(X,U)
```

It should be noted, however, that this doubling is only a rule of thumb. One could use $t(i) = (3/2)^i$, or 4^i, or just i and the algorithm will still return a perfect sample, no matter how big t' is. Only the number of steps (on average) taken by the algorithm could change.

For a more sophisticated block size method designed to give greater concentration of the running time around the mean, see Section 11.6.

3.2.2 Recursion as history

There is another way of viewing CFTP that can make more intuitive sense, especially to those already familiar with Markov chain Monte Carlo. Suppose that a Markov chain is run forward in time yielding $X_0, X_1, X_2, \ldots$. Then as the number of steps grows towards infinity, Theorem 1.3 tells us that the state will converge towards stationarity.

Now suppose that the Markov chain steps are not just nonnegative integers, but all integers. So the process is $\ldots, X_{-2}, X_{-1}, X_0, X_1, X_2, \ldots$ and not only runs infinitely forward in time, but also infinitely backwards in time as well. Then by time 0, intuitively the process has *already* been running for an "infinite" number of steps. Therefore X_0 should already be stationary.

This intuitive idea can be made precise using the idea of the reverse process for a Markov chain (see Section 5.2.1,) but for now let us just follow where this intuition leads.

Suppose CFTP is run with $t = 1$. Then ϕ_t can be viewed as the step from X_{-1} to X_0. If coalescence occurs, the algorithm quits and returns X_0. Otherwise, recursion occurs, and t is doubled to 2. These two steps can be viewed as the step from X_{-3} to X_{-2} and the step from X_{-2} to X_{-1}

If coalescence occurs, then one more step in the Markov chain is taken, and recall this was the step from X_{-1} up to X_0. The two blocks together have moved the chain from step X_{-3} up to step X_0.

If coalescence does not occur. then now $t = 4$, and steps $X_{-6} \to X_{-5} \to X_{-4} \to X_{-3}$ are taken. Then as the recursion unpacks, the two steps from $X_{-3} \to X_{-2} \to X_{-1}$ are taken, finally followed by the step $X_{-1} \to X_0$.

No matter how far back the recursion goes, the final output of the procedure (from this point of view) is $X_0 \sim \pi$. This way of viewing CFTP becomes very important for designing perfect simulation algorithms for spatial point processes (see Chapter 7.)

3.3 Monotonic CFTP

To use CFTP requires an update function that is stationary with respect to π, and a set A such that for $U \in A$, $\phi(\Omega, U) = \{x\}$. The stationary requirement is is usually easy, as it can be done by constructing a Markov chain whose stationary distribution is π. The hard part is creating the sets $A_{t(i)}$ together with a method for determining if $U \in A_{t(i)}$.

It is this difficulty that prevents CFTP from being used for every Monte Carlo application. In the remainder of this chapter, different methods for finding such $A_{t(i)}$ will be discussed, together with applications. The first method considered is the use of a monotonic update function.

Monotonicity was used by Johnson [74] to detect the speed of convergence for the Gibbs chain for the Ising model, although the term monotonicity was not used there. Propp and Wilson [112] saw that this same idea could be used as the basis of a CFTP-type algorithm for perfect simulation.

The idea is as follows. Suppose that the state space Ω admits a partial order. Recall Definition 1.24, which states that the binary relation $\preceq$ is a partial order if it is reflexive (so $s \preceq s$), antisymmetric (so $s \preceq t$ and $t \preceq s$ gives $s = t$), and transitive (so $s \preceq t$ and $t \preceq r$ gives $s \preceq r$.) But unlike a total order (such as $\leq$ over $\mathbb{R}$), there could be elements s and t for which neither $s \preceq t$ or $t \preceq s$ holds.

As an example, consider two Ising configurations x and y. Say that $x \preceq y$ if for all nodes v, $x(v) \leq y(v)$. So if there are four nodes, $(-1,-1,1,1)$ precedes $(-1,-1,-1,1)$. However, the states $(-1,-1,1,1)$ and $(1,1,-1,-1)$ are incomparable; neither one precedes the other.

Suppose further that Ω has a largest element $x_{\max}$, and a smallest element $x_{\min}$. Continuing the Ising model example, the configuration of all -1 values is the smallest element, and the configuration of all 1 values is the largest element.

Definition 3.3. *Let ϕ be an update function for a single step of a Markov chain. If ϕ has the property that for all $x \preceq y$ and $u \in [0,1]$, it holds that $\phi(x,u) \preceq \phi(y,u)$, then call ϕ* monotonic.

Consider running two Markov chains $\{X_t\}$ and $\{Y_t\}$ where $X_0 = x_0$ and $Y_0 = y_0$ with $x_0 \preceq y_0$. Suppose at each step, $X_{t+1} = \phi(X_t, U_t)$ and $Y_{t+1} = \phi(Y_t, U_t)$, where $U_1, U_2, \ldots$ is an iid stream of $\text{Unif}([0,1])$ random variables.

Then a simple induction gives that $X_{t+1} \preceq Y_{t+1}$ for all values of t. Moreover, suppose $x_0 = x_{\min}$ and $y_0 = x_{\max}$. Then any other state w_0 satisfies $x_0 \preceq w_0 \preceq y_0$, and so building a W_t chain in similar fashion gives that $X_t \preceq W_t \preceq Y_t$ for all t.

Suppose that $X_t = Y_t$. The properties of a partial order imply that $X_t = W_t = Y_t$ as well. The starting state of W_t was arbitrary, so for any starting state $w_0 \in \Omega$, $\phi_t(w_0, u) = X_t = Y_t$.

In other words, if after taking a fixed number of steps, the largest state chain and the smallest state chain (also referred to as the upper and lower chain) have reached the same state, every other state in the chain has been trapped in between them, and also reached the same state. This gives rise to one of the most important CFTP variants, monotonic CFTP.

```
Monotonic_coupling_from_the_past    Input: t    Output: Y

1)  Draw U ← Unif([0,1]^t)
2)  Let X_t ← φ_t(x_max, U), and Y_t ← φ_t(x_min, U).
3)  If X_t = Y_t
3)      Return X_t
4)  Else
5)      Draw X ← Monotonic_coupling_from_the_past(2t)
6)      Return φ_t(X, U)
```

3.3.1 The Ising model

Consider the `Ising_Gibbs` update from Section 1.8.1. In this update, a node v was chosen uniformly at random, as was $U \sim \text{Unif}([0,1])$. Then N_c was the number of neighbors of v labeled c. If $U < \exp(N_1\beta)/[\exp(N_1\beta) + \exp(N_{-1}\beta)]$, then the node was labeled 1, otherwise it was labeled 0.

In order to use this update for perfect simulation, it needs to be written as an update function. That is, the current state and the random choices become input parameters.

```
Ising_Gibbs    Input: x ∈ {−1,1}^V, u ∈ [0,1], v ∈ V    Output: x ∈ {−1,1}^V

1)  Let N_1 be the number of neighbors of v labeled 1 in x
2)  Let N_{-1} be the number of neighbors of v labeled -1 in x
3)  If U < exp(N_1 β)/[exp(N_1 β) + exp(N_{-1} β)]
4)      x(v) ← 1
5)  Else
6)      x(v) ← −1
```

When $u \sim \text{Unif}([0,1])$ and $v \sim \text{Unif}(V)$, this is the random scan Gibbs update from Section 1.8.1.

Fact 3.1. *The* `Ising_Gibbs` *update is monotonic when* $\beta > 0$.

Proof. Let $x \preceq y$. Let v be any node, and $N_c(z)$ denote the number of neighbors of v labeled c in configuration z. Then any neighbor of v in x labeled 1 is also labeled 1 in y, so $N_1(x) \leq N_1(y)$. Similarly $N_{-1}(x) \geq N_{-1}(y)$.

Let $f(n_1,n_{-1}) = \exp(N_1(x)\beta)/[\exp(N_1(x)\beta) + \exp(N_{-1}\beta)]$. Since $\beta > 0$, $f(n_1,n_{-1})$ is increasing in n_1, and decreasing in n_2. Hence if $U < f(N_1(x),N_{-1}(x))$, then certainly $U < f(N_1(y),N_{-1}(y))$. That is, if $x(v)$ is changed to 1, then certainly y is as well.

Similarly, if $U > f(N_1(y),N_{-1}(y))$ so that $y(v)$ has its label changed to 0, then $U > f(N_1(x),N_{-1}(x))$ and $x(v)$ is changed to 0 as well.

Therefore no matter what the choice of v and U is in the update, if $x \preceq y$ then $\phi(x,v,U) \preceq \phi(y,v,U)$. □

This means that monotonic CFTP can be used to generate a random sample example from the stationary distribution.

`Monotonic_Ising_Gibbs` *Input: t* *Output: X*

Draw $(U_1,\ldots,U_t) \leftarrow \text{Unif}([0,1]^t)$, draw $(V_1,\ldots,V_t) \leftarrow \text{Unif}(V^t)$
Let $x_{\max} \leftarrow (1,1,\ldots,1)$, $x_{\min} \leftarrow (0,0,\ldots,0)$
For i from 1 to t
    $x_{\max} \leftarrow$ `Ising_Gibbs`$(x_{\max}, U_i, V_i)$
    $x_{\min} \leftarrow$ `Ising_Gibbs`$(x_{\min}, U_i, V_i)$
If $x_{\max} \neq x_{\min}$
    Draw $x_{\max} \leftarrow$ `Monotonic_Ising_Gibbs`$(2t)$
    For i from 1 to t
        $x_{\max} \leftarrow$ `Ising_Gibbs`$(x_{\max}, U_i, V_i)$
$X \leftarrow x_{\max}$

3.3.2 The hard-core gas model on bipartite graphs

Consider the `HCGM_Gibbs` update from Section 1.8.5. This update function is not monotonic using the simple partial order $x \preceq y \Leftrightarrow (\forall v \in V)(x(v) \leq y(v))$.

Therefore either a new partial order or a new update is needed. The problem is that the chain is repulsive—neighbors in the graph cannot both be given the label 1. However, when the graph is bipartite, it is possible to build a new partial order that is monotonic.

Definition 3.4. *A graph is* bipartite *if the vertex set V can be partitioned into V_1 and V_2 so that for every edge e, one node of e belongs to V_1 and one node belongs to V_2.*

For configurations on a bipartite graph, the partial order is similar to that of the previous section:

$$x \preceq y \Leftrightarrow (\forall v_1 \in V_1)(\forall v_2 \in V_2)(x(v_1) \leq y(v_1) \wedge x(v_2) \geq y(v_2)). \tag{3.12}$$

Fact 3.2. `HCGM_Gibbs` *is monotonic with respect to* (3.12).

Proof. If $x \preceq y$ and node $i \in v_2$ is chosen to update, then if any neighbor w of i has $x(w) = 1$ then $y(w) = 1$ as well. So both $x(i)$ and $y(i)$ will always stay 0. The only way there is a chance that $y(i)$ is changed to 1 is if $y(w) = 0$ for all neighbors w of i. In this case all $x(w) = 0$ as well. $x(i) = y(i) = \mathbb{1}(U < \lambda/(\lambda+1))$.

When $i \in v_1$ the analysis is similar. □

The largest state in this graph has all nodes in v_2 labeled 1 and all nodes in v_1 labeled 0. The smallest state flips all these labellings.

3.3.3 Monotonicity and mixing times

One of the reasons monotonicity is important is that monotonic CFTP cannot take much more time to run than the mixing time of the Markov chain.

Definition 3.5. *A* chain of length ℓ *in a partial order is a subset of* $\ell+1$ *elements where* $x_0 \prec x_1 \prec x_2 \prec x_3 \prec \cdots \prec x_\ell$.

An important fact about total variation distance is that it allows a representation of the two random variables using a third common random variable.

Fact 3.3. *Suppose* $d_{TV}(X,Y) = d$, *where* X *and* Y *respectively have densities* f_X *and* f_Y *with respect to measure* μ *over* Ω. *Then there exist random variables* W_1, W_2, W_3, *and* $B \sim \text{Bern}(d)$ *independent so that*

$$X \sim (1-B)W_1 + BW_2$$
$$Y \sim (1-B)W_1 + BW_3.$$

Proof. Let $g(s) = \min(f_X(s), f_Y(s))$. If $\int_\Omega g\,d\nu = 0$, then f_X and f_Y disagree on a set of measure 0, so $d = 0$ and give W_1 the common distribution of X and Y. (W_2 and W_3 can be assigned any distribution.)

If $\int_\Omega g\,d\nu \neq 0$, give W_1 density $g/\int_\Omega g\,d\nu$, W_2 density $[f_X(s) - g(s)]/\int_\Omega (f_X(r) - g(r))\,d\nu(r)$, and W_3 density $[f_Y(s) - g(s)]/\int_\Omega (f_Y(r) - g(r))\,d\nu(r)$.

In turns out that $\int_\Omega g\,d\nu = 1-d$. To see why, recall that the total variation distance is $d = \sup_D |\mathbb{P}(Y \in D) - \mathbb{P}(X \in D)|$. Because $|\mathbb{P}(Y \in D) - \mathbb{P}(X \in D)| = |\mathbb{P}(Y \in D^C) - \mathbb{P}(X \in D^C)|$, the supremum in the total variation distance can be taken over sets D such that $\mathbb{P}(Y \in D) \geq \mathbb{P}(X \in D)$.

Let $C = \{s \in \Omega : f_X(s) \leq f_Y(s)\}$. Let D be any other measurable set with $\mathbb{P}(Y \in D) \geq \mathbb{P}(X \in D)$. For any random variable W, $\mathbb{P}(W \in D) = \mathbb{P}(W \in C) - \mathbb{P}(W \in CD^C) + \mathbb{P}(W \in C^C D)$. Let $h(W) = \mathbb{P}(W \in CD^C) - \mathbb{P}(W \in C^C D)$. Then

$$\mathbb{P}(Y \in D) - \mathbb{P}(X \in D) = \mathbb{P}(Y \in C) - \mathbb{P}(X \in C) - h(Y) + h(X).$$

Now

$$-h(Y) + h(X) = [\mathbb{P}(X \in CD^C) - \mathbb{P}(Y \in CD^C)] + [\mathbb{P}(X \in C^C D) - \mathbb{P}(Y \in C^C D)].$$

By the way C was chosen, both terms in brackets above are nonpositive, so

$$0 \leq \mathbb{P}(Y \in D) - \mathbb{P}(X \in D) \leq \mathbb{P}(Y \in C) - \mathbb{P}(X \in C).$$

Since D was an arbitrary set satisfying $\mathbb{P}(Y \in D) \geq \mathbb{P}(X \in D)$, the total variation distance is $d = \mathbb{P}(Y \in C) - \mathbb{P}(X \in C)$. Now consider the normalizing constant of W_2.

$$\begin{aligned}\int_\Omega f_X(s) - g(s)\,d\nu(s) &= \int_C f_X(s) - g(s)\,d\nu(s) + \int_{C^C} f_X(s) - g(s)\,d\nu(s)\\ &= 0 + \int_{C^C} f_X(s) - f_Y(s)\,d\nu(s)\\ &= \mathbb{P}(X \in C^C) - \mathbb{P}(Y \in C^C) = \mathbb{P}(Y \in C^C) - \mathbb{P}(X \in C^C) = d.\end{aligned}$$

So the density of $(1-B)W_1+BW_2$ is

$$(1-d)\frac{\min(f_X(s),f_Y(s))}{1-d}+d\frac{f_X(s)-\min(f_X(s),f_Y(s))}{d}=f_X(s),$$

the same as that of X. A similar result holds for Y. □

Theorem 3.4 (Propp and Wilson [112]). *For a discrete set Ω, let ℓ be the length of the longest chain in the partial order of Ω. Let p denote the probability that a call to* `Monotonic_coupling_from_the_past`(t) *must recursively call itself. Let $d_t = \sup_{x\in\Omega} d_{TV}([X_t|X_0=x],\pi)$. Then*

$$\frac{p}{\ell}\leq d_t \leq p. \tag{3.13}$$

Proof. That $d_t \leq p$ is a well-known result called the Coupling Lemma (see [29, 1]).

For an element $x\in\Omega$, let $h(x)$ denote the length of the longest chain whose largest element is x. If X_t is the chain started at $x_{\min}$ and Y_t the chain started at $x_{\max}$, then always $X_t \preceq Y_t$. Note if $X_t \prec Y_t$ then $h(X_t)+1\leq h(Y_t)$, but if $X_t=Y_t$ then $h(X_t)=h(Y_t)$. Hence $h(Y_t)-h(X_t)$ is a nonnegative, integer valued random variable that is at least 1 whenever $X_t\neq Y_t$. So Markov's inequality gives

$$p=\mathbb{P}(X_t\neq Y_t)\leq \mathbb{E}[h(Y_t)-h(X_t)]$$

From the last fact, there exist W_1,W_2,W_3, and $B\sim \textsf{Bern}(d_t)$ independent such that $X_t\sim(1-B)W_1+BW_2$ and $Y_t\sim(1-B)W_1+BW_3$. So

$$p\leq \mathbb{E}[h(Y_t)-h(X_t)]=\mathbb{E}[B(h(W_2)-h(W_3))]=d_t\ell.$$

□

The total variation distance tends to decrease very quickly because it is an example of a submultiplicative function.

Lemma 3.2. *The total variation distance d_t satisfies $d_{t+s}\leq d_t d_s$.*

The proof involves the use of the Coupling Lemma alluded to in the proof of Theorem 3.4. This property allows us to show that the block size in monotonic CFTP needed for coalescence is not too much larger than the time needed for the Markov chain to start to mix.

Lemma 3.3. *Suppose that $d_t\leq e^{-1}$. Then for a block of length $t[k+\ln(\ell)]$, the probability p that a single block does not coalesce is at most e^{-k}.*

Proof. By the submultiplicative property, $d_{t[k+\ln(\ell)]}\leq [e^{-1}]^{k+\ln(\ell)}=\ell^{-1}e^{-k}$. By the previous theorem this makes $p\leq \ell\ell^{-1}e^{-k}=e^{-k}$. □

Lemma 3.4. *Suppose that in monotonic CFTP run for a block of length t, the chance that $\mathbb{P}(X_t\neq Y_t)\leq e^{-1}$. Then the expected number of steps in the Markov chain used by* `Monotonic_coupling_from_the_past`(t) *is at most* $19.2t$.

Proof. A call to `Monotonic_coupling_from_the_past` with input s requires at most $2s$ steps in the Markov chain, s steps when first determining if $X_t = Y_t$, and then s more steps after the recursive call if $X_t = Y_t$.

Let R_k denote the event that the kth call to `Monotonic_coupling_from_the_past` does not have $X_t = Y_t$. If the first call uses one step and the number of steps doubles after each call, then after $\lceil \log_2(t) \rceil$ calls, the next call has parameter at least t, and so only fails with probability at most e^{-1}. Hence

$$\mathbb{E}[2S] = \sum_{k=1}^{\infty} 2 \cdot 2^k \mathbb{P}(R_1 R_2 \cdots R_{k-1}) \leq 2(2t) + \sum_{k=\lceil \log_2(t) \rceil + 1}^{\infty} 2^{k+1} (e^{-1})^{k-1-\lceil \log_2(t) \rceil}$$
$$\leq 4t + 4t/(1 - 2/e) \leq 19.2t.$$

□

For example, consider the Ising model. The maximum height of a chain is $\#V$, the number of nodes in the graph. Therefore, if t is the time needed for the total variation distance to drop below e^{-1}, then $O(t \ln(\#V))$ steps are needed on average to produce a perfect simulation from the model. On the other hand, $\Omega(t)$ steps are also needed to produce a perfect simulation: CFTP does not make the Markov chain converge any faster than it normally does.

The important part about CFTP is that knowledge of the value of t where the total variation is at most e^{-1} is not necessary to use the procedure. When the algorithm stops, the result is always guaranteed to be a sample exactly from the correct distribution.

3.4 Slice samplers

The slice sampler was mentioned briefly as an auxiliary variable method in Section 1.8.4. Mira, Møller, and Roberts [94] realized that a slice sampler for a bounded density that is only nonnegative over a bounded set could be used to create a monotonic update function that can be used with monotonic CFTP.

`Slice_Sampler_update_step`	*Input:* x	*Output:* new state x
1) $y \leftarrow f_X(x)$		
2) Draw $U \leftarrow \mathsf{Unif}([0,1])$		
3) $A \leftarrow \{x' : f_X(x') \geq U \cdot y\}$		
4) $x \leftarrow \mathsf{Unif}(A)$		

The partial order on elements of Ω is: $x' \preceq x$ iff $f_X(x') \leq f_X(x)$. To make this a partial order, it is necessary to work with equivalence classes where x and x' are equivalent if $f_X(x') = f_X(x)$. The bottom state is then the set of equivalent x that minimizes f_X, and the top state is the set of x that maximizes f_X.

Now suppose that $A_1 \subseteq A_2$, and the goal is to draw $U_1 \sim \mathsf{Unif}(A_1)$ and $U_2 \sim \mathsf{Unif}(A_2)$ in such a way that $\mathbb{P}(U_1 = U_2)$ is as high as possible. This can be accomplished as follows. First draw U_2 uniformly from the larger set A_2. Conditional on

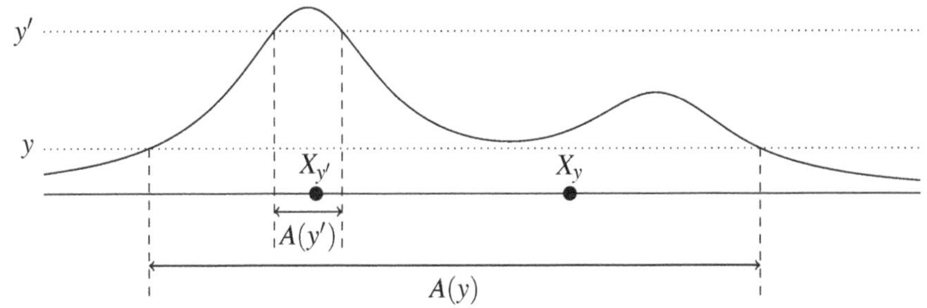

Figure 3.2 *The slice sampler dividing up the state space. Note that since $y < y'$, it holds that $A(y') \subseteq A(y)$. In this example, the draw $X_y \sim$ Unif$(A(y))$ did not land in $A(y')$, so an independent choice of $X_{y'} \sim$ Unif$(A(y'))$ was simulated.*

$U_2 \in A_1$, it holds that U_2 is uniform over A_1. If $U_2 \notin A_1$, then just draw U_1 uniformly from A_1 independently of U_2.

With this approach, $\mathbb{P}(U_1 = U_2)$ equals the measure of A_1 divided by the measure of A_2. This is the highest possible probability, since if $U_2 = U_1$, then $U_2 \in A_1$, and $\mathbb{P}(U_2 \in A_1)$ equals the measure of A_1 divided by the measure of A_2.

Next apply this idea to $A(y) = \{x : f_X(x) \geq y\}$. Note that if $y \leq y'$, then $A(y') \subseteq A(y)$. To draw $(X_{y'}, X_y)$ such that $X_{y'} \sim \text{Unif}(A(y'))$, draw $X_y \sim \text{Unif}(A(y))$ first. If $X_y \in A_{y'}$ as well, then set $X_{y'}$ equal to X_y. Otherwise, draw $X_{y'}$ independently from $A(y')$. This is illustrated pictorially in Figure 3.2.

This idea can be extended from just two values of y to give a way of updating every state $x \in \Omega$ simultaneously. The method is called the slice sampler. Start with $y = 0$ so $A(y) = \Omega$. Next draw X_1 uniformly from $A(y)$. Then for every $y' \geq 0$ such that $X_1 \in A(y')$, accept X as a uniform draw from $A(y')$. Let $y = \inf\{y' : A(y') \subseteq A(y)\} = f_X(X)$. Then start over, drawing X_2 uniformly from $A(y)$ and accepting as a draw for all $y' \geq y$ such that $X_2 \in A(y')$. Repeat until every $A(y')$ has been assigned some point X_i that is uniform over $A(y')$.

`Slice_Sampler_update_step_theoretical` *Input:* x *Output:* new state x

$y_0 \leftarrow 0, N \leftarrow 0$
Repeat
  $N \leftarrow N + 1$
  Draw $x_N \leftarrow \text{Unif}(A(y_{N-1}))$
  $y_N \leftarrow f_X(x_N)$
Until $y_N > f_X(x)$
Draw $U \leftarrow \text{Unif}([0,1])$
$i \leftarrow \min\{i : y_i > U f_X(x)\}, x \leftarrow x_i$

For a draw $A \sim f_X$, assume that $\mathbb{P}(f_X(A) > f_X(x)) > 0$. Then the result is a set of $0 < y_1 < y_2 < \cdots y_N$ values where N is finite with probability 1, and a corresponding set $\{x_i\}$ of points in Ω with $x_i \sim \text{Unif}(A(y_i))$. Since the y_i are increasing, $x_1 \preceq x_2 \preceq$

$\cdots \preceq x_N$. That makes the overall update function monotonic: if $f_X(x) \leq f_X(x')$ then the i chosen at the last step is lower for x than x', which makes the new state for x precede the new state for x' in the partial order.

The above method is a theoretical update that typically will not be used in this form. This is because to use CFTP it is not necessary to update the state over all possible starting states. It is only necessary to update the state from the minimum state, and the maximum state. Since this update is monotonic, every other state is unimportant. So all other steps can be skipped. This gives the following much simpler method for updating the lower and upper process.

`Slice_Sampler_update_step`	*Input:* $x_{\text{lower}}, x_{\text{upper}}$	*Output:* new $x_{\text{lower}}, x_{\text{upper}}$

1) Draw $U \leftarrow \mathsf{Unif}([0,1])$
2) Draw $x_{\text{lower}} \leftarrow \mathsf{Unif}(A(U f_X(x_{\text{lower}})))$
3) If $f_X(x_{\text{lower}}) \geq U f_X(x_{\text{upper}})$
4) $\quad x_{\text{upper}} \leftarrow x_{\text{lower}}$
5) Else
6) $\quad$ Draw $x_{\text{upper}} \leftarrow \mathsf{Unif}(A(U f_X(x_{\text{upper}})))$

This gives an algorithm that reinforces the message of the Fundamental Theorem of Monte Carlo. If it is possible to simulate uniformly from the level sets of a density, then it is possible to sample from the density.

Of course, this method only works when it is possible to maximize the density f_X of interest, and when there is a bounded space for the x values so that it is possible to sample from $A(0)$.

3.5 Drawbacks to CFTP

While CFTP is a powerful tool for creating perfect simulation algorithms, it does have some flaws. The main two problems are that basic CFTP is noninterruptible and read twice.

3.5.1 *Noninterruptibility*

As mentioned earlier, an algorithm is interruptible if it can be terminated and restarted (immediately or later) without biasing the results. This is such a common feature of most algorithms that users seldom even stop to consider it. If a sorting algorithm is interrupted any time in the middle of a run, no one worries that if it is begun again later starting from the original unsorted data, the result will be anything but a sorted set of data.

However, some perfect simulation algorithms are set up so that if the user aborts a run of the simulation because it is taking too long, or *simply has positive probability of aborting the algorithm because it could take too long*, the result is no longer a perfect simulation algorithm. An algorithm that can be stopped and restarted at any point is known as interruptible.

Definition 3.6. *A perfect simulation algorithm that uses random variables*

$X_1,X_2,\ldots,X_T$ *to produce* $f_T(X_1,\ldots,X_T) \sim \pi$ *is* interruptible *if* $f_T(X_1,\ldots,X_T)$ *and* T *are independent. Otherwise* T *is* noninterruptible.

As an example, consider generating a geometric random variable by letting $B_1,B_2,\ldots$ be iid $\mathsf{Bern}(p)$ random variables, then setting $T = \inf\{t : B_t = 1\}$ and $f_T(B_1,\ldots,B_T) = T$. Then $T \sim \mathsf{Geo}(p)$. Of course, in this instance, T and $f_T(B_1,\ldots,B_T)$ are not independent, they have the same value. Hence this algorithm is noninterruptible.

Now consider this from the perspective of a simulator. Perhaps the simulator wishes to generate a geometric random variable, and intends to run the above algorithm exactly as written. But lurking in the subconscious of the simulator is an ultimatum: if the algorithm actually takes longer than five million steps, the simulator will become impatient and abort the algorithm.

This unacknowledged piece of the algorithm means that the simulator does not obtain samples drawn exactly from $\mathsf{Geo}(p)$, but rather from $\mathsf{Geo}(p)$ conditioned to lie in $\{1,\ldots,5\cdot 10^6\}$. In this case, it is a minor point: unless p is very small, the algorithm is very unlikely to reach such high values of T. But the potential is there, and the willingness of the simulator to possibly interrupt the algorithm if the running time is too long makes this no longer a perfect simulation algorithm.

Next consider the acceptance/rejection algorithm that draws from measure ν over A given the ability to draw from ν over B such that A is contained in B.

Lemma 3.5. *Suppose that* ν *is a finite measure over* B*, and* $A \subseteq B$ *has* $\nu(A) > 0$*. For iid* $X_1,X_2,\ldots \sim \nu(B)$ *and* $T = \inf\{t : X_t \in B\}$ *it holds that* T *and* X_T *are independent random variables. That is,* AR *is an interruptible algorithm.*

Proof. Let C be a measurable subset of A and $i \in \{1,2,\ldots\}$. Then

$$\begin{aligned}
\mathbb{P}(X_T \in C, T = i) &= \mathbb{P}(X_1,\ldots,X_{i-1} \notin C, X_i \in C) \\
&= (1-\nu(A)/\nu(B))^{i-1}(\nu(C)/\nu(B)) \\
&= (1-\nu(A)/\nu(B))^{i-1}(\nu(A)/\nu(B))(\nu(C)/\nu(A)) \\
&= \mathbb{P}(X_1,\ldots,X_{i=1} \notin C, X_i \in A)\mathbb{P}(X_T \in C) \\
&= \mathbb{P}(T = i)\mathbb{P}(X_T \in C),
\end{aligned}$$

where the fact that $\mathbb{P}(X_T \in A) = \nu(A)/\nu(B)$ is just Theorem 2.1. □

The advantage of an interruptible perfect simulation algorithm is that a user can interrupt the algorithm as any moment and restart the algorithm without changing the output. Another way to say this is follows: even if the use intends to abort the algorithm after a fixed number of steps, then conditioned on the algorithm succeeding, the distribution of the output is unaffected.

Lemma 3.6. *Suppose that* T *is a stopping time with respect to* $X_1,X_2,\ldots$ *and the family* $\{f_t\}_{t=1}^{\infty}$ *satisfies* $f_T(X_1,\ldots,X_T) \sim \pi$*, where* T *and* $f_T(X_1,\ldots,X_T)$ *are independent. Then for a fixed time* t_{stop}*,* $[f_T(X_1,\ldots,X_T) | T \leq t_{stop}] \sim \pi$*.*

Proof. Let C be any measurable set, and break the probability into pieces:

$$\begin{aligned}\mathbb{P}(f_T(X_1,\ldots,X_T)\in C|T\le t_{\text{stop}}) &= \frac{\mathbb{P}(f_T(X_1,\ldots,X_T)\in C, T\le t_{\text{stop}})}{\mathbb{P}(T\le t_{\text{stop}})} \\ &= \frac{\sum_{t=1}^{t_{\text{stop}}}\mathbb{P}(f_T(X_1,\ldots,X_T)\in C, T=t)}{\sum_{t=1}^{t_{\text{stop}}}\mathbb{P}(T=t)} \\ &= \frac{\sum_{t=1}^{t_{\text{stop}}}\mathbb{P}(f_T(X_1,\ldots,X_T)\in C)\mathbb{P}(T=t)}{\sum_{t=1}^{t_{\text{stop}}}\mathbb{P}(T=t)} \\ &= \frac{\sum_{t=1}^{t_{\text{stop}}}\pi(C)\mathbb{P}(T=t)}{\sum_{t=1}^{t_{\text{stop}}}\mathbb{P}(T=t)} \\ &= \pi(C).\end{aligned}$$

□

The benefit of using interruptible algorithms is that the practitioner does not have to worry about any known or unknown limitations on the running time of the algorithm affecting the distribution of the output.

Here is the bad news: in general, CFTP is noninterruptible.

Recall that CFTP effectively writes the target distribution π as a mixture of two other distributions. Let A be the event that $\phi(x,U)=\{y\}$ for all $x\in\Omega$. Then for $Y\sim\pi$ and any measurable C,

$$\mathbb{P}(Y\in C)=\mathbb{P}(Y\in C|A)\mathbb{P}(A)+\mathbb{P}(Y\in C|A^C)\mathbb{P}(A^C).$$

When A occurs, $\phi(x,U)=\{y\}$, and y is output as the answer. When A^C occurs, recursion is used to draw an $X\sim\pi$, and then $\phi(X,U)$ gives $[Y|A^C]$.

Now suppose a simulator does not have time to do the recursive step. Then the result does not come from $Y\sim\pi$, but rather from the distribution of $[Y|A]$.

As an example, consider drawing uniformly from $\{1,2,3\}$ using the Markov chain whose update function is

$$\pi_1(x,U)=x+\mathbb{1}(x<3,U>1/2)-\mathbb{1}(x>1,U\le 1/2).$$

This is a simple symmetric random walk with partially reflecting boundaries.

Suppose that $\phi_2(x,U_1,U_2)=\phi_1(\phi_1(x,U_1),U_2)$, that is to say, ϕ_2 takes two steps in the Markov chain. When does $\phi_2(\Omega,U_1,U_2)=\{y\}$? Only in the case that both U_1 and U_2 fall in $(1/2,1]$, or if they both fall in $[0,1/2]$. In the first case, $\phi_2(\Omega,U_1,U_2)=\{3\}$, otherwise $\phi_2(\Omega,U_1,U_2)=\{1\}$. In no case does $\phi_2(\Omega,U_1,U_2)=\{2\}$, so the resulting draw is not uniform over $\{1,2,3\}$.

In reality, known reasons (researcher time, computer time) and unknown reasons (lightning strike on a computer, power outage) mean that virtually every simulation is run with a fixed or random upper limit on the stopping time. This makes the noninterruptible property of CFTP troubling. If there is an unknown limit to the amount of time it can be run, it ceases to be an exact simulation algorithm.

That being said, the bias introduced into the samples by noninterruptibility is bounded by the chance that the algorithm is interrupted. With sufficient time given, this chance can be made arbitrarily low. Still, perfect simulation is the name of the game here, and in later chapters methods are described that do not suffer from this noninterruptible flaw.

3.5.2 Read twice

The other problem with CFTP lies in the overall structure. When random variables U are generated to be used by the simulation, if recursion occurs, those variables must be used again. Therefore, in theory they must be stored, which can impose a sizable memory burden in running the algorithm.

There are two ways to ameliorate this effect. First, truly random variables are never used in a simulation, instead pseudorandom numbers that follow a deterministic sequence are used instead. This sequence of numbers is initialized by what is called a seed.

Therefore, instead of saving the entirety of the random choices made by the algorithm, it instead suffices to save the seed used to create those pseudorandom choices.

Even better, it is possible to write CFTP in such a way that the random choices need not be stored. This variant is known as read-once coupling from the past, and will be discussed in Chapter 5.

Chapter 4

Bounding Chains

It is often safer to be in chains than to be free.

Franz Kafka

Monotonicity is a valuable property to have, but unfortunately, not every Markov chain update function has it. From among our examples, the Ising model when $\beta < 0$, the Potts model with $k \geq 4$, the hard-core gas model on nonbipartite graphs, weighted permutations, sink-free orientations, the antivoter model, and self-organizing lists all have nonmonotonic Gibbs samplers. (While the slice sampler is monotonic, for these examples it cannot be employed efficiently in its pure form.) While Gibbs samplers are often monotonic, Metropolis-Hastings, chains on permutations, and auxiliary variable chains (like Swendsen-Wang) are often not.

The first move away from monotonic Markov chains was the notion of antimonotonicity, introduced in [44, 96, 46]. Bounding chains, introduced independently in [62] and [44], and further refined in [63, 64, 65, 66, 51], are a generalization of antimonotonicity that apply to a wide variety of Markov chains.

4.1 What is a bounding chain?

Suppose an underlying chain state has nodes, each of which has a label. Then a bounding chain keeps track of a set of possible labels at each node. It bounds the underlying chain if it is possible to ensure that the actual label of each node falls into the subset of labels in the bounding chain.

Let $\Omega = C^V$ be the state space of the underlying chain. Call C the set of colors, values, or labels, and V the set of dimensions, positions, coordinates, nodes, or vertices. For instance, in the Ising model $C = \{-1, 1\}$ and V is the vertex set of the graph. In the autonormal model $C = [0, 1]$. For Brownian motion both C and V are $\mathbb{R}$. For permutations on n items, both C and V equal $\{1, 2, \ldots, n\}$.

A bounding chain wraps around the underlying chain C^V. At each node of V, the bounding chain keeps a subset of C that represents the possible colors permitted for the underlying chain. So a bounding chain for the Ising model could have either $\{-1\}, \{1\}$, or $\{-1, 1\}$ as the label of a vertex. The bounding part comes when steps in the Markov chain are made. Let X_t be the underlying chain state and Y_t the bounding chain state. If $X_t(v) \in Y_t(v)$ for every v, then it should also be true that $X_{t+1}(v) \in Y_{t+1}(v)$. Let 2^C denote the set of all subsets of the set C.

Definition 4.1. *Say that* $y \in (2^C)^V$ bounds $x \in C^V$ *if for all* $v \in V$, $x(v) \in y(v)$.

Definition 4.2. *Say that the Markov process* $\{Y_t\}$ *over* $(2^C)^V$ *is a* bounding chain *for the Markov chain* $\{X_t\}$ *over* C^V *if there exists a coupling between* $\{Y_t\}$ *and* $\{X_t\}$ *such that if* Y_t *bounds* X_t, *then* Y_{t+1} *bounds* X_{t+1}.

The simplest way to create such a coupling is by using an update function for the Markov chain.

Theorem 4.1 (Building a bounding chain). *Suppose* $X_{t+1} = \phi(X_t, U_t)$ *for all* t, *where* $U_0, U_1, U_2, \ldots$ *are iid Unif*$([0,1])$. *Create* $y' = \phi_t^{BC}(y,u)$ *as follows. For all* $v \in V$, *set*

$$y'(v) = \{c : \exists x \text{ bounded by } y \text{ such that } \phi_t(x,u) \text{ has color } c \text{ at vertex } v\}.$$

Then ϕ^{BC} *gives a bounding chain for* $\{X_t\}$.

In other words, consider all the possible states x that are bounded by the bounding state y. Take a step in each of those chains, and let the resulting set of labels for v comprise the new bounding state at v.

Once a bounding chain update has been created, CFTP can be run as follows.

```
Bounding_chain_cftp    Input: t    Output: X

Draw U_1, U_2, ..., U_t
For all v ∈ V, set Y(v) = C
For i from 1 to t
    Y ← Bounding_chain_update(Y, U_i)
If #Y(v) > 1 for some v
    X ← Bounding_chain_cftp(2t)
    For all v, Y(v) ← {X(v)}
    For i from 1 to t
        Y ← Bounding_chain_update(Y, U_i)
    For all v, let X(v) be the single element of Y(v)
```

As usual, when writing something like `Bounding_chain_cftp`(Y, U_i), the U_i stands for all the random choices needed to be made inside the bounding chain step. So it can be multiple random variables, or even an iid sequence if necessary.

With the general framework in place, now consider some specific examples.

4.2 The hard-core gas model

Consider again the hard-core gas model (Definition 1.20) where each node is labeled either 0 or 1, two adjacent nodes cannot both be labeled 1, and the weight of a configuration is proportional to λ raised to the number of nodes labeled 1. When the graph was bipartite, it was possible to cast the Gibbs chain as a monotone chain with the proper partial order. Now consider a graph (such as the triangular lattice) that is not bipartite.

In the Gibbs update step from Section 1.8.5, a vertex is chosen uniformly at random along with $U \sim \text{Unif}([0,1])$. If $U < \lambda/(1+\lambda)$ and no neighbors are labeled 0, then the node is relabeled 1. Otherwise, it stays at 0.

Now consider the bounding chain. Nodes are now labeled with nonempty subsets of $\{0,1\}$, so each node is labeled either $\{0\}$, $\{1\}$, or $\{0,1\}$. If a node v has a neighbor w that is labeled $\{1\}$, then any x bounded by v has $x(w) = 1$, so the node v must be given the label $\{0\}$. If all neighbors of v are labeled $\{0\}$, then any x bounded by v has all nodes labeled 0, and so U determines whether $y(v)$ is $\{0\}$ or $\{1\}$.

But if some neighbors of v are labeled $\{0\}$, and some neighbors are labeled $\{0,1\}$, and $U < \lambda/(1+\lambda)$, then it is uncertain how to proceed. There is a state x bounded by y with $x(w) = 1$, in which case $x(v) = 0$, but another state x' could have all neighbors labeled 0, giving $x'(v) = 1$. Hence $y(v) = \{0,1\}$ at the next state. The following pseudocode formalizes this idea.

`HCGM_bounding_chain_Gibbs` *Input:* $y, v \in V, U \in [0,1]$ *Output:* y

1) Let $N_{\{1\}}$ be the number of neighbors of v labeled 1 in y
2) Let $N_{\{0,1\}}$ be the number of neighbors of v labeled $\{0,1\}$ in y
3) If $U > \lambda/(1+\lambda)$ or $N_{\{1\}} > 0$
4) $\quad y(v) \leftarrow 0$
5) Else if $N_{\{0,1\}} = 0$
6) $\quad y(v) \leftarrow \{1\}$
7) Else if $N_{\{1\}} = 0$ and $N_{\{0,1\}} > 0$
8) $\quad y(v) \leftarrow \{0,1\}$

This bounding chain can be used with CFTP to obtain perfect samples. It can also be a useful aid in analyzing the mixing time of the Markov chain. To understand how long CFTP takes, it is necessary to bound the expected time needed for all of the $\{0,1\}$ labels to disappear.

Lemma 4.1. *Let ϕ_t^{BC} denote t steps in* `HCGM_bounding_chain_Gibbs`. *For $\lambda < 1/(\Delta-1)$, the chance that the bounding chain bounds more than one state is at most*

$$\#V \exp\left(-t(\#V)^{-1}(1-\lambda\Delta/(1+\lambda))\right).$$

In particular, setting

$$t = \#V \ln(2\#V)\left(\frac{1+\lambda}{1-\lambda(\Delta-1)}\right)$$

gives an expected running time of at most $4t$ to generate one sample with CFTP.

Proof. Let $W_t = \{v : Y_t(v) = \{0,1\}\}$ denote the number of positions where the underlying state x is not known completely. When a node v labeled $\{0,1\}$ is selected, its label is wiped away, reducing W_t by 1. Since there are W_t nodes labeled $\{0,1\}$, the chance that the chosen node has this label is $W_t/\#V$.

After v is chosen, consider the value of U. If $U < \lambda/(1+\lambda)$ then the node might end the step with the label $\{0,1\}$. In order for this to happen, it must be true that v is adjacent to a node already labeled $\{0,1\}$. There are at most $W_t\Delta$ nodes labeled $\{0,1\}$ in the graph.

The chance that the v picked is adjacent to a node labeled $\{0,1\}$ is at most

$W_t\Delta/\#V$. Hence

$$\mathbb{E}[W_{t+1}|W_t] \leq W_t - \frac{W_t}{\#V} + \frac{W_t\Delta}{\#V} \cdot \frac{\lambda}{1+\lambda} = W_t\left[1 - \frac{1}{\#V}\left(1 - \frac{\lambda\Delta}{1+\lambda}\right)\right].$$

Let $\gamma = (\#V)^{-1}(1 - \lambda\Delta/(1+\lambda))$. Note $\mathbb{E}[W_t] = \mathbb{E}[\mathbb{E}[W_t|W_{t-1}]] \leq \mathbb{E}[\gamma W_{t-1}] = \gamma\mathbb{E}[W_{t-1}]$. So a simple induction gives $\mathbb{E}[W_t] \leq \gamma^t\mathbb{E}[W_0]$. For $\alpha > 0$, $1-\alpha \leq \exp(-\alpha)$ so $\gamma^t \leq \exp(-t(\#V)^{-1}(1-\lambda\Delta/(1+\lambda)))$.

Since W_t is nonnegative and integral, and $W_0 = \#V$, Markov's inequality gives $\mathbb{P}(W_t > 0) \leq \mathbb{E}[W_t] \leq \#V\exp(-t(\#V)^{-1}(1-\gamma\Delta/(1+\lambda))$.

The bound on the expected running time for CFTP then follows from Lemma 3.1.

□

4.2.1 The Dyer-Greenhill shifting chain

Now consider the more effective chain for the hard core gas model of Dyer and Greenhill from Section 1.8.5, called `HCGM_shift`. Recall that in this chain when an attempt is made to label node v with a 1 and exactly one neighbor is preventing that from happening, there is a Bernoulli $S \sim \mathsf{Bern}(p_{\text{swap}})$. When $S = 1$, the value $x(v) = 1$ and $x(w) = 0$. So when $S = 1$, the value of $x(w)$ gets shifted over to $x(v)$.

For the bounding chain, this raises many cases to deal with. For instance, suppose that $U < \lambda/(1+\lambda)$, $S = 1$, all but one neighbor of v has label $\{0\}$, and exactly one neighbor w has label $\{0,1\}$. In the original chain, this meant that $y(v) = \{0,1\}$.

But now consider how the shifting chain works. If that $\{0,1\}$ neighbor w is hiding a 0 in the underlying state, then $y(v)$ should be 1. If the $\{0,1\}$ label at w is hiding a 1 in the underlying state, then $y(v)$ should still be 1, since the neighboring 1 would get shifted to v anyway! Moreover, the value of $y(w)$ should be changed to $\{0\}$, since if there was a 1 there, it has been shifted away!

Then the underlying state x might have $x(w) = 0$ or $x(w) = 1$. But if $U < \lambda/(1+\lambda)$ and $S = 1$, then if $x(w) = 0$, the new value of $x(v)$ is 1. And if $x(w) = 1$, then the new value of $x(v)$ is still 1, and the new value of $x(w) = 0$. So $y(v) = \{1\}$ and $y(w) = \{0\}$ for all underlying states.

So for all underlying states, $y(v) = \{1\}$ and $y(w) = \{0\}$ in the bounding chain.

That is a good move: the $\{0,1\}$ label at the node w has been removed! Now suppose that that all but two neighbors of v have label $\{0\}$, one neighbor w_1 has label $\{0,1\}$, and one neighbor w_2 has label $\{1\}$. Then for $x(w_1) = x(w_2) = 1$, whether or not $S = 1$ or $S = 0$, no swap occurs, and $x(v) = 0, x(w_1) = x(w_2) = 1$.

However, if $x(w_1) = 0, x(w_2) = 1$, then a swap would occur, so $x(v) = 1, x(w_1) = 0, x(w_2) = 0$. So in the bounding chain, $y(v) = y(w_1) = y(w_2) = \{0,1\}$. This is a bad move for the bounding chain, as two new $\{0,1\}$ labels have been added!

To break down the bounding chain into cases, let N_ℓ denote the number of neighbors of the chosen node v given label ℓ. When $N_{\{0,1\}} = 1$, let $w_{\{0,1\}}$ denote the neighbor of v with label $\{0,1\}$, and when $N_{\{1\}} = 1$, let $w_{\{1\}}$ be the neighbor of v with label $\{1\}$. The complete bounding chain update is given in Table 4.1. In this table, $\deg(v)$ denotes the degree of v, and $*$ is a wildcard character that indicates any value.

So why go to the extra trouble of executing an update which is far more complex

Table 4.1 *Cases for updating the bounding chain for the swap chain for the hard-core gas model. A $*$ indicates that there is no restriction on that component of the vector. When $N_q = 1$, let w_q be the neighbor of v with $y(w_q) = q$.*

Case	$(N_{\{0\}}, N_{\{1\}}, N_{\{0,1\}})$	Is $U < \lambda/(1+\lambda)$?	S	new bounding state
1	$(*,*,*)$	No	0 or 1	$y(v) = \{0\}$
2	$(*,0,0)$	Yes	0 or 1	$y(v) = \{1\}$
3	$(*,0,1)$	Yes	0	$y(v) = \{0,1\}$
4	$(*,0,1)$	Yes	1	$y(v) = \{1\}, y(w_{\{0,1\}}) = 0$
5	$(*,1,*)$	Yes	0	$y(v) = \{0\}$
6	$(*,1,0)$	Yes	1	$y(v) = \{1\}, y(w_{\{1\}}) = \{0\}$
7	$(*,1,\geq 1)$	Yes	1	$y(v) = \{0,1\}, y(w_{\{1\}}) = \{0,1\}$
8	$(*,\geq 2,*)$	Yes	0 or 1	$y(v) = \{0\}$

than that for the simple Gibbs model? Because it is possible to prove that this method runs in polynomial time over a wider range of λ.

Lemma 4.2. *Let ϕ_t^{BC} denote t steps in the bounding chain for the swap chain with $p_{swap} = 1/4$. If $\lambda < 2/(\Delta - 2)$, then the chance that the bounding chain bounds more than one state is at most*

$$\#V \exp\left(-t(\#V)^{-1}(1 - \lambda\Delta/[2(1+\lambda)])\right).$$

In particular, setting

$$t = \#V \ln(2\#V)\left(\frac{2(1+\lambda)}{2 - \lambda(\Delta - 2)}\right)$$

gives an expected running time of at most $4t$ to generate one sample with CFTP.

Proof. As in the proof of the previous lemma, let $W_t = \{v : Y_t(v) = \{0,1\}\}$ denote the number of positions labeled $\{0,1\}$ in the bounding chain.

Use the case numbers in Table 4.1. When node v is first selected, its label is removed. If its label had been $\{0,1\}$, then W_t is reduced by 1. Then, depending on the case, W_t is increased by either 1 or 2. For example, in case 3, which occurs with probability $p_{\text{swap}}\lambda/(1+\lambda)$, W_t increases by 1. Case 3 increases W_t by 1, case 4 actually decreases W_t by 1, and case 7 increases W_t by 2. Let n_{34} denote the number of nodes that activate either case 3 or 4 (so $(N_{\{0\}}, N_{\{1\}}, N_{\{0,1\}}) = (*,0,1)$), and n_7 the number of nodes where $(N_{\{0\}}, N_{\{1\}}, N_{\{0,1\}}) = (*,1,\geq 1)$.

Then

$$\mathbb{E}[W_{t+1} - W_t | W_t, n_{34}, n_7] = \frac{1}{\#V}\left[-W_t + \frac{\lambda}{1+\lambda}\left(n_{34}(1 - p_{\text{swap}}) - n_{34}p_{\text{swap}} + 2n_7 p_{\text{swap}}\right)\right].$$

Let $p_{\text{swap}} = 1/4$, so $n_{34}(1 - p_{swap}) - n_{34}p_{\text{swap}} = n_{34}/2$, and $2n_7 p_{\text{swap}} = n_7/2$. Then

$$\mathbb{E}[W_{t+1} - W_t | W_t, n_{34}, n_7] = \frac{1}{\#V}\left[-W_t + \frac{1}{2}\cdot\frac{\lambda}{1+\lambda}(n_{34} + n_7)\right].$$

Then the number of nodes adjacent to nodes labeled $\{0,1\}$ is at most $\Delta \cdot W_t$, which means that $n_{34} + n_7 \leq \Delta \cdot W_t$. That gives

$$\mathbb{E}[W_{t+1}|W_t] \leq W_t \left[1 - (\#V)^{-1} \frac{\lambda\Delta}{2(1+\lambda)}\right].$$

Taking the expectation of both sides and setting $\alpha = (\#V)^{-1}\lambda\Delta/(2(1+\lambda))$ gives

$$\mathbb{E}[W_{t+1}] = \mathbb{E}[W_t](1-\alpha).$$

An easy induction then gives $\mathbb{E}[W_t] = \mathbb{E}[W_0](1-\alpha)^t$. Since $W_0 \leq \#V$ and $(1-\alpha) \leq \exp(-\alpha)$, that completes the proof of the bound on the probability.

The bound on the expected running time of CFTP then follows from Lemma 3.1. □

In the proofs of Lemma 4.2, a p_{swap} value of $1/4$ was used to get the best analytical result. Of course, the algorithm is correct for any value of p_{swap} in $[0,1]$, and so in practice a user would run experiments to determine which value of p_{swap} gave the fastest running time.

4.2.2 Bounding chains for the Strauss model

The Strauss model (Definition 1.21) allows two adjacent nodes to both be labeled 1, but at the cost of assigning a penalty factor $\gamma < 1$ to the probability of the state. (In the hard-core gas model, $\gamma = 0$ as any adjacent nodes labeled 1 immediately reduce the probability to 0.) Both the bounding chain for the original Gibbs sampler as well as for the shift chain can easily be modified to handle this more general model.

4.3 Swendsen-Wang bounding chain

The Swendsen-Wang chain of Section 1.8.3 begins with an Ising configuration x. Every edge $\{i,j\}$ receives a value $y(\{i,j\})$ chosen uniformly from $[0, \exp(\beta \cdot \mathbb{1}(x(i) = x(j)))]$. Then the configuration is discarded. Let $A \subseteq E$ be the set of edges with $y(\{i,j\}) > 1$. Then A partitions the nodes of the graph into a set of connected components. Each node in these components must have the same label. The label for the components is chosen uniformly from $\{-1,1\}$. The exact value of $y(\{i,j\})$ is unimportant—what really matters is if $y(\{i,j\})$ is greater than or at most 1. Let $w(\{i,j\}) = \mathbb{1}(y(\{i,j\}) > 1)$. These are the values on edges that form the underlying chain to be bounded.

The first perfect simulation algorithm for Swendsen-Wang was given in [66], and utilized a bounding chain approach. In the bounding chain, some $w_{\text{BC}}(\{i,j\})$ are $\{0\}$ (indicating the edge is always out), $\{1\}$ (indicating the edge is always in), and $\{0,1\}$ (indicating the edge might be in or out.)

Then the edges labeled $\{1\}$ divide the graph into connected components where it is sure that each node in the component receives the same color.

With edges labeled $\{0,1\}$, that edge might or might not be in the graph. If it joins two connected components that are otherwise unjoined, then the labels for one component are known, but the labels for the other component could either be the same or a new choice. The resulting method looks like this. (Assume the vertex set is numbered $\{1, 2, \ldots, \#V\}$.)

`Swendsen_Wang_bounding_chain`
Input: $y \in (2^{\{-1,1\}})^V$, $(U_1, \ldots, U_{\#E}) \in [0, \exp(\beta)]^{\#E}$, $c_1, \ldots, c_{\#V} \in \{-1, 1\}^V$
Output: y

---

1) For each edge $\{i, j\} \in E$
2)   If $U_{\{i,j\}} \leq 1$ then $w(\{i, j\}) = \{0\}$
3)   Else if $y(i)$ or $y(j)$ is $\{-1, 1\}$ then $w(\{i, j\}) = \{0, 1\}$
4)   Else $w(\{i, j\}) = \mathbb{1}(y(i) = y(j))$.
5) $A_{\{1\}} \leftarrow \{e : w(e) = \{1\}\}$, $A_{\{0,1\}} \leftarrow \{e : w(e) = \{0, 1\}\}$
6) For every connected component D in $A_{\{1\}}$
7)   Let v be the smallest numbered vertex in D
8)   For all $w \in D$, let $x(w) \leftarrow c_v$
9) For every node $v \in V$
10)   Let V_A be all nodes reachable from v using edges in $A_{\{0,1\}}$ or $A_{\{1\}}$
11)   Let $y(v) = \cup_{w \in V_A} \{x(w)\}$

Fact 4.1. *Suppose* $\Delta \geq 3$ *and* $p = 1 - \exp(-\beta) < 1/(\Delta - 1)$. *Then set*

$$\gamma = p\left[1 + \frac{2p(\Delta - 1)}{1 - p(\Delta - 1)}\right]. \tag{4.1}$$

The probability that the Swendsen-Wang bounding chain allows more than one underlying state after t *states is at most* $\#V \cdot \gamma^t$.

Proof. Let D_t denote the number of edges in the set $A_{\{0,1\}}$ after t steps of the chain. Consider the value of D_{t+1} conditioned on D_t. Note that any edges in $A_{\{0,1\}}$ at time $t+1$ had to either be an edge in $A_{\{0,1\}}$ at the previous time step, or had to be connected to an edge in $A_{\{0,1\}}$ at the previous time step using edges in $A_{\{0,1\}}$ or $A_{\{1\}}$.

Now, the chance that an edge survives being labeled $\{0\}$ in the bounding chain during a step is $(\exp(\beta) - 1)/\exp(\beta)$. So let $p = 1 - \exp(-\beta)$ denote this chance of survival.

Then starting with D_t edges at time t in $A_{\{0,1\}}$, on average only pD_t edges survive the culling at line 2 of the algorithm.

Suppose that a particular edge e survives this part of the step. Now consider how many edges are connected to e. Consider one endpoint of the edge. At most $(\Delta - 1)p$ nodes are connected to this endpoint (on average) by edges still labeled $\{1\}$ or $\{0, 1\}$. So the total number of edges (on average) connected to this endpoint is at most

$$p(\Delta - 1) + [p(\Delta - 1)]^2 + \cdots = \frac{p(\Delta - 1)}{1 - p(\Delta - 1)}. \tag{4.2}$$

The original edge had two endpoints. Hence

$$\mathbb{E}[D_{t+1} | D_t] = D_t p\left[1 + \frac{2p(\Delta - 1)}{1 - p(\Delta - 1)}\right] = D_t \gamma, \tag{4.3}$$

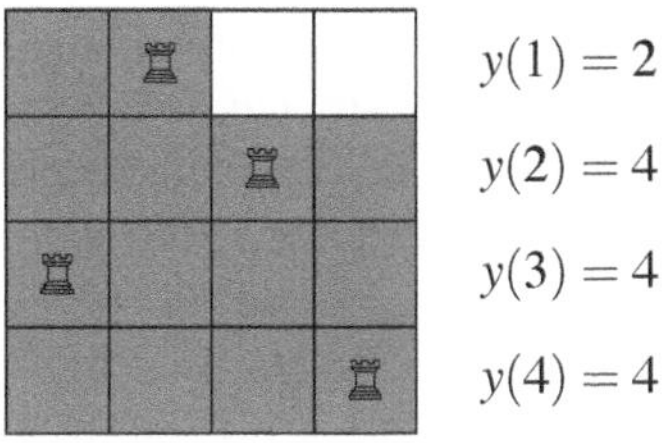

Figure 4.1 *The permutation* $x = (3,1,2,4)$ *(so* $x^{-1} = (2,3,1,4)$*) illustrated as a rook placement. The bounding state* $y = (2,4,4,4)$ *is shown as gray shaded squares.*

and a simple induction completes the proof. □

4.4 Linear extensions of a partial order

Formally, a bounding chain labels each node with a nonempty subset of possible labels. If there are q labels, there are $2^q - 1$ different states. Often, not all subsets are possible, which can lead to great savings in memory and in the time needed for an update function. For instance, often the subsets are intervals of form $\{x : a \leq x \leq b\}$. In this case there are only q^2 possible states. Monotonic CFTP is a special case of this situation, but there are also bounding chains of this form where the update function is not monotonic.

An example of this is the bounding chain for the problem of linear extensions of a partial order from Section 1.6.2. Recall that a permutation σ is a linear extension of a partial order on $\{1,\ldots,n\}$ if $\sigma(i) \preceq \sigma(j)$ implies $i < j$. Without loss of generality, assume that $(1,2,\ldots,n)$ is a valid linear extension of the partial order.

A permutation can be viewed as a rook placement on a chessboard. A rook placement is when n rooks are placed onto an n by n chessboard so that no two rooks fall into the same row or column. Say that $x(j) = i$ if a rook is placed into row i and column j. Figure 4.1 illustrates how the permutation $x = (3,1,2,4)$ looks as a rook placement.

The bounding state then is the set of squares upon which it is possible to place rooks. In Figure 4.1 permitted locations are shaded grey. To keep the set of bounding states y as simple as possible, row i will have permitted squares from 1 up to $y(i)$. This is equivalent to saying that y bounds x if and only if $x^{-1}(i) \leq y(i)$ for all i. The bounding state can have multiple rows i with $y(i) = n$, but if $y(i) < n$, then there does not exist $j \neq i$ with $y(j) = y(i)$.

Moreover, the partial order is respected by the bounding state in the sense that $i \preceq j$ implies that $y(i) \leq y(j)$. With these bounding states, a natural initial state is $(n,n,\ldots,n)$, since that bounds every possible linear extension.

For example, consider the bounding state $(3,2,4,4)$. The 3 in the first component indicates that any underlying permutation is of the form $(1,*,*,*)$, $(*,1,*,*)$, or $(*,*,1,*)$. That is, the 1 has to be in position 1, 2, or 3. Similarly, the 2 has to be in the first 2 positions, so underlying states have the form $(2,*,*,*)$ or $(*,2,*,*)$.

The 3 and 4 can appear anywhere in the permutation, as the 4 in the bounding state indicates that they are essentially unrestricted.

The Markov chain that will be considered here is due to Karzinov and Khachian [77]. Their chain can be viewed as a Metropolis-Hastings style chain where j is chosen uniformly from $\{1,\ldots,n-1\}$, and with probability $1/2$ the proposed state is the state σ with the values of position j and $j+1$ transposed, denoted $\sigma_{j\leftrightarrow j+1}$. The transposed state is accepted if and only if it is not true that $\sigma(j) \preceq \sigma(j+1)$. That is, the new state is accepted if the transposition does not violate the partial order. Otherwise the state does not change. (As mentioned earlier, a transposition that only swaps values at adjacent positions is called an adjacent transposition.)

In Figure 4.1, this corresponds to switching columns j and $j+1$ (with probability 1/2) as long as the resulting rook placement does not violate the partial order.

For instance, suppose the partial order is $1 \preceq 4, 2 \preceq 4$, $y = (2,4,4,4)$ and columns 2 and 3 are being considered for swapping. Then rows 1 and 2 are active. Since there is no restriction of the form $1 \preceq 2$, with probability $1/2$, the swap occurs.

If the swap occurs, note that in the first row, the rook is now in the third column. So it would be necessary to make $y(1) = 3$ so that it was still bounding. The value of $y(2)$ could remain unchanged.

As another example, suppose that $y(2) = 3$ and $y(4) = 4$, and columns 3 and 4 are being considered for swapping. Suppose the roll is made to swap columns 3 and 4 if possible. Should $y(2)$ and $y(3)$ be changed?

The answer is no. The only way $y(2) = 3$ could cease to be a true bound on the underlying state with such a swap is if $x^{-1}(2) = 3$. Since $2 \preceq 4$, if $x^{-1}(2) = 3$ then $x^{-1}(4) > 3$. The bounding chain says $x^{-1}(4) \leq 4$, so $x^{-1}(4) = 4$. The partial order means that those two never swap. So y would also remain unchanged in this case.

In general, the bounding chain behaves much like the original underlying chain; if $a \prec b$, then $y(a)$ must stay smaller than $y(b)$. This is encapsulated in the following bounding chain. In this pseudocode, $n+1$ is a dummy position that is treated as though it is incomparable with every element of $\{1,\ldots,n\}$, and $y(n+1)$ is a dummy coordinate of the bounding state that affects nothing.

`Karzanov_Khachiyan_bounding_chain`
Input: y, $j \in \{1,\ldots,n-1\}, c \in \{0,1\}$
Output: y

If $c = 1$ then
    $a \leftarrow n+1, b \leftarrow n+1$
    If $j = n-1$ then $b \leftarrow \inf\{b' : y(b') = n\}$
    Else if $\exists b'$ such that $y(b') = j+1$, then $b \leftarrow b'$
    If $\exists a'$ such that $y(a') = j$, then $a \leftarrow a'$
    If $a \not\prec b$ then
        $y(a) \leftarrow j+1$
        $y(b) \leftarrow j$

Note that $a \not\prec b$ could mean that $b \prec a$, or it could be that a and b are incomparable elements of the partial order.

In order for this to be a bounding chain, it is necessary to find an update function for the underlying chain such that if the underlying state is bounded, then after one step in the underlying and bounding chain, the state is still bounded.

Let $\phi_{BC}(y,j,c)$ denote the output of `Karzanov_Khachiyan_bounding_chain` given the input y, j, and c. Then the goal is to find an update function $\phi(x,j,c)$ for the original Karazanov-Khachiyan chain such that if y bounds x, then $\phi_{BC}(y,j,c)$ bounds $\phi(x,j,c)$ for all values of j and c.

Consider

$$\phi_1(x,j,) = (1 - c\mathbb{1}(x(j) \prec x(j+1)))x + c\mathbb{1}(x(j) \not\prec x(j+1))x_{j\leftrightarrow(j+1)}. \tag{4.4}$$

Unpacking the notation, ϕ_1 says that if $c = 0$ or $x(j) \prec x(j+1)$, then the chain stays where it is. Otherwise, the values at positions j and $j+1$ are swapped. Therefore, this update function simulates the Karzinov-Khachian chain as long as i is chosen uniformly over $\{1,\ldots,n-1\}$ and c is $\mathsf{Bern}(1/2)$.

Of course if $c \sim \mathsf{Bern}(1/2)$, then $1-c \sim \mathsf{Bern}(1/2)$ as well, so an equally valid update function is

$$\phi_2(x,j,c) = \phi_1(x,j,1-c). \tag{4.5}$$

Note that since c and $1-c$ have the same distribution, it is possible to alter the update function dependent on properties of the state x and position j and still get a valid update function. For example, let

$$\phi_3(x,j,c) = \phi_1(x,j,c)\mathbb{1}(x(j) = 1) + \phi_2(x,j,c)\mathbb{1}(x(j) \neq 1) \tag{4.6}$$

is a valid update function. This flexibility between using ϕ_1 or ϕ_2 at each step is essential to constructing an update function for which ϕ_{BC} is a bounding chain.

With these preliminaries, here is the update function that works for the bounding chain update ϕ_{BC}.

$$\phi(x,j,c) = \phi_1(x,j,c)\mathbb{1}(y(x^{-1}(j)) \neq j+1) + \phi_2(x,j,c)\mathbb{1}(y(x^{-1}(j)) = j+1) \tag{4.7}$$

Lemma 4.3. *Suppose y bounds x. Then $\phi(x,j,c)$ is bounded by $\phi_{BC}(y,j,c)$ for all choices of j and c.*

Proof. Consider the current permutation x. Then item $x^{-1}(j)$ is in position j, and item $x^{-1}(j+1)$ is currently in position $j+1$. The move in the Markov chain might (or might not) swap the positions of item $x^{-1}(j)$ and $x^{-1}(j+1)$.

Since those are the only two positions affected by the move, showing the result is equivalent to showing that the items $x^{-1}(j)$ and $x^{-1}(j+1)$ are properly bounded by the new bounding state.

There are three ways $\phi_{BC}(y,j,c)$ could fail to bound $\phi(x,j,c)$. One, $y(x^{-1}(j)) = j$, so the bound on item $x^{-1}(j)$ was tight, and then x executed the transposition but y did not. Second, $y(x^{-1}(j)) = j+1$, so the bound on the position of item $x^{-1}(j)$ was $j+1$, and both x and y executed the transposition: this would move item $x^{-1}(j)$ to position $j+1$, but change the bound on the item to j. Third $y(x^{-1}(j+1)) = j+1$, and y executed the transposition and x did not. This would keep item $x^{-1}(j+1)$ in position $j+1$, but make the bound of the item j, which is incorrect.

These are the only three possible ways for the new bounding state not to contain the new underlying state, so consider each in turn.

Case 1: Suppose $y(x^{-1}(j)) = j$, and x does the transposition. Then it must be that $c = 1$ and $x^{-1}(j) \not\prec x^{-1}(j+1)$. Consider the value of b (as found in `Karzanov_Khachian_bounding_chain`). If b is $n+1$ then y does the transposition. If $b = x^{-1}(j+1)$, y does the transposition. If b is neither of those things, then $x^{-1}(j+1) \neq b$, which means that $x(b) \notin \{j, j+1\}$, so $x(b) < j$. Hence $x^{-1}(j) \not\prec b$, as otherwise the bounding state y would be out of order. So y executes the transposition as well and the new bounding state continues to bound the new underlying state.

Case 2: Suppose $y(x^{-1}(j)) = j+1$, and x executes the transposition. Then by the definition of ϕ in (4.7), $c = 0$, which means that y does not execute the transposition. So the new bounding state still has $y(x^{-1}(j)) = j+1$, and it continues to bound the new underlying state.

Case 3: Suppose $y(x^{-1}(j+1)) = j+1$, and x does not do the transposition. When $c = 0$, y also does not do the transposition. When $c = 1$, this must be because $x^{-1}(j) \prec x^{-1}(j+1)$. Because it is a bounding state, $y(x^{-1}(j)) \geq j$, but because $x^{-1}(j) \prec x^{-1}(j+1)$, $y(x^{-1}(j)) < y(x^{-1}(j+1)) = j+1$. Hence $y(x^{-1}(j)) = j$, and in the `Karzanov_Khachian_bounding_chain`, $a = x^{-1}(j) \prec b = x^{-1}(j+1)$. So y does not make the transposition either.

Since in all three possibilities for failure, y continues to bound x after the Markov chain move, it is a valid bounding chain. □

Let Y_t be the bounding chain process. Note that the number of items c for which $Y_t(c) = n$ either stays the same or decreases by one at each step. Once there remains only a single element c with $Y_t(c) = n$, all of the labels of Y_t are unique because they are all different. That means there is a unique state x with $x^{-1} \leq Y_t$, and that is $x = Y_t^{-1}$. For instance, if $Y_t = (3,2,4,1)$, then the underlying state must have $X_t^{-1}(4) = 1$, which in turn means $X_t^{-1}(2) = 2$, which implies $X_t^{-1}(1) = 3$, which gives the final piece of $X_t^{-1}(3) = 4$.

The probability that the bounding chain allows for more than one state goes down exponentially in t.

Fact 4.2. *The probability that the bounding chain for linear extensions has more than one underlying state is at most*

$$2n^2 \exp(-t(\pi^2/8)n^{-3}). \tag{4.8}$$

Proof. When there is no item a such that $y(a) = j$, say that there is a *hole* at position j. The analysis proceeds by tracking the behavior of the holes. As steps are taken in the bounding chain, the holes move around. For instance, if at bounding state y there is a hole at position j, then in $\phi_{BC}(y, j, 1)$ the transposition will move the hole to position $j+1$.

When $j = n-1$, the hole disappears after moving to n, since there is always at least one a such that $y(a) = n$.

Similarly, if there is a hole at $j+1$, $\phi_{BC}(y, j, 1)$ moves this hole to position j. A hole already at position 1 cannot be moved any farther to the left.

If there is no hole at j or $j+1$, then after the move $\phi_{BC}(y, j, 1)$, there will still be no hole at j or $j+1$.

So the holes can only be destroyed or moved around, never created. There are many ways to analyze the behavior of the holes. Following [53], set

$$\phi(i) = \frac{\sin(Ci)}{\sin(C)}, \quad C = \frac{\pi}{2(n-1)}. \tag{4.9}$$

Let W_t be the distance that a particular hole is away from position n. Suppose $W_{t-1} < n$. For the hole to move, either the position of the hole or the spot to the left of the hole must be chosen, and it must be true that $c = 1$. Therefore, with probability at least $1 - 1/(n-1)$ the hole stays where it is, with probability at most $(1/2)/(n-1)$ the hole moves left (when the hole is in position 1 the chance it moves left is 0), and with probability $(1/2)/(n-1)$ the hole moves to the right. Hence

$$\mathbb{E}[\phi(W_t)|W_{t-1}] = \frac{1}{2(n-1)}\phi(W_t - 1) + \left(1 - \frac{1}{(n-1)}\right)\phi(W_t) + \frac{1}{2(n-1)}\phi(W_t + 1). \tag{4.10}$$

This can be written more compactly with the forward difference operator defined as $\Delta\phi(i) = \phi(i+1) - \phi(i)$. (This use of Δ as an operator should not be confused with the use of Δ to indicate the maximum degree in a graph from earlier.) This makes the second forward difference operator $\Delta^2\phi(i) = \Delta\phi(i+1) - \Delta\phi(i) = \phi(i+2) - 2\phi(i+1) + \phi(i)$. Therefore (4.10) can be written

$$\mathbb{E}[\phi(W_t)|W_{t-1}] \geq \phi(W_{t-1}) + (1/2)(n-1)^{-1}\Delta^2\phi(W_t - 1). \tag{4.11}$$

Using the sum of angles formula for sine gives

$$\Delta^2[\sin(CW_{t-1})] = \sin(CW_{t-1})[2\cos(C) - 2]. \tag{4.12}$$

Note $(1/2)(n-1)^{-1} = \pi^{-1}C$, so

$$\mathbb{E}[\phi(W_t)|W_{t-1}] = \phi(W_{t-1}) + \pi^{-1}C\phi(W_{t-1})(2\cos(C) - 2) \leq \phi(W_t)\exp(-\pi^2/[8n^3]) \tag{4.13}$$

where the last inequality follows from the first three terms in the Taylor series expansion of cosine. When $W_{t-1} = 0$, $W_t = 0$ as well, so (4.13) holds even for $W_{t-1} = 0$. Then using the expectational conditioning rule:

$$\mathbb{E}[\phi(W_t)] = \mathbb{E}[\mathbb{E}[\phi(W_t)|W_{t-1}]] \leq \phi(W_{t-1})\exp(-\pi^2/[8n^3]).$$

At this point a simple induction gives $\mathbb{E}[\phi(W_t)] \leq \phi(W_0)\exp(-t\pi^2/[8n^3])$.

Now let H_t denote the sum over the holes of the ϕ function applied to the distance from each hole to n. Then linearity of expectations gives $\mathbb{E}[H_t] \leq H_0\exp(-t\pi^2/[8n^3])$.

As long as any holes remain, $H_t \geq 1$. By Markov's inequality, $\mathbb{P}(H_t > 0) \leq \mathbb{E}[H_t]$. It is straightforward to verify that $1/\sin(C) \leq 2n/\pi$, which makes $H_0 \leq 2n^2/\pi$ and the result follows directly. □

4.5 Self-organizing lists

Consider the problem of designing a self-organizing list for a database. Like linear extensions, this application also involves a distribution on permutations. A linked list is a permutation of objects where the objects can only be accessed in order. So if the permutation on $\{1,2,3,4,5\}$ is 43512, and the list attempts to access object 1, then objects 4, 3, 5, and then 1 must be accessed to find the 1. In other words, the time to access an object is just equal to its location in the permutation.

A simple probabilistic model is that each access of an object occurs independently, and object i is chosen to be accessed with probability p_i. Then the goal is to develop a simple way of updating the list after each access to ensure that the more popular (high p_i) items are located near the front of the permutation, giving lower access times on average.

Two simple schemes are move to front (MTF), and move ahead 1 (MA1). In MTF, when an item is accessed, it is immediately moved directly to the front of the permutation. In MA1, when an item is accessed, it moves forward one place in the permutation. So with permutation 43512, after object 1 is accessed, in MTF the new permutation would be 14352, and using MA1, the new permutation is 43152.

Suppose that $p_i > 0$ for all i. Then in either MTF or MA1, every possible permutation can be reached, and there is a chance that the permutation remains unchanged after the access. Hence MTF and MA1 give irreducible, aperiodic Markov chains on the set of permutations. That means that they each have a stationary distribution, and the limiting distribution of the chain converges to the stationary distribution.

Call π_{MTF} and π_{MA1} the stationary distributions for the MTF and MA1 process, respectively.

Definition 4.3. *For a self-organizing list with stationary distribution π, suppose $X \sim \pi$. Then the* long-term average access time *is*

$$\sum_{i=1}^{n} p_i \sum_{j=1}^{n} j\mathbb{P}(X(j) = i). \tag{4.14}$$

This is essentially the average time to access the next object given that the self-organizing list is a draw from the stationary distribution of the Markov chain. To study the long-term average access time through a Monte Carlo approach, a Monte Carlo approach needs to be able to draw samples from this stationary distribution.

A bounding chain for the MTF rule works as follows. Let $Q = \{1,2,\ldots,n\}$. Then the bounding state begins at $(Q,Q,\ldots,Q)$ indicating that the position of each object in the permutation is unknown. Suppose object 3 is chosen to be moved to the front. Then the bounding chain becomes $(3,Q,Q,\ldots,Q)$. If 5 is chosen next, the bounding chain becomes $(5,3,Q,Q,\ldots,Q)$.

It is easy to see that the bounding chain never loses track of an object that has been chosen. Therefore, the probability that the bounding chain bounds a unique state after t steps is exactly the probability that all elements from 1 to n have been chosen at least once during the t steps. In fact, it is possible to determine the long-term average access time for the MTF chain, so it is not surprising that the easiest Markov chain works so well with perfect simulation.

The MA1 distribution is a tougher nut to crack. It is currently unknown how to efficiently sample from the MA1 distribution for all access vectors $\{p_i\}$. Still, something can be said. First, note that (unlike the MTF chain) the MA1 Markov chain is reversible. This gives a means to compute the stationary distribution up to an unknown normalizing constant:

$$\pi_{\text{MA1}}(x) \propto \prod_{j=1}^{n} p_{x(j)}^{n-j}. \tag{4.15}$$

Now that the distribution is known (up to an unknown normalizing constant), a new chain can be built that has the same distribution. The idea is to use random transpositions together with a Metropolis Hastings approach.

First, choose i and j iid uniformly over $\{1,2,\ldots,n\}$. Then propose transposing the objects at positions i and j. Note the proposal density divided by the original density is $p_{x(i)}^{j-i}/p_{x(j)}^{j-i}$. That makes the Metropolis-Hastings chain as follows.

```
Transposition_chain_for_MA1
1)  Draw i, j uniformly (iid) from {1,2,...,n}
3)  If U < p_{y(i)}^{j-i} / p_{y(j)}^{j-i}
4)      x ← x_{i↔j}
```

Now to turn this Metropolis-Hastings setup into a bounding chain step. Call the bounding state y. Then each $y(i)$ will either have size 1 (refer to such elements as *known*) or size n (refer to such elements as *unknown*.) Of course, initially $y(i) = \{1,2,\ldots,n\}$ for all i. Similarly, if there exists j such that $y(j) = j'$, refer to object j' as *known*, and otherwise it is *unknown*.

Consider the positions i and j that are to be swapped. There are three different cases for i and j that need to be considered.

First, both i and j are known. Then the bounding chain just proceeds as in the original chain. Second, i could be known and j could be unknown. This presents a problem, since that makes the acceptance ratio unknown. Find the smallest and largest that the acceptance ratio could be, and proceed from there. Third, both i and j could be unknown.

This actually presents an opportunity. In order to learn more about the state, it helps to treat j as being an unknown object rather than a position. In other words, in this case swap position i with what position contains object j. Since each unknown object is equally likely to be picked in this instance, this does not change the transition probabilities for the underlying Markov chain.

The reason for this change in how i and j work is as follows. If i is the leftmost unknown position, and j is the highest-probability unknown object, then the move of j to position i always happens. This means that moving forward, $y(i) = \{j\}$.

Here is how to convert j from a uniform unknown position to a uniform unknown object. Let $a_1 < a_2 < \cdots < a_k$ be the set of unknown positions, and $b_1 < b_2 < \cdots < b_k$ be the set of unknown objects. Then let j' be the value such that $a_{j'} = j$. Then make b_j the unknown object.

`Bounding_Transposition_chain_for_MA1`

Input: $y, i, j, U \in [0,1]$, *Output:* y

If i and j are both known positions
    If $U < p_{y(i)}^{j-i}/p_{y(j)}^{j-i}$ then $y \leftarrow y_{i\leftrightarrow j}$
Else if i is a known position but j is an unknown position
    $k_1 \leftarrow \arg\max_{k \text{ unknown}}\{p_k\}$, $k_2 \leftarrow \arg\min_{k \text{ unknown}}\{p_k\}$
    If $i < j$ then $r_{\min} \leftarrow (p_{y(i)}/p_{k_2})^{i-j}$, $r_{\max} \leftarrow (p_{y(i)}/p_{k_1})^{i-j}$
    Else $r_{\min} \leftarrow (p_{y(i)}/p_{k_1})^{i-j}$, $r_{\max} \leftarrow (p_{y(i)}/p_{k_2})^{i-j}$
    If $U < r_{\min}$ then $y \leftarrow y_{i\leftrightarrow j}$
        Else if $U < r_{\max}$ then $y(i) \leftarrow \{1,2,\ldots,n\}$
Else if i and j are both unknown positions
    Let $a_1 < a_2 < \cdots < a_k$ be the unknown positions
    Let $b_1 < b_2 < \cdots < b_k$ be the unknown objects
    Let j' be such that $a_{j'} = j$, then set $j \leftarrow b_{j'}$
    If $i = a_1$, and $p_j = \max_{\text{object } \ell \text{ unknown}}\{p_\ell\}$ then $y(i) \leftarrow \{j\}$

Now suppose that the p_i values are decreasing geometrically, so that there exists $\gamma \in [0,1)$ such that $p_{i+1} \leq \gamma p_i$. Then one can show that the bounding chain converges quickly.

Lemma 4.4. *When $\gamma < 1/2\ldots$, the expected number of moves to reach a state that has $\#y(i) = 1$ for all i is bounded above by $n^3(1-\gamma^2/(1-\gamma)^2)^{-1}$.*

Proof. Given the bounding chain state Y_t after t steps, let

$$K_t = \max\{k : \{1,2,\ldots,k\} \text{ are known positions in } Y_t\}. \tag{4.16}$$

After zero steps, $K_0 = 0$, but if $K_t = n$, then that means that there does not exist i with $\#y(i) > 1$. So let us examine how K_t evolves on average.

Given $K_t = k$, $K_{t+1} \in \{0,1,\ldots,k+1\}$. In order for $K_{t+1} = k+1$, i must be position $k+1$, and j must be the highest-probability unknown object. These two choices occur with probability

$$r_1 = \frac{1}{n^2}. \tag{4.17}$$

In order for $K_{t+1} = r \leq k$, the position i must be chosen so that $y(i) = r$, and an unknown value of j must be picked. Since p_r is greater than $p_{i'}$ for any unknown object i', the transposition always occurs when $j < i$, and accepted with probability at most $(p_r/p_{i'})^{-a}$ when $i < j$, where a is the minimum distance from i to an unknown position.

The unknown with the largest probability has $p_{i'} \leq p_r \gamma^{k+1-r}$, and subsequent probabilities get smaller by a factor of γ. Hence the chance that object r becomes unknown is bounded above by

$$r_i = \left(\frac{1}{n^2}\right)\left(\gamma^{k+1-r} + (\gamma^2)^{k+1-r} + \cdots\right) = \frac{\gamma^{k+1-r}}{n^2(1-\gamma^{k+1-r})}. \tag{4.18}$$

Combining these gives

$$\mathbb{E}[K_{t+1}|K_t] = K_t + \frac{1}{n^2} - \sum_{r=0}^{K_t} \frac{(K_t - r)}{n^2} \frac{\gamma^{K_t+1-r}}{1-\gamma^{K_t+1-r}} \leq K_t + \frac{1}{n^2}\left[1 - \frac{\gamma^2}{(1-\gamma)^2}\right]. \quad (4.19)$$

So K_t is changing on average by a small amount α, which is positive when $\gamma < 1/2\ldots$ (the solution to $\gamma^2/(1-\gamma)^2 = 1$.) Standard methods for martingales with drift (see [31] then can be used to show that the expected time for K_t to reach n is n/α, which gives the desired result. □

Interestingly, it is known that for any probability vector $(p_1, \ldots, p_n)$, the MA1 rule has a lower long-run expected access time than the MTF rule [114], but Monte Carlo simulation is necessary to discover exactly how much lower it is.

4.6 Using simple coupling with bounding chains

For many problems, finding an initial bounding state is easy, but it is unclear how to move any from that initial state.

An approach to this type of situation is to partition the state space into pieces where in each piece, some information about the labeling is known. For instance, if a node v is labeled from $\{0,1,2\}$, then the set of configurations can be partitioned into three sets Ω_0, Ω_1, and Ω_2, where Ω_i consists of those configurations where node v is labeled i.

This gives the bounding chain some information to start with, and allows for every state within each Ω_i to coalesce down to a single state. That leaves a single state coming from each Ω_i, for three states total.

Now these three states can be run forward in time using an update function with the hope that these final three then coalesce down to a single state, allowing the application of CFTP.

4.6.1 *Sink-free orientations*

This technique was used to give the first polynomial time perfect simulation algorithm for sampling uniformly from the sink-free orientations of a graph. In an undirected graph $G = (V,E)$, the edges are size two subsets of the set of nodes where order does not matter. The edge $\{i,j\}$ and $\{j,i\}$ are identical. To orient an edge is to give it one of two directions, either from i to j, or from j to i.

Definition 4.4. *An* orientation *of a graph is a labeling where each edge $\{i,j\}$ is labeled either (i,j) or (j,i). An edge labeled (i,j) is said to be* directed *from i to j. Call i the* tail *of the edge and j the* head.

Definition 4.5. *An orientation is said to have a* sink *at node i if every edge of the form $\{i,j\}$ is oriented (j,i). An orientation where none of the nodes are a sink is said to be* sink-free.

If the degree of a node is 1, then there is only one way to orient the outgoing edge to avoid creating a sink. That node and edge can then be removed from consideration, and repeat the process on any new degree 1 nodes created by the removal. So assume without loss of generality that all the nodes of the graph have degree at least 2.

Similarly, if a graph is disconnected, then each component will have to be oriented separately, which means that again without loss of generality it is possible to only consider connected graphs.

Consider the set of sink-free orientations of a graph. A simple Gibbs sampler chain chooses an edge uniformly at random, and then an orientation for the edge is chosen uniformly from among the orientations that do not create a sink. Pseudocode for this update function looks like this.

Sink_Free_Orientation_Gibbs_chain	
Input: x, $\{i,j\} \in E$, $B \in \{0,1\}$	*Output:* x
1)	If $B = 1$ then $a \leftarrow (i,j)$, else $a \leftarrow (j,i)$
2)	Let $n_{\text{out}} \leftarrow \#\{w : x(\{a(2),w\}) = (a(2),w)\}$
3)	If $n_{\text{out}} \geq 1$ then $x(\{i,j\}) \leftarrow a$

In other words, if the chosen orientation for $\{i,j\}$ leaves at least one other edge leaving the head of the edge (so the change does not create a sink), then make the change in orientation.

Bubley and Dyer showed that this chain mixes rapidly as long as the graph was connected and contained more than one cycle. More specifically, they showed that if two chains X and X' are passed through the pseudocode above using the same choice of edge and coin B, then after $O(\#E^3)$ steps there is a large chance that the two chains have come together.

Lemma 4.5 ([17]). *Suppose $e_1, e_2, \ldots$ are iid uniform over E, and $B_1, B_2, \ldots$ are iid Bernoulli with mean 1/2. Then consider using the same update for X_t and Y_t so $X_{t+1} =$ Sink_Free_Orientation_Gibbs_chain(X_t, e_{t+1}, B_{t+1}) and also $Y_{t+1} =$ Sink_Free_Orientation_Gibbs_chain(Y_t, e_{t+1}, B_{t+1}). Then no matter the values of X_0 and Y_0, $\mathbb{P}(X_t \neq Y_t) \leq (1/2)^{t/\#E^3}$.*

Using the coupling lemma, this means that starting at about $\#E^3$ number of steps, the total variation distance from the Markov chain state distribution and the stationary distribution starts declining exponentially.

So it is possible to bring two states together. However, to use CFTP it is necessary to bring all the states together. The problem is that for this chain, it is impossible to use the bounding chain technique. Each edge $\{i,j\}$ is oriented either (i,j) or (j,i), so initially each bounding chain state is $\{(i,j),(j,i)\}$. But then taking steps in the Markov chain does not help: no matter what edge is chosen to flip or not flip, the fact that all the adjacent edges have unknown orientation means that the state of the edge will still be $\{(i,j),(j,i)\}$ at the next step.

Here is the solution (a revision of an idea first presented in [64]): first bring all the states down to two states, then use the Bubley-Dyer coupling to bring those two states together.

Pick an edge $\{i,j\}$. Let Ω_1 consist of those orientations where $\{i,j\}$ is oriented (i,j), and Ω_2 be those orientations with $\{i,j\}$ oriented (j,i). Then the Gibbs chain will choose uniformly at random any edge other than $\{i, j\}$, and then uniformly choose from the orientations of that edge that do not contain a sink.

Since $\{i, j\}$ can never be picked, a bounding chain for Ω_1 can now make progress, since the orientation of $\{i, j\}$ is fixed at (i, j), so i will always be sink-free regardless of how its adjacent edges are oriented.

Similarly, at the same time these moves can make progress on a bounding chain for Ω_2. After $\Theta(\#E^3)$ steps, it is likely that each of these two bounding chains have reduced the set of possible states to just one apiece. Then the regular coupling of Bubley and Dyer can be used to bring these two states together down to just one state, thereby allowing the use of CFTP.

As with the regular Gibbs step, the input to the bounding chain will be the edge $\{i, j\}$ whose orientation is trying to change, and B the coin flip that determines if the orientation to be attempted is (i, j) or (j,i).

Suppose the orientation is (i, j). Then this could make j a sink. But suppose that there is a k, such that $y(\{j,k\}) = \{(j,k)\}$. Then it is known that j already has an edge leaving it, so it is okay to change the orientation to (i, j). On the other hand, if no edge is known to be leaving j, and some edges have unknown orientation, then it is unknown whether the flip can be accomplished or not. This procedure is given by the following pseudocode.

Sink_Free_Orientation_bounding_chain
Input: y, $\{i, j\} \in E, B \in \{0,1\}$ *Output:* y

---

1) If $B = 1$ then $a \leftarrow (i, j)$, else $a \leftarrow (j,i)$
2) Let $n_{\text{out}} \leftarrow \#\{w : y(\{a(2), w\}) = \{(a(2), w)\}$
3) Let $n_{\text{unknown}} \leftarrow \#\{w : y(\{a(2), w\}) = \{(a(2), w), (w, a(2))\}$
4) If $n_{\text{out}} > 1$ then $y(\{i, j\}) \leftarrow \{(a(1), a(2))\}$
5) Else if $n_{\text{out}} = 0$ and $n_{\text{unknown}} > 0$, $y(\{i, j\}) \leftarrow y(\{i, j\}) \cup \{(a(2), a(1))\}$

Lemma 4.6. *Suppose there exists an edge e such that $\#Y_0(e) = 1$, $e_1, e_2, \ldots, e_t$ are iid uniform over $E \setminus \{e\}$, and $B_1, \ldots, B_t$ are iid uniform over $\{0,1\}$. Then if $Y_{t+1} =$* Sink_Free_Orientation_Gibbs_chain(Y_t, e_{t+1}, B_{t+1}), *the chance that any edge e' exists with $\#Y_t(e') > 1$ is at most*

$$\exp\left(-t\left[\frac{1}{2\#E^3} + \frac{1}{24\#E^4}\right]\right). \tag{4.20}$$

In particular, to get the chance of any edge remaining unknown to be at most δ requires $(2\#E^3 + O(\#E^1))\log(\delta^{-1})$ steps.

Proof. Let $D_t = \#\{e : \#Y(e) > 1\}$. Then $D_0 = \#E - 1$. At step t, the value at edge e_{t+1} could change, so $|D_{t+1} - D_t| \leq 1$.

How can D_t increase? Suppose $Y_t(\{i, j\}) = \{(i, j)\}$, and $e_{t+1} = \{i, j\}$. Then if $B = 1$ the edge just keeps its current orientation, so consider $B = 0$.

Then $\#Y_{t+1}(\{i, j\}) = 2$ only if there is at least one edge $\{i,w\}$ with $\#Y_{t+1}(\{i,w\}) > 1$, and no other edge $\{i,w'\}$ with $Y_{t+1}(\{i,w'\}) = \{(i,w')\}$. That means that for the edge $\{i,w\}$, the edge $\{i, j\}$ is the unique edge leaving i that can be made to increase D_t by 1.

But since edge $\{i, j\}$ is known to be oriented (i, j), if edge $\{i, w\}$ had been selected and attempted to be oriented (w, i), that would have been accepted, and D_t would be reduced by 1.

This argument means that the chance that D_t increases by 1 exactly equals the chance that D_t decreases by 1. Let that chance that it goes up 1 or down 1 be p_t. Then if $f(a) = \sin(a/\#E)$, then

$$\begin{aligned}\mathbb{E}[f(D_{t+1})|\mathscr{F}_t] &= f(D_{t+1}+1)p_t + f(D_{t+1}-1)p_t - f(D_{t+1}) \\ &= f(D_t)[1-2p_t(1-\cos(1/\#E)],\end{aligned}$$

where the last line follows from the sum of angles formula for sine. Since at least one edge is always known, there is at least a $(1/2)(1/\#E)$ (the $1/2$ comes from the choice of B) chance of D_t changing, hence $2p_t \geq \#E^{-1}$. An easy induction (and the Taylor series bound on cosine) then gives

$$\mathbb{E}[f(D_t)] \leq f(D_0)[1 - \#E^{-1}((\#E^{-2}/2) - (\#E^{-4}/24))]^t.$$

Of course $f(D_0) \leq 1$, and when $\mathbb{E}[f(D_t)] \leq \sin(\gamma/\#E)$, by Markov's inequality, $\mathbb{P}(D_t) > 0) \leq \gamma$. Using $\sin(1/\#E) \geq \#E^{-1} - \#E^{-3}/6$ completes the proof. □

Chapter 5

Advanced Techniques Using Coalescence

> Time has been transformed, and we have changed; it has advanced and set us in motion; it has unveiled its face, inspiring us with bewilderment and exhilaration.
>
> *Khalil Gibran*

In the previous chapter, coupling from the past was introduced, which remains the most widely applicable protocol for creating perfect simulation algorithms. However, CFTP has two important drawbacks. It requires that the random variables generated be used twice, and it is noninterruptible. There are two variants of the algorithm that each deal with one of these issues.

5.1 Read-once coupling from the past

First consider acceptance/rejection. Recall that AR generates $X_i \sim \nu(B)$ until first encountering $X_T \in A$. Denote the event that $X_i \in A$ by S (for success) and $X_i \notin A$ by F (for failure.)

Then a sequence of AR consists of a sequence of failures followed by a single success. So for instance, $FFFFFFFS$, or FS, or $FFFS$, or S are all valid possible sequences of blocks. That is, a valid sequence consists of a finite number of F blocks followed by an S block. The total number of blocks is a geometric random variable with parameter equal to the probability of an S block.

Suppose that multiple samples $X \sim \nu(A)$ are desired. Then simply consider the infinite stream of blocks:

$$FFFSFSSFFFFFSFFFSFSSFFFS\ldots$$

Every state at the end of an S block represents a sample exactly from the target distribution.

Now consider homogeneous CFTP. Let S denote the event that $U \in A$, so it is possible to determine if $\phi(\Omega, U) = \{y\}$ for some y. For instance, when the update function is monotonic, then S is the event that $\phi(x_{\max}, U) = \phi(x_{\min}, U)$.

Then CFTP can be summarized as follows. Generate U randomly. If S occurs, then report the state at the end of the S block. Otherwise, the value of U chosen results in an F block, and recursion occurs to generate the stationary sample, which is then run through the F block.

Suppose, for instance, three levels of recursion are necessary. Then the structure of the output is $S_3F_2F_1F_0$. The zeroth level (the original level) is an F, the first is an F, and the second is an F, which have been labeled with subscripts for their appropriate levels. Only the third level of recursion gives an S, which is then fed through the second level F_2 block, which is fed through the first level F_1 block, and finally the F_0 block.

So in general the form of homogeneous CFTP is $SFF\cdots F$, where the final answer is the state at the end of the final F block, and where the total number of blocks is a geometric random variable with parameter equal to the probability of an S block.

Now suppose blocks are just run forward in time. Then the result looks very similar to an AR run:

$$FSFFFSFFFSSFFSFSFFFFFFFFSFFF\ldots.$$

Notice that after the first S block, there is a geometric number of blocks until the next S block occurs. Hence the first S block is followed by a geometric number (minus 1) of F blocks, exactly as in the original homogeneous CFTP!

Therefore, the state at the end of the F block that precedes the first S block must have the target distribution. In fact, the state that precedes the second S block, the third S block, etc., must all have the target distribution.

This is the essence of read-once coupling from the past (ROCFTP) as created by Wilson [129]. Run the blocks forward until the first S block. Then keep running forward, and before each subsequent S block, the state will come from the target distribution.

As before, let A be a set for an update function ϕ_t such that if $U \in A$, then $\phi_t(\Omega, U)$ contains only a single element.

```
Read_once_coupling_from_the_past
Input: k, t     Output: y_1, ..., y_k ~ π iid

i ← 0, x ← an arbitrary element of Ω
Repeat
    y_i ← x
    Draw U ← Unif([0,1]^t)
    x ← φ_t(x, U)
    i ← i + 1(U ∈ A)
Until i > k
```

Unlike the earlier formulations of perfect simulation, there is no explicit recursion. As with AR, the above code could be reformulated as a recursive function since repeat loops can always be written recursively. From a practical perspective, using a repeat loop rather than recursion speeds up the process.

When $i = 0$, the repeat loop draws from $\phi_t(\Omega, U)$ until a success is obtained. This is wasted effort, as these draws are then thrown away by the algorithm, and y_0 is not part of the output. So to generate k iid draws from π, ROCFTP requires $k+1$ different success blocks, while CFTP only requires k of them.

However, CFTP needs to evaluate each step twice, therefore to generate k samples requires $2k$ evaluations of the ϕ_t function, while ROCFTP only requires $k+1$. For $k = 1$ there is no gain, but for larger numbers of samples this is a tremendous speedup.

This improvement requires that lines 4, 5, and 6 be written in such a way that the state updates and the test that $U \in A$ can be accomplished by looking at $U_1, \ldots, U_t$ as a stream of choices. For instance, when the update is monotonic, this can be accomplished by applying the update function to both the upper, lower, and middle state.

`Monotonic_Read_once_coupling_from_the_past`
Input: k, t *Output:* $y_1, \ldots, y_k \sim \pi$ iid

---

1) $i \leftarrow 0$, $x \leftarrow$ an arbitrary element of Ω
2) Repeat
3) $\quad y_i \leftarrow x$
4) $\quad x_{\text{max}} \leftarrow$ largest element of Ω $x_{\text{min}} \leftarrow$ smallest element of Ω
5) $\quad$ For t' from 1 to t
6) $\quad\quad$ Draw $U_t \leftarrow \text{Unif}([0,1])$
7) $\quad\quad x \leftarrow \phi(x, U_t)$, $x_{\text{max}} \leftarrow \phi(x_{\text{max}}, U_t)$, $x_{\text{min}} \leftarrow \phi(x_{\text{min}}, U_t)$
8) $\quad i \leftarrow i + \mathbb{1}(x_{\text{max}} = x_{\text{min}})$
9) Until $i > k$

The downside is that in the repeat loop, the state x needs to be saved at each step before $\phi(\Omega, U)$ is calculated. Therefore, the memory requirements to hold the configuration has doubled.

Since this extra memory requirement is usually small compared to the storage of the U variable, generally ROCFTP comes out ahead of CFTP in the memory game. Furthermore, it can be written without the overhead of a random number of recursions, which in many computer languages are very slow compared to repeat loops. Therefore ROCFTP is typically preferred in practice to CFTP.

Lemma 5.1. `Read_once_coupling_from_the_past` *generates* $Y_1, \ldots, Y_k \sim \pi$ *that are iid.*

Proof. The fact that a geometric number of blocks where the first is an S block and the remainder are F blocks gives a state in the target distribution is just basic CFTP where t is the same at each level of recursion. The correctness of this then follows as a corollary of Theorem 3.1.

So the Y_i are identically distributed. Are they independent? Well, the block that follows after a state Y_i is an S block. The output of this S block is independent of the state Y_i that precedes it, since this is a Markov chain. Hence all $Y_{i'}$ with $i' > i$ are independent of Y_i. $\square$

5.1.1 Example: ROCFTP for the Ising model

Now look at how this method works for the Ising model. The easiest way to use ROCFTP for monotonic systems is to first create an update function that takes as input the current state and all randomness needed to complete the step.

```
Ising_Gibbs_update_function
Input: x ∈ Ω, v ∈ V, U ∈ [0,1]    Output: x(v) ∈ {−1,1}

1)  Let n_1 be the number of neighbors of v labeled 1
2)  Let n_{-1} be the number of neighbors of v labeled -1
3)  x(v) ← −1 + 2·𝟙(U ≤ exp(βn_1)/[exp(βn_1) + exp(βn_{-1})]
```

This pseudocode actually takes into account how computers work: given that only the label on state v changes at each step, it does not make sense for the update function to return the entire new state (although that is the formal definition). Instead, only the new value of $x(v)$ is returned. The calling function, of course, needs to be aware of this behavior in order to utilize this function properly.

The next thing needed for monotonic CFTP is code that runs a single block forward, reports the new state of the chain, and also reports if the block coalesced or not. The output variable B will be true if the block coalesced, and false otherwise.

```
Ising_Gibbs_block
Input: x ∈ Ω, t    Output: x ∈ Ω, B ∈ {TRUE, FALSE}

1)  x_max ← (1,1,...,1), x_min ← (−1,−1,...,−1)
2)  For t' from 1 to t
3)      Draw v uniformly from V, U uniformly from [0,1]
4)      x(v) ← Ising_Gibbs_update_function(x, v, U)
5)      x_max(v) ← Ising_Gibbs_update_function(x_min, v, U)
6)      x_max(v) ← Ising_Gibbs_update_function(x_max, v, U)
7)  B ← (∀v ∈ V)(x_min(v) = x_max(v))
```

Note that as soon as the call to this routine is over, the random variables used are gone: as the name says, the random variables in read-once coupling from the past do not need to be stored. With the block algorithm, the general algorithm can be built.

```
ROCFTP_Ising_Gibbs    Input: k, t    Output: y_1,...,y_k

1)  i ← 0, x ← (1,...,1)
2)  Repeat
3)      y_i ← x
4)      (x, B) ← Ising_Gibbs_block(x, t)
5)      i ← i + 𝟙(B)
6)  Until i > k
```

This example illustrates a strength of ROCFTP: the random variable generation can occur inside the block structure instead of outside it. This can greatly simplify the code for problems where the randomness used to generate the Markov chain step is complex and uses a random number of uniform random variables.

5.2 Fill, Machida, Murdoch, and Rosenthal's method

So the two problems with basic CFTP is that it is read twice and noninterruptible. ROCFTP is read once but still noninterruptible.

Fill [35] was the first to introduce a variant of CFTP that was interruptible, but sadly is still read twice. Møller and Schladitz [103] extended his method to anti-monotone chains, and then in [37], Fill, Machida, Murdoch, and Rosenthal generalized Fill's algorithm for general update functions. This last algorithm will be referred to here as FMMR.

To use ROCFTP, first you need an S block, followed by a number of F blocks, then followed by a second S block. The stationary state is the state at the beginning of the second S block. An S block is an update such that $\phi(\Omega, U) = \{x\}$ for some state x in Ω.

Suppose we first fix an $x \in \Omega$. Now call a block an S_x block if $\phi(\Omega, U) = \{x\}$. Otherwise call the block an F block. Then the same argument for why the output of ROCFTP is stationary means that if we generate an S_x block, followed by some number of F blocks, then a second S_x block, the state at the beginning of the S_x block will be stationary.

Suppose that we could just generate an S_x block, and it was possible to run the chain backwards in time, to figure out what the state at the beginning of the block y was given that it ended at state x. Then $y \sim \pi$.

Here is the central FMMR idea: start with state $X_t = x$. Run X_t backwards in time to get $X_{t-1}, X_{t-2}, \ldots, X_0$. Now run the chain forward in time conditioned on $X_0, X_1, \ldots, X_t$. If $\phi(\Omega, U) = \{x\}$, then accept this as a draw from an S_x block. Otherwise reject and start over.

This is similar to CFTP, but in AR form! So unlike CFTP it is interruptible. This makes FMMR the first widely applicable interruptible perfect simulation algorithm. (It was not the last, however, see Chapters 8 and 9.)

To see what is meant by running the Markov chain backwards in time, it helps to have an example. Consider the biased random walk on $\{0,1,2\}$, with update function

$$\phi(x,U) = x + \mathbb{1}(x < 2, U > 2/3) - \mathbb{1}(x > 0, U \leq 2/3).$$

Hence the chain adds one to the state with probability 1/3, and subtracts one from the state with probability 2/3 (unless the move would take the state out of $\{0,1,2\}$.)

The stationary distribution of this chain is $\pi(\{0\}) = 4/7$, $\pi(\{1\}) = 2/7$, $\pi(\{2\}) = 1/7$. Suppose $X_4 \sim \pi$, $X_5 \sim \pi$, and the goal is to figure out how to simulate $[X_4 | X_5]$.

If $X_5 = 1$, then X_4 is either 0 or 2. By Bayes' rule

$$\begin{aligned}\mathbb{P}(X_4 = 0 | X_5 = 1) &= \mathbb{P}(X_4 = 0)\mathbb{P}(X_5 = 1 | X_4 = 0)/\mathbb{P}(X_5 = 1) \\ &= \pi(\{0\})(1/3)/\pi(\{1\}) = (4/7)(1/3)/(2/7) = 2/3.\end{aligned}$$

In fact, it turns out that $\mathbb{P}(X_4 = i | X_5 = j) = \mathbb{P}(X_5 = i | X_4 = j)$, the transition probabilities for the Markov chain look exactly the same whether the chain is being run forward or backward in time. In general, this situation holds when $\pi(dx)\mathbb{P}(X_{t+1} \in$

$dx|X_t = x) = \pi(dy)\mathbb{P}(X_{t+1} \in dx|X_t = y)$, which is why in Definition 1.34 such chains were labeled reversible.

Okay, so now we know how to run the chain backwards in time. Suppose in our simulation, $X_5 = 0$, $X_4 = 1$, $X_3 = 2$, $X_2 = 1$, $X_1 = 0$, and $X_0 = 0$. Now look at how the chain can be run forward in time conditioned on the path $(0,0,1,2,1,0)$ using the updates $X_t = \phi(X_{t-1}, U_t)$.

In the first step $X_1 = 0$ given $X_0 = 0$. That means that $U_1 \leq 2/3$. So generate U_1 uniformly from $[0,2/3]$. Similarly, since $X_1 = 0$ and $X_2 = 1$, $U_2 > 2/3$. The rest of the U_i can be bounded similarly.

Note that regardless of the value of x_0, setting $x_t = \phi(x_{t-1}, U_t)$ using $U_1 \leq 2/3$, $U_2 > 2/3$, $U_3 > 2/3$, $U_4 \leq 2/3$, and $U_5 \leq 2/3$ gives $x_5 = 0 = X_5$. Hence this gives an S_x block.

On the other hand, suppose $(X_0, X_1, X_2, X_3, X_4, X_5) = (1,0,1,0,1,0)$. Then generating the U_i gives a final state of 1 if $x_0 = 1$ and 0 if $x_0 = 0$. So that is an example of an F block.

In practice, it is not necessary to retain the entire path $(X_0, \ldots, X_t)$. As the pair $(X_{t'-1}, X_{t'})$ the value of $U_{t'}$ can be imputed and stored. So only the $(U_0, \ldots, U_{t'})$ need be retained.

5.2.1 Reversing a Markov chain

In the random walk on $\{0,1,2\}$ example, it is was easy to determine how the reversed walk process behaved. For instance, if $X_{t+1} = X_t + 1$, then in the reverse process $X_t = X_{t+1} - 1$, and so on. Now the general method for reversing a Markov chain is discussed.

Recall Definition 1.27 of a Markov kernel K requires that for all $x \in \Omega$, $K(x,\cdot)$ is a probability distribution, and that for all measurable A and $a \in [0,1]$, the set of states x such that $K(x,A) \leq a$ is also measurable.

In other words, what the kernel K gives us is a family of probability distributions. Given the current state x, $K(x,\cdot)$ is the probability distribution for the next state y. (The second part of the definition is necessary for technical reasons.)

Now the notion of a reverse kernel with respect to a distribution π can be created.

Definition 5.1. *Let K be a kernel from $(\Omega, \mathcal{F})$ to itself, and π a probability measure on $(\Omega, \mathcal{F})$. Then K_{rev} is a* reverse kernel associated with π *if for all bounded measurable functions f defined on $\Omega \times \Omega$*

$$\int_{(x,y)\in\Omega\times\Omega} f(x,y)\nu(dx)K(x,dy) = \int_{(x,y)\in\Omega\times\Omega} f(x,y)\nu(dy)K_{rev}(y,dx).$$

Note that in general the reverse kernel might not even exist, and if it does, is not always uniquely defined!

While the definition above is presented in an abstract fashion, the notion that it is capturing is more straightforward. Suppose $X_t \sim \pi$, and then X_{t+1} comes from taking one step in the Markov chain. Now suppose the value of X_t is lost. What is the new distribution of X_t given X_{t+1}? This is what the reverse kernel gives us.

As was seen in the previous section, for the biased simple random walk on

$\{0,1,2\}$, the kernel is also the reverse kernel. Fortunately, the basic toolkit for building Markov chains with a target stationary distribution often gives us the reversed kernel alongside the original one!

Theorem 5.1. *Both Gibbs samplers and Metropolis-Hastings chains are reversible.*

For more complicated chains, a reversible kernel can be constructed as long as the distribution of the next state can be written using a density with respect to π.

Theorem 5.2. *Suppose that for all $x \in \Omega$, there exists a density $p(x,\cdot)$ with respect to π such that $K(x,B) = \int_B p(x,y)\pi(dy)$. Then a reverse kernel exists where*

$$K_{rev}(y,dx) = \frac{p(x,y)\pi(dy)}{\int_{z\in\Omega} p(z,y)\pi(dz)}.$$

The general proof of this theorem requires the use of a tool from measure theory called Fubini's Theorem, but when the density is with respect to the counting measure things are simpler. In this case $\pi(dy) = \pi(dz) = 1$ for all y and z, and $\int_{z\in\Omega} p(z,y)\pi(dz) = \sum_{z\in\Omega} \mathbb{P}(X_1 = y, X_0 = x)$, so this is saying

$$\mathbb{P}(X_0 = x | X_1 = y) = \frac{\mathbb{P}(X_0 = x, X_1 = y)}{\sum_z \mathbb{P}(X_0 = z, X_1 = y)},$$

just as in our earlier discrete example.

5.2.2 General FMMR

Now that we have a precise notion of how to reverse a Markov chain, the FMMR procedure can be written as follows. Let ϕ_{rev} denote the update function for the reverse kernel of the Markov chain. As earlier, let $\phi_t(x_0, U_1, \ldots, U_t)$ be the state X_t conditioned on $X_0 = x_0$.

FMMR *Input:* $x \in \Omega, t$ *Output:* $X_0 \sim \pi$
1) Repeat
2) $X_t \leftarrow x$
3) For i from $t-1$ to 0
4) Draw $U \leftarrow \textsf{Unif}([0,1])$
5) $X_i \leftarrow \phi_{\text{rev}}(X_{i+1}, U)$
6) Draw U_{i+1} uniformly from $[0,1]$ conditioned on $\phi(X_i, U_{i+1}) = X_{i+1}$
7) Until $\phi_t(\Omega, U) = \{x\}$

Lemma 5.2. *Suppose that for $x \in \Omega$ and $X_0 \sim \pi$, $\mathbb{P}(\phi_t(\Omega, U) = \{x\}) > 0$. Then the output of* `FMMR` *comes from π.*

Proof. Fix a state x in Ω. Suppose $X_0 \sim \pi$. Generate the forward uniforms $U = U_1, \ldots, U_t$ until $\{U \in A\} = \{\phi(\Omega, U) = \{x\}\}$. Then X_0 and $\{U \in A\}$ are independent of each other, so $[X_0 | U \in A] \sim \pi$.

From AR theory, another way to generate $[X_0 | U \in A]$ is to draw X_0 and generate U until $\{U \in A\}$ occurs. When this happens, the sample path $X_t = \phi(X_{t-1}, U_t)$ will end at $X_t = x_0$. Therefore, by only working with the draws of U where the sample paths end at $X_t = x_0$, the last accepted draw of U_t remains from the correct distribution. □

The memory requirement here is two states of the configuration and recording all of the $U_1, \ldots, U_t$ for use in the until condition. Therefore, FMMR is a read twice, interruptible perfect simulation algorithm.

5.2.3 *Example: FMMR for the Ising model*

To see this approach on a nontrivial problem, consider perfect simulation from the Ising model of Section 1.6.1 using the `Ising_Gibbs` algorithm from Section 1.8.1. Since this is a Gibbs sampler style chain, the same kernel works for both the forward process and the reverse process.

Let X_{t+1} be $\phi(X_t, V_{t+1}, U_{t+1})$ whose value is the output of the function `Ising_Gibbs_update_function` from Section 5.1.1. Suppose the states x_t and x_{t+1} are known, and the question is how can the values of V_{t+1} and U_{t+1} for the forward step be simulated?

First consider the case that $X_t \neq X_{t+1}$. Since ϕ can only change the value at a single node, there exists a unique v such that $X_t(v) \neq X_{t+1}(v)$. So $V_{t+1} = v$.

As in `Ising_Gibbs_update_function`, let n_1 denote the number of neighbors of v labeled 1 in X_t, and n_{-1} the number of neighbors labeled -1. Set $r = \exp(\beta n_1)/[\exp(\beta n_1) + \exp(\beta n_{-1})]$. Then if $X_{t+1}(v) = -1$, then U_{t+1} is uniform over $[0, r]$, and if $X_{t+1}(v) = 1$, then U_{t+1} is uniform over $(r, 1]$.

Now suppose that $X_{i+1} = X_i$. Then the goal is to choose (V_{t+1}, U_{t+1}) uniformly from the set (v, U) such that $\phi(X_{i+1}, v, U) = X_i$. But that was already done! The (v, U) that resulted in moving from X_{t+1} to $X_t = X_{t+1}$ was exactly such a draw. Hence no additional work is needed.

This gives the following method.

`FMMR_Ising_Gibbs` *Input:* x, t *Output:* $X \sim \pi$

$X \leftarrow x$
Repeat
    For i from $t-1$ to 0
        Draw $(V_{i+1}, U_{i+1}) \leftarrow \mathsf{Unif}(V \times [0,1])$
        $c \leftarrow$ `Ising_Gibbs_update_function`(X, V_{i+1}, U_{i+1})
        For $d \in \{-1, 1\}$, let n_d be the # of neighbors of V_{i+1} labeled d in X
        $r \leftarrow \exp(\beta n_1)/[\exp(\beta n_1) + \exp(\beta n_{-1})]$
        If $X(V_{i+1}) = 1$ and $c = -1$ then $U_{i+1} \leftarrow \mathsf{Unif}([0,r])$, $X(V_{i+1}) \leftarrow c$
        Else if $X(v_{i+1}) = 1$ and $c = 1$ then $U_{i+1} \leftarrow \mathsf{Unif}((r,1])$, $X(V_{i+1}) \leftarrow c$
    $x_{\min} \leftarrow (-1, \ldots, -1)$, $x_{\max} \leftarrow (1, \ldots, 1)$
    For i from 0 to $t-1$
        $x_{\min}(V_{i+1}) \leftarrow$ `Ising_Gibbs_update_function`$(x_{\min}, V_{i+1}, U_{i+1})$
        $x_{\max}(V_{i+1}) \leftarrow$ `Ising_Gibbs_update_function`$(x_{\max}, V_{i+1}, U_{i+1})$
Until $x_{\max} = x_{\min}$

5.2.4 Doubling time

Note that in the proof of Lemma 5.2, the size of t is not important: the distribution of X_0 conditioned on coalescence detected in the block is π.

That means that it is perfectly legal to double the value of t used at each step of the process, in the same way as in the original CFTP. In this way it is not necessary to worry about for what value of t the backwards coalescence is likely to occur.

5.3 Variable chains

In a bounding chain, a subset of labellings was updated at each step. A more general approach is to keep track of the state as a function of the initial values of nodes. This type of chain is called a *variable chain*, and can be used to understand how changes in the initial state propagate through the system.

In a variable chain, each node is labeled not with a set of labellings, but rather with a function of variables $x_1, \ldots, x_{\#V}$. Initially, node v is given the label x_v. As the chain evolves, each node will always have a label that is a function of $(x_1, \ldots, x_{\#V})$. The variables that appear in $(x_1, \ldots, x_{\#V})$ indicate where changes in the initial state would have changed the current value of a node.

5.3.1 Antivoter model

Variable chains were introduced in [61] in order to generate samples from the antivoter model stationary distribution.

The antivoter model is a sequence of states that gives voter preferences that change through time based on the preferences of the local neighbors.

The state space for a graph $G = (V, E)$ is $\Omega = \{0, 1\}^V$, so that each node is labeled either 0 or 1. The Markov chain then chooses a node v uniformly at random, and then a neighbor w of node v uniformly from the neighbors of v. Finally, $x(v)$ is set to $1 - x(w)$.

In the voter model, if v then w is chosen $x(v)$ is set to be equal to $x(w)$. On a finite graph, the voter model eventually reaches the all 0's state or the all 1's state and stays there.

For the antivoter model, the behavior is more interesting. When the graph G is regular with degree $r \geq 3$ and is not bipartite, Donnelly and Welsh [30] showed that this Markov chain has a unique stationary distribution that is uniform over every state except the all 0's and all 1's states.

In the variable chain, each node v begins with a label $x_{\text{BC}}(v) = x_v$. When the step in the chain chooses v and then w to modify, the label on v becomes $1 - x_v$. Suppose that edge $(3,4)$ in the chain was chosen. Then $x_{\text{VC}}(4)$ is set to be $1 - x_3$. Note that at this point, no node has an x_4 in its label. Since the only way a node can get a variable in its label is if it is next to a label with that variable, there is no longer any way for a node to be labeled x_4. That variable is gone forever from the state of the variable chain.

The update then looks as follows.

```
Antivoter_model_variable_chain
```
Input: old state x_{VC} *Output:* new state x_{VC}

1) Draw v uniformly from V
2) Draw w uniformly from the neighbors of v
3) Set $x_{VC}(v)$ to be $1 - x_{VC}(w)$

It is easy to show that eventually, only one variable remains.

Fact 5.1. *Consider a connected graph G that is regular of degree $r \geq 3$, and has a minimum cut of size $c \geq 1$. Then starting from state $x_{VC}(v) = x_v$ for all $v \in V$, after t steps, the chance that more than one variable remains in the node labels is at most* $\exp(-ct((2/\pi^2)(n-1)^{-2}2 + (1/(6\pi))(n-1)^{-4})\#E^{-1})$.

Proof. Let $n(z) = \#\{v : x_{VC}(v) \in \{x_z, 1-x_z\}\}$ be the number of nodes that are labeled either x_z or $1 - x_z$. (At time 0, $n(z) = 1$ for all $z \in V$.) Let $W_t = \max_z n(z)$ after t steps, $v_{\max} = \arg\max_z n(z)$, and $V_t = \{v : x_{VC}(v) \in \{x_{v_{\max}}, 1 - x_{v_{\max}}\}\}$. Suppose there are k edges connecting nodes in V_t to nodes in $V \setminus V_t$. Then the chance of choosing one of these edges at the first step is $2k(1/\#V)(1/r)$. If $v \in V_t$ and $w \notin V_t$ then W_t increases by 1. If $v \notin V_t$ and $w \in V_t$ then W_t decreases by at most 1. (It could be less: for instance, if $W_t = 4$ using x_7 or x_6, and an edge removed an x_7 instance, then x_6 takes over as the most common label and W_{t+1} is still 4.)

Let $W_t' = n - W_t$. Then $W_t = n$ when $W_t' = 0$. Use the same potential function as in the proof of Lemma 4.2: $\phi(i) = \sin(Ci)/\sin(C)$, with $C = \pi/(2(n-1))$. Then as in (4.10),

$$\begin{aligned}
\mathbb{E}[\phi(W_t')|W_{t-1}'] &\leq \frac{k}{2\#E}\phi(W_{t-1}' - 1) + \left(1 - \frac{k}{\#E}\right)\phi(W_{t-1}') + \frac{k}{2\#E}\phi(W_{t-1}' + 1) \\
&= \phi(W_{t-1}') + k(2\#E)^{-1}\Delta^2\phi(W_{t-1}' - 1) \\
&= \phi(W_{t-1}') + k(2\#E)^{-1}\phi(W_{t-1}')(2\cos(C) - 2) \\
&\leq \phi(W_{t-1}')\gamma,
\end{aligned}$$

where $\gamma = 1 - ct(1 - \cos(C))\#E^{-1}$. (Since c is the size of the minimum cut, $k \geq c$.) Now a simple induction gives $\mathbb{E}[\phi(W_t')] \leq W_0'\gamma^t$.

When $W_t' > 0$ it holds that $W_t' \geq 1$, so by Markov's inequality $\mathbb{P}(W_t' > 0) = \mathbb{P}(\phi(W_t') > 0) \leq \mathbb{E}[\phi(W_t')]$. Recall that $1 - x \leq \exp(-x)$ for all real x. Hence $\gamma \leq \exp(-ct(1 - \cos(\pi/(2(n-1))))\#E^{-1})$. Using $1 - \cos(x) \geq x^2/2 - x^4/24$ completes the proof. □

Once the variable chain has been reduced to a configuration where every node is labeled x_z or $1 - x_z$, there are only two possible states, the one where $x_z = 0$ and the one where $x_z = 1$. Since the stationary distribution assigns equal weight to all configurations, simply decide uniformly which of these two states to pick to complete the perfect simulation algorithm.

With the antivoter model, the problem was that the basic bounding chain did not have any traction to get started. In order to learn about the value of a node, at least one node already needed to be known. The variable chain sidestepped that problem by keeping track of how the values of states percolate through the chain. (Lubetzky would later refer to this as information percolation, which he used to study cutoff phenomena in the mixing time for the Ising model [87].)

5.4 Dealing with infinite graphs

So far, the algorithms used to generate from the Markov random fields have only applied when the graph in question V is finite. In this section it is shown how to draw from a projection of the infinite graph onto a finite subset.

Let V denote a possibly infinite set of vertices, and C be the finite set of labellings for a node.

Definition 5.2. *Let $\partial S = \{w : \exists s \in S, \{s,w\} \in E\}$ be the neighbors of the set S. Then a random configuration X with distribution ν is a* Markov random field *if ν has conditional probabilities such that for all finite $S \subseteq V$, all $x \in C^S$ and $y \in C^{V \setminus S}$,*

$$\mathbb{P}_\nu(X(S) = x | X(V \setminus S) = y) = \mathbb{P}_\nu(X(S) = x | X(\partial S) = y(\partial S)). \tag{5.1}$$

(So the distribution of X over S only depends on the values on the boundary of S, and not on further distant nodes.) Call the set of conditional distributions a specification.

In the case that V is finite, a specification is enough to uniquely determine the distribution ν. When V is infinite, there will be at least one ν with a given specification, but there also might be more than one such distribution.

Definition 5.3. *Suppose that a node has a fixed distribution given the values on the neighboring nodes, and that there exists a unique measure on the state space satisfying these conditional distributions. Then call the unique measure a* Gibbs measure.

The question of when the Gibbs measure is unique is a central issue in Markov field theory. Nonuniqueness occurs at values of the parameter known as a phase transition.

Definition 5.4. *The* Ising model Gibbs measure *has a specification given by*

$$\mathbb{P}(X(S) = x | X(\partial S) = y) \propto \exp(-\beta H(x(S \cup \partial S))), \tag{5.2}$$

where

$$H(x) = -\sum_{\{i,j\}} \mathbb{1}(x(i) = x(j)). \tag{5.3}$$

When the distribution of a single node is "sufficiently" independent of the values of its neighbors, say that the specification satisfies a high-noise requirement. Typically the higher the noise, the easier it becomes to show there is a unique Gibbs measure, and to generate samples from the measure.

There have been several different high-noise requirements, such as Dobrushin uniqueness [28] or a condition of van den Berg and Maes [126]. The one that will be of use here is due to Häggström and Steif [45], and has the nice feature that not only does it guarantee the existence of a unique Gibbs measure, it also guarantees the existence of a linear time method for drawing a sample perfectly from the measure.

Definition 5.5. *For $c \in C$, define a parameter of independence of the distribution of a node from its neighbors as follows:*

$$\gamma(c) = \min_{v \in V} \min_{y \in C^{\partial\{v\}}} \mathbb{P}(X(v) = c | X(\partial\{s\}) = y), \tag{5.4}$$

$$\gamma = \sum_{c \in C} \gamma(c). \tag{5.5}$$

Say that a specification is HS-high noise *if for Δ the maximum degree in the graph,*

$$\gamma > \frac{\Delta - 1}{\Delta}. \tag{5.6}$$

At each step of the process, a subset of nodes must be chosen so that no two nodes in the subset are connected by an edge (this is known as an *independent set* of nodes.) The labels on these nodes will simultaneously be updated. Because none are connected, they can all be updated simultaneously.

One way to do this suggested in [45] is to independently flip a p-coin for every node in V. If a node comes up heads and all of its neighbors are tails, then the node becomes part of the updating set. The probability that a particular node becomes part of the set is then at least $p(1-p)^\Delta$, which is maximized by taking $p = (\Delta+1)^{-1}$.

For every node in the updating set, choose a new label for the node conditioned on the values of the neighbors. If the value of one or more neighbors are unknown, then draw U uniformly from $[0,1]$, and set aside a subinterval of length $\gamma(c)$ for each $c \in C$. If U falls in the subinterval associated with color c, then label the node c. Otherwise label the node "unknown."

To be precise, label a node v with $Z_i(v) = 1$ if it is unknown and $Z_i(v) = 0$ if it is known after i steps in this process.

Lemma 5.3 (Proposition 2.1 of [45]). *Suppose there is a subset of nodes S with $Z_0(v) = 1$ for all $v \in \lambda$. Let $p^* = p(1-p)^\Delta$. Then using the updating scheme given in the previous paragraph, for a node v that is at least k distance in the graph from ∂S, and γ as in* (5.5).

$$\mathbb{P}(Z_k(v) = 1) \leq \left(1 - \frac{\Delta^\Delta(1-\Delta(1-\gamma))}{(\Delta+1)^{\Delta+1}}\right)^k. \tag{5.7}$$

Proof. The proof is by induction on k. The base case of $k = 0$ is trivially true. Assume it holds for k. Let x be any node at least $k+1$ distance from the boundary of S.

Let $b(i) = (1 - (\Delta^\Delta(1-\Delta(1-\gamma)))/(\Delta+1)^{\Delta+1})^i$. Then any node adjacent to x is at least k distance from the boundary of S. Let Y_k denote the set of nodes chosen to update at each step. As before, let $p(x) = \mathbb{P}(x \in Y_k)$. Then if x is not updated, use the induction hypothesis to bound the chance $Z_{k+1}(x) = 1$. If it is updated, then the chance $Z_{k+1}(x) = 1$ is at most $1-\gamma$ if any neighbor w of x has $Z_k(w) = 1$. The union bound makes this chance at most $\Delta b(i)$. Hence

$$\begin{aligned}
\mathbb{P}(Z_{k+1}(x) = 1) &= \mathbb{P}(x \in Y_k)\mathbb{P}(Z_{k+1}(x) = 1 | x \in Y_k) + \mathbb{P}(x \notin Y_k)\mathbb{P}(Z_{k+1}(x) = 1 | x \notin Y_k) \\
&\leq p(x)(1-\gamma)\Delta \cdot b(i) + (1-p(x))b(i) \\
&= b(i)\left[(1-\gamma)\Delta \cdot p(x) + (1-p(x))\right].
\end{aligned}$$

For $(1-\gamma)\Delta < 1$, the right-hand side is maximized when $p(x)$ is minimized, and $p(x) \geq p^* = (\Delta+1)^{-1}$. Simplifying this expression then completes the induction. □

Let $N(v)$ denote the set of nodes w such that $\{v,w\} \in E$. This then gives the following method:

```
Finitary_coding    Input: W

k ← 0
Repeat
   k ← k + 1
   Let W_{-k} be all points within distance k of W
   For all v ∈ W_{-k}, let Z_{-k}(v) = 1
   For t from −k to −1
      Draw Y_t ⊆ W_{-k} to update
      For all v ∈ Y_t
         Draw U ← Unif([0,1])
         If U ≤ γ
            Z_{t+1}(v) ← 0, set X_{t+1}(v) by which subinterval for c that U is in
         Else
            If there is a neighbor w of v with Z_t(w) = 1
               Z_{t+1}(v) = 1
            Else
               Z_{t+1}(v) = 0, set X_{t+1}(v) given X_t(N(v)) and U > γ
Until Z_0(v) = 0 for all v ∈ W
```

Theorem 5.3 (Theorem 2.2 of [45]). *For any finite W, the* `Finitary_coding` *terminates with probability 1, and outputs an unbiased sample X from the distribution over the observation window W.*

Proof. From Lemma 5.3, the chance a node $v \in W$ has $Z_0(v) = 1$ is at most

$$\left(1 - \frac{\Delta^{\Delta}(1-\Delta(1-\gamma))}{(\Delta+1)^{\Delta+1}}\right)^k. \tag{5.8}$$

So the chance that any of the elements in W has $Z_0(v) = 1$ is at most $\#W$ times this bound. This tends to 0 as $k \to \infty$, so the algorithm terminates with probability 1.

The proof that the output comes from the correct distribution is similar to earlier versions of CFTP. Let T denote the minimum value of k needed for the algorithm to terminate. Then as shown above T is finite with probability 1.

For each k, let R_k^* be a draw from the distribution over the observation window W_{-k}. Then running this forward k steps gives a result R_k^0, which is still stationary. Thus the set of variables $R_1^0, R_2^0, R_3^0, \ldots$ all have the correct distribution. But R_k^0 converges to X, the output of the algorithm with probability 1. The only way this can happen is if X has the stationary distribution. □

5.5 Clan of ancestors

So far the running time for generating samples has been nearly linear in the number of nodes in the sample, $O(\#V \ln(\#V))$. In this section it is shown that by restricting carefully the nodes that are considered in an update, this can be reduced to time linear in the number of nodes.

The idea is as follows. Consider the finitary coding of Häggström and Steif [45], and think about what the value of a specific node v at time 0 is. To learn about the value of the node, it is necessary to go back in time until the last time that the node was updated.

At this time in the past, there is a γ chance that the value of v does not depend in any way on the values of the neighbors. On the other hand, if a U is drawn for v at this point that indicates that the values of the neighbors are important, then the neighbors of v need to be added to the clan of ancestors of v.

Each of these neighbors, then, needs its value to be determined, which can be done recursively. Some of these neighbors might have had $U \leq \gamma$ when it was updated, or their neighbors might go on the list to be determined.

The result is similar to a branching process, and certainly has a tree structure. It is not exactly a branching process because of the lack of independence between "children." However, ideas from branching processes can be used to bound the overall size of the cluster. Because the nodes of the cluster are present for the birth of later nodes, call these nodes the *clan of ancestors* of the original node.

Lemma 5.4. *Suppose $\gamma > (\Delta - 1)/\Delta$. Then the expected size in the clan of ancestors of v (including v) is at most $(1 - \Delta(1-\gamma))^{-1}$.*

Proof. Let $C_0 = \{v\}$, C_1 denote the first generation of ancestors, C_2 the second, and so on. Then the expected number of nodes in generation $i+1$ is at most $\Delta(1-\gamma)$ times the expected number of nodes in generation i, since each node in i has only a $1-\gamma$ chance of adding at most Δ nodes to the next generation.

Hence $\mathbb{E}[\#C_{i+1} | \#C_i] \leq \Delta(1-\gamma)C_i$, and taking the expected value of both sides gives $\mathbb{E}[\#C_{i+1}] \leq \Delta(1-\gamma)\mathbb{E}[C_i]$. An easy induction proof then gives $\mathbb{E}[\#C_i] \leq [(\Delta)(1-\gamma)]^i$, and so the expected size of the clan of ancestor is at most

$$\mathbb{E}[\#C_0 + \#C_1 + \cdots] \leq \sum_{i=0}^{\infty} [(\Delta)(1-\gamma)]^i = \frac{1}{1-\Delta(1-\gamma)}.$$

□

Fact 5.2. *The expected time to use the clan of ancestors method to find the values on a finite window is $\#W(1-\Delta(1-\gamma))^{-1}$.*

Proof. The resolution of each clan of ancestors gives the value of exactly one node. Since each clan of ancestors has a constant (in expectation) number of nodes, the overall algorithm to find the values of a finite window W is just $\#W$ times the bound from the last fact on the expected size of the clan of ancestors. □

5.6 Deciding the length of time to run

While the doubling time method works for both CFTP, and FMMR, when many samples are being taken it does involve some wasted effort.

A better method in this case is to use the doubling method to initially find a value of t such that the probability of coalescence is at least $1/2$, and then use that method throughout. In this case the number of steps taken by ROCFTP will be at most $2t$ on average, and for FMMR $4t$ (since it is read twice.)

One approach is to run the chain forward starting from the initial bounding state k times, and let $T_1, \ldots, T_k$ be the times needed for coalescence to occur. Then the median value of $\{T_1, \ldots, T_k\}$ will be a good estimate of the time needed for a block to coalesce roughly half of the time. Let this median be t_{med}. The average running time per sample using ROCFTP will then approach $2t_{\text{med}}$ as the number of samples gets large.

Of course, at this point it might be possible to do much better. For instance, if there is a value t that is only 10% greater than the median, but for which coalescence occurs 80% of the time, then using this value for the size of a block in ROCFTP will approach a time of $(1.1/0.8)t_{\text{med}}$.

Because coalescence often has the kind of cutoff phenomena where for t less than a critical threshold coalescence is unlikely, but past that threshold coalescence becomes very likely, there is usually a point somewhere past the median that will minimize the running time of the algorithm.

Chapter 6

Coalescence on Continuous and Unbounded State Spaces

The self is not something ready-made, but something in continuous formation through choice of action.

John Dewey

The algorithms of Chapters 3 and 5 are not restricted in theory to discrete state spaces, but the examples presented therein were all discrete. This is because the problem of designing a chain that coalesces down to a single state can be trickier when Ω is a continuous state space. Trickier, but not impossible, and in this chapter several methods for dealing with continuous state spaces are considered.

6.1 Splitting chains

Suppose that $X_0, X_1, X_2, \ldots$ is a Markov chain over a continuous state space, and that given $X_t = x_t$, the distribution of the next state X_{t+1} is given by the transition kernel $K(x_t, \cdot)$.

The transition kernel usually depends on the starting state x_t. That is, the distribution $K(x_t, \cdot)$ varies as x_t varies. However, it can sometimes be the case that there is an $\varepsilon > 0$ and kernels K_1 and K_2, such that $K(x_t, \cdot) = K_1(x_1, \cdot)\varepsilon + K_2(x_t, \cdot)(1 - \varepsilon)$ with the property that $K_1(x_1, \cdot)$ does not depend on x_1!

For example, suppose that X_t is a Markov chain conditioned to lie in $[0,1]$ that updates as follows. With probability $1/2$, X_{t+1} given X_t is uniform over $[0,1]$, and with probability $1/2$, $[X_{t+1}|X_t] \sim \textsf{Unif}(\{X_t - \delta, X_t + \delta\})$, where $\delta = \min\{1 - X_t, X_t\}$. Figure 6.1 shows the density of X_{t+1} given that $X_t = 0.3$.

The idea then to take a move is given X_t, draw $Y_1 \sim \textsf{Unif}([0,1])$, and $[Y_2|X_t] \sim \textsf{Unif}([X_t - \delta, X_t + \delta])$ (where $\delta = \min\{1 - X_t, X_t\}$). Last, draw $B \sim \textsf{Bern}(1/2)$. Set

$$X_{t+1} = Y_1 B + Y_2(1 - B) \tag{6.1}$$

so that $X_{t+1} = Y_1$ if $B = 1$ and $X_{t+1} = Y_2$ otherwise. Note that Y_1 did not depend on X_t, and so no matter what the value of X_t, the next state X_{t+1} will be Y_1 if $B = 1$. So the chain couples if $B = 1$.

Since the chain has been split into a random variable that depends on X_t and one

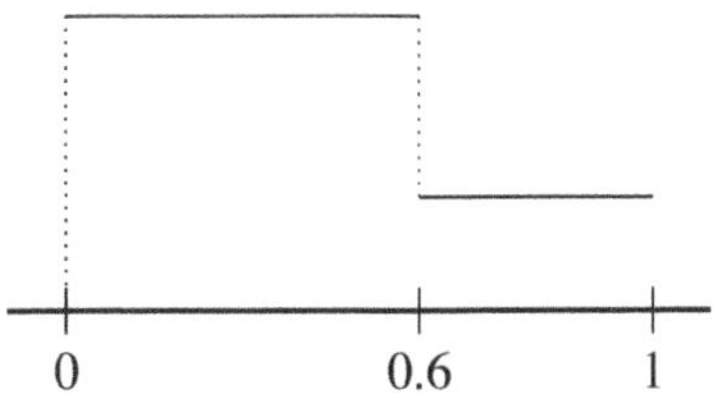

Figure 6.1 *The density of* X_{t+1} *given that* $X_t = 0.3$, *which is* $(1/2)\mathbb{1}(x \in [0,1]) + (1/2)\mathbb{1}(x \in [0,0.6])$.

that does not, this is called a *splitting chain* [109]. If $B \sim \textsf{Bern}(p)$ for the distribution, on average $1/p$ steps are needed before the chain coalescences by choosing the random variable that does not depend on the current state.

6.2 Multigamma coupling

The multigamma coupler of Murdoch and Green [107] takes this splitting chain idea and uses it to build an update function. Previously, the splitting chain [109] had been used by Lindvall [85] to create a coupling for two processes to bound the mixing time. Since an update function couples together multiple processes (usually an infinite number), Murdoch and Green called this process the multigamma coupler.

The idea is to explicitly write the update function into a mixture of a piece that couples completely, and a piece that does not. That is, suppose that for all $x \in \Omega$, the chain can be written as

$$\phi(x,U,B) = \phi_1(U)B + \phi_2(x,U)(1-B) \tag{6.2}$$

where $U \sim \textsf{Unif}([0,1])$ and $B \sim \textsf{Bern}(\rho)$ for some $\rho > 0$.

At each step of this update function, there is a ρ chance that $\phi(\Omega,U,B) = \phi_1(U)$. Hence such a chain quickly coalesces. But can this be done in practice?

Suppose the update function $\phi(x,U,B)$ has unnormalized density $f(y|x)$, and there exists f_1 so that $0 \leq f_1(y) \leq f(y|x)$ for all x and y. Then let $f_2(y|x) = f(y|x) - f_1(y|x)$.

With probability $\int f_1(y)\,dy / \int f(y|x)\,dy$, draw the next state from f_1, otherwise draw from f_2.

`Multigamma_update` *Input:* current state x, *Output:* next state y
1) Draw $C \leftarrow \textsf{Bern}(\int f_1(y)\,dy / \int f(y|x)\,dy)$
2) If $C = 1$
3)   Draw y from unnormalized density f_1
4) Else
5)   Draw y from unnormalized density $f_2(\cdot|x)$

To employ multigamma coupling in density form, it is necessary to obtain a uniform lower bound on the density of the next state.

6.2.1 *Example: hierarchical Poisson/gamma model*

In statistical inference, the goal is to learn from the data more about parameters of a statistical model. In Bayesian statistical inference, the unknown parameter is treated as a random variable. That means that it is modeled initially using a distribution called the prior that is set before any data is taken.

Then given the data and the statistical model, the distribution of the parameters can be updated using Bayes' Rule. The resulting distribution is called the posterior distribution.

In a hierarchical model, the parameter of interest has a prior that itself depends on unknown parameters that usually are given by a prior with high variance.

As an example, consider a dataset on pump reliability, with a model originally given by Gelfand and Smith [39], and studied further by Reutter and Johnson [113].

In this case the data is the number of failures in ten pump systems at a nuclear power plant. Let s_i denote the number of failures of pump i, and t_i be the time that pump i has been operating.

A common model is to treat the failures of a pump i as occurring according to a Poisson process with failure rate λ_i. In such a model, the number of failures has a Poisson distribution with mean equal to the failure rate times the length of time of operation. That is, $s_i \sim \mathsf{Pois}(\lambda_i t_i)$ for each $i \in \{1,\ldots,10\}$.

In order for the prior to not overwhelm the data, each failure rate is given a high-variance distribution for its prior, say $[\lambda_i|\alpha,\beta] \sim \mathsf{Gamma}(\alpha,\beta)$. The value of $\alpha = 1.802$ comes from a method of moments argument. The inverse scale β is taken from another high-variance prior $\beta \sim \Gamma(\gamma,\delta)$, where the values $\gamma = 0.01$ and $\delta = 1$ were used in [113].

The reason for employing Gamma distributions is their special nature as a conjugate prior. When the λ_i have gamma distributions and are used to generate a Poisson random variable, then conditioned on the Poisson random variable, they still have a Gamma distribution with slightly altered parameters. To be precise:

$$[\lambda_i|\beta,s_i] \sim \mathsf{Gamma}(\alpha+s_i,\beta+t_k). \tag{6.3}$$

Similarly, the distribution of β given the failure parameters is also gamma:

$$[\beta|\lambda_1,\ldots,\lambda_{10},s_1,\ldots,s_{10}] \sim \mathsf{Gamma}\left(\gamma+10\alpha,\delta+\sum\lambda_i\right). \tag{6.4}$$

It is important to note here that for this particular model, perfect simulation is unnecessary. The variable β is an example of a *linchpin variable* meaning that once it is known, the distribution of the remaining variables is also known. So what initially looks like a ten-dimensional problem actually reduces to a one-dimensional problem that can be integrated numerically.

Still, there are many hierarchical models that do not split apart so easily, and this particular model serves as a nice illustration of the multigamma coupling method.

In order to do that, lower bounds on gamma densities are needed.

Lemma 6.1. *Say $X \sim \mathsf{Gamma}(a,b)$ if X has density $f_X(x) = x^{a-1}b^a\exp(-xb)\mathbb{1}(x \geq 0)/\Gamma(a)$. Then if $b \in [b_0,b_1]$, and $r(x) = x^{a-1}b_0^a\exp(-xb_1)\mathbb{1}(x>0)/\gamma(a)$, then $r(x) \leq f_X(x)$ for all x, and $\int_{x\in\mathbb{R}} r(x)\,dx = (b_0/b_1)^a$.*

Proof. Since $a > 0$ and $x > 0$, $b_0^a \leq b^a$ and $\exp(-xb) \leq \exp(-xb_1)$, so $r(x) \leq f_X(x)$. Also $r(x)(b_1/b_0)^a$ is a gamma density, and so integrates to 1. □

To use this for the pump model, consider taking one step in the Gibbs sampler Markov chain, where first β is sampled conditioned on the λ_i, and then the λ_i are sampled conditioned on β.

Since $[\beta|\lambda_1,\ldots,\lambda_{10}] \sim \mathsf{Gamma}(1.803, 1+\sum\lambda_i)$, then if $\sum\lambda_i$ was bounded, it would be possible to use the previous lemma to write the distribution as the sum of a draw from density r plus a draw from whatever remained.

But in this case, the λ_i are unbounded! There are two ways to solve this issue:

1. Alter the prior slightly to impose the restriction $\sum\lambda_i < L$. When L is large enough, this alters the prior very little.
2. Employ a technique developed by Kendall and Møller [82] known either as *dominated coupling from the past* or *coupling into and from the past.* This idea is explained in Section 6.7.

The first approach is much easier. Murdoch and Green found that after a 10,000-step run of the Markov chain, the largest $\sum\lambda_i$ value that was seen was 13.6, and so they employed $L = 20$ as their maximum value [107]. Plugging into the lemma gives $[b_0, b_1] = [1, 21]$, and $\int r(x)\,dx = (1/21)^{18.03} < 10^{-24}$. Therefore, the chance that the multigamma coupler coalesces across the entire space simultaneously is very small.

One solution is not to try to couple the entire space simultaneously, but rather to partition the possible $\sum\lambda_i$ values into intervals $0 = L_0 < L_1 < L_2 \cdots < L_m = L$. Then all the values of $\sum\lambda_i \in [L_k, L_{k+1}]$ can be updated simultaneously using the multigamma coupler.

If every one of these couplers coalesce, then there is now a finite set of β values to bring forward. The problem has been reduced to updating a finite set. By setting the L_k so that $[(1+L_k)/(1+L_{k+1})]^{18.03}$ is always the same number, the chance of coalescence in each interval will be the same. By making this ratio sufficiently close to 1, the overall chance of every interval coupling will be high.

A more effective approach for this particular problem was developed by Møller [96]. This approach takes advantage of the following well known property of the gamma distribution.

Lemma 6.2. *Say $X \sim \mathsf{Gamma}(a,b)$. Then for $c > 0$, $X/c \sim \mathsf{Gamma}(a,cb)$.*

Therefore, given $1+\sum\lambda_i \in [1,21]$, it is possible to generate $X \sim \gamma(18.03,1)$, and then say $\beta \in [X/21, X/1]$. These bounds can then be used to bound each λ_i in a similar fashion, and then keep going back and forth until the upper and lower bound on $\sum\lambda_i$ is close enough together that there is a reasonable chance that the multigamma coupler coalescences.

6.3 Multishift coupling

The partitioned multigamma coupler reduces a continuous space down to a finite number of states. Wilson developed another method called the layered multishift coupler [130] that accomplishes this for a wide variety of target distributions. Moreover, the multishift coupler is monotone, thus allowing the use of monotonic CFTP.

To illustrate how multishift coupling works, consider the Markov process that is simple symmetric random walk on the real line, where if the current state is x, the next state is chosen uniformly from $[x-1,x+1]$.

One update function is $\phi_1(x,U) = x+2U-1$. For $U \sim \text{Unif}([0,1])$, $2U-1 \sim \text{Unif}([-1,1])$, and so this update function has the correct kernel. It is also monotone, since for any $u \in [0,1]$ and $x \leq y$, $\phi_1(x,u) \leq \phi_1(y,u)$.

It is, however, extraordinarily bad at coalescence. If $x \neq y$, then $\phi_1(x,u) \neq \phi_1(y,u)$ for all $u \in [0,1]$. A different update is needed in order to make states comes together.

Multishift coupling does the following. Consider the set of numbers $S = \{\ldots,-6,-4,-2,0,2,4,6,\ldots\}$. For any real α, let $S+\alpha$ denote the set $\{s+\alpha : s \in S\}$. Hence $S+0.5 = \{\ldots,-5.5,-3.5,-1.5,0.5,\ldots\}$. Again let $U \sim \text{Unif}([0,1])$, and consider the set $S+2U$.

Then for any $x \in \mathbb{R}$, exactly one point of $S+2U$ falls in the interval $[x-1,x+1)$. (Note that the point $x+1$ has been removed from the interval, but this does not change the kernel since this only occurs with probability 0 anyway.)

So the new update function is

$$\phi_{\text{multishift}}(x,U) \text{ is the unique element of } (S+2U)\cap[x-1,x+1). \quad (6.5)$$

Some pertinent facts about this update function:

Fact 6.1. *Let $\phi_{multishift}(x,U) = \min((S+2U)\cap[x-1,x+1))$. Then for $U \sim \text{Unif}([0,1])$,*

1. *The update function has $\phi_{multishift}(x,U) \sim \text{Unif}([x-1,x+1))$.*
2. *The update function is monotonic.*

$$(\forall x)(\forall y \geq x)(\forall u \in [0,1])(\phi_{multishift}(x,u) \leq \phi_{multishift}(y,u)). \quad (6.6)$$

Proof. Let $a \in [x-1,x+1)$. Then $s = 2\lfloor a/2 \rfloor$ is the largest element of S that is at most a. Then the event that $\phi_{\text{multishift}}(x,U) \leq a$ is just the event that $s+2U \in [x-1,a)$, which occurs with probability $(a-(x-1))/(s+2-s) = (a-(x-1))/2$. Hence $\phi_{\text{multishift}}(x,U) \sim \text{Unif}([x-1,x+1))$.

Let $x < y$ and $u \in [0,1]$. Then if an element of $S+2u$ falls in $[x-1,x+1)\cap[y-1,y+1)$, then $\phi_{\text{multishift}}(x,u) = \phi_{\text{multishift}}(y,u)$. Otherwise $\phi_{\text{multishift}}(x,u) < y-1$ and $\phi_{\text{multishift}}(y,u) > x+1$, so monotonicity is trivially obtained. □

Suppose that the current state of the chain is bounded between $x_{\min}$ and $x_{\max}$. Then after taking one step using $\phi_{\text{multishift}}$, the number of points to be coupled will have been reduced to $(x_{\max}-x_{\min})/2$.

There was nothing special about the width 2 on the interval. In general, given a width w between elements of S and upper and lower bounds $x_{\min}$ and $x_{\max}$, after one step in the Markov chain there will remain at most $(x_{\max}-x_{\min})/w$ points.

6.3.1 *Random walk with normals*

Now examine a different random walk, one where $\phi_2(x,Z) = x + Z$, where Z is a standard normal random variable. At first glance it does not appear that the same technique can be applied, but in fact adding Z can be transformed into a process where a uniform is added.

The basic idea is to write the normal distribution as a mixture of uniform random variables. That is, build random variables A and B so that if $[X|A,B] \sim \textsf{Unif}([A,B])$, then $X \sim \textsf{N}(0,1)$.

The easiest way to do this is to begin by generating $Z \sim \textsf{N}(0,1)$. Then given X, create the auxiliary variable $[Y|Z] \sim \textsf{Unif}([0,\exp(-Z^2/2)])$. At this point (Z,Y) are uniform over the region under the unnormalized density $f_X(x) = \exp(-x^2/2)$. Now turn it around and generate $[W|Y] \sim \textsf{Unif}(x' : 0 \leq y \leq \exp(-x'^2/2))$. Then $W \sim \textsf{N}(0,1)$. But before that happens, $[W|Y] \sim \textsf{Unif}([-\sqrt{-2\ln(y)}, \sqrt{-2\ln(y)}])$. See Figure 6.2 for an illustration.

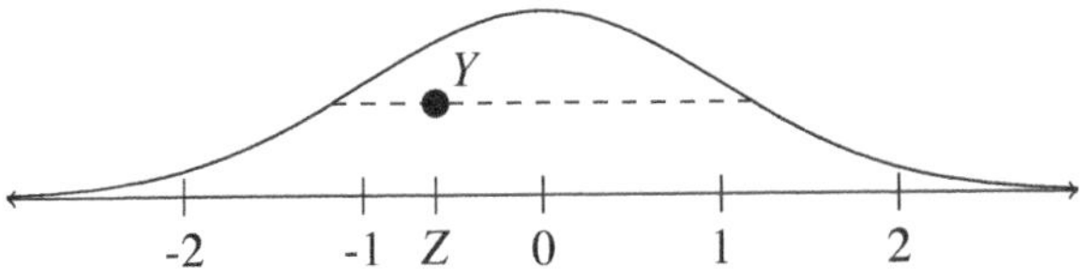

Figure 6.2 *First generate $Z \sim f$, then $[Y|Z] \sim \textsf{Unif}([0, f(Z)])$. Finally, draw $W \sim \textsf{Unif}(\{x : f(x) \geq Y\})$. Then $W \sim Z$. In the figure, the state space for the draw W is the dashed line.*

So the W that is being added to the current state is uniform, and so multishift coupling can be used. Pseudocode for this procedure is as follows.

`Multishift_update_normal`   *Input:* current state x *Output:* next state y

1) Draw $Z \leftarrow \textsf{N}(0,1)$
2) Draw $Y \leftarrow \textsf{Unif}([0,\exp(-Z^2/2)])$
3) $w \leftarrow 2\sqrt{-2\ln(Y)}$
4) Draw $U \leftarrow \textsf{Unif}([0,1])$
5) $W \leftarrow wU - w/2$
6) $y \leftarrow w\lfloor x/w \rfloor + W$

Although this was written for a standard normal being added, to deal with a normal with mean μ and variance σ^2 being added, simply multiply by σ and add μ to X' at step 5.

When this update is applied to an interval $[x_{\min}, x_{\max}]$, the result is $(x_{\max} - x_{\min})/w$ different points returned. This can be a problem as w can be arbitrarily close to 0.

Wilson [130] solved this issue by taking the normal distribution density, and "flipping over" the right-hand side. That is, instead of generating (X,Y) uniformly over

$\{(x,y) : 0 \leq y \leq \exp(-x^2/2)\}$, generate (X,Y) uniformly over

$$\{(x,y) : x \leq 0, 0 \leq y \leq \exp(-x^2/2)\} \cup \{(x,y) : x > 0, 1 - \exp(-x^2/2) \leq y \leq 1\}. \tag{6.7}$$

See Figure 6.3 for an illustration.

Given x, y is still uniform over an interval of length $\exp(-x^2/2)$. and so X is still normally distributed. This leads to the following algorithm.

`Multishift_update_normal_flipped` *Input: x* *Output: y*
1) Draw $X \leftarrow \mathsf{N}(0,1)$
2) If $X \leq 0$ then draw $Y \leftarrow \mathsf{Unif}([0, \exp(-X^2/2)])$
3) Else draw $Y \leftarrow \mathsf{Unif}([1 - \exp(-X^2/2), 1])$
4) $a \leftarrow -\sqrt{-2\ln(Y)}$, $b \leftarrow \sqrt{-2\ln(1-Y)}$, $w \leftarrow b - a$
5) Draw $U \leftarrow \mathsf{Unif}([0,1])$
6) $X' \leftarrow wU + a$
7) $y \leftarrow w\lfloor x/w \rfloor + X'$

Unlike the first method, here $w \geq 2\sqrt{\ln(4)} > 2.35$. Therefore, when applied to the interval $[x_{\min}, x_{\max}]$, this multishift coupler results in moving to at most $(x_{\max} - x_{\min})/2.35$ different points.

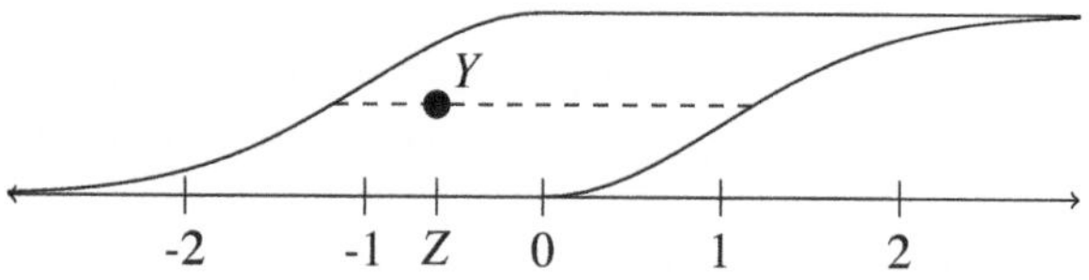

Figure 6.3 *Inverting the right half of the normal density. Z is still drawn from f, now [Y|Z] is uniform over the area between the two lines with horizontal coordinate Z. Then W is uniform over the shaded area with vertical coordinate Y. Now the state space of W is an interval of length at least 2.35.*

6.4 Auto models

Multishift coupling is especially useful in auto models. Recall that an auto model is a Markov random field where the density can be written as a product of functions on the nodes and functions over labels on endpoints of edges.

In particular, in an auto model the distribution of $X(v)$ for a single node $v \in V$ only depends on the immediate neighbors, and not on more distant nodes. That is, $[X(v)|X(V \setminus v)] \sim [X(v)|X(\{w : \{v,w\} \in E\})]$.

When the distribution of $X(v)$ conditioned on its neighbors always comes from the same distribution family, the model can be named as auto followed by the name of the distribution family.

For instance, suppose $x \in \{0,1\}^V$ is a labeling of the nodes with either 0 or 1.

Then trivially the distribution of x conditioned on its neighbors is Bernoulli since it must be either 0 or 1. Hence this is an autobernoulli model.

In the hierarchical Poisson/gamma model from Section 6.2.1, consider the star graph with node v_β at the center, and nodes $v_1,\ldots,v_{10}$ each connected only to v_β. Label node v_β with the value β, and v_i with λ_i.

Then this forms an autogamma model, since conditioned on the neighbors of each node, the distribution of the node will come from the gamma family of distributions.

6.4.1 *Example: autonormal model*

In [9], Besag considered a statistical model for the output from a camera. Each pixel of the camera returns numerical values based on the color of light that hits it, with larger numbers denoting brighter values. In a standard photo, the color space is three dimensional, with one dimension giving the red, one the blue, and one the green. In a grayscale photo, the color space is one dimensional. For simplicity, only the one-dimensional case will be considered here, but this easily extends to higher dimensions.

The model is that at each pixel of the camera, the returned value is normally distributed with mean equal to the actual entering value, and some fixed variance.

Then the prior distribution on the picture entering the camera (the data) assigns a penalty for neighboring pixels that goes down exponentially in the square of the difference between the pixel values.

If the observed picture is y, then the posterior on the true picture X has density

$$f_X(x) = Z^{-1}\exp(-H(x,d))\mathbb{1}(x \in [0,1]^V) \tag{6.8}$$

$$H(x,y) = -\frac{1}{2\sigma^2}\sum_i (x(i)-y(i))^2 - \frac{\gamma^2}{2}\sum_{\{i,j\}\in E}(x(i)-x(j))^2, \tag{6.9}$$

where σ and γ are parameters of the model.

Now suppose a Gibbs sampler is implemented for this problem. A random scan Gibbs sampler chooses a node uniformly at random, then updates the value of the node conditioned on the rest of the neighbors.

Fact 6.2. *For the autonormal Gibbs sampler,*

$$[X(v)|X(\text{neighbors of } v)] \propto \exp(-(x-a(i))^2/[2b(i)]),$$
$$b(i) = (\sigma^{-2}+\deg(i)\gamma^2)^{-1},$$
$$a(i) = \left[\sigma^{-2}y(i)+\gamma^2\sum_{j:\{i,j\}\in E} X(j)\right]b(i),$$

where $\deg(i)$ *is the number of neighbors of i in the graph. That is,* $X(v)$ *conditioned on the values of its neighbors is normally distributed with mean* $a(i)$ *and variance* $b(i)$.

So the target distribution is a normal whose variance is independent of the neighbors, but whose mean is an increasing function of the neighboring values.

This is a situation where multishift coupling could be applied, since the goal is to couple normally distributed random variables with means over an interval from the smallest value of $a(i)$ to the largest value of $a(i)$.

6.5 Metropolis-Hastings for continuous state spaces

Gibbs sampler Markov chains can often be made monotonic, but require special approaches such as multigamma coupling or multishift coupling. Metropolis-Hastings Markov chains are usually not monotonic, but can make continuous spaces coalesce down to a single point. To get the best from these two approaches, create an update function that consists of several steps in a Gibbs chain to bring the upper and lower states close together, and then use a Metropolis-Hastings step to completely reduce the image down to a single state.

6.5.1 Example: truncated autonormal

The multishift coupler was used previously to deal with the unrestricted autonormal model, but in fact for image processing usually values are conditioned to lie in the real interval from 0 (solid black) to 1 (solid white).

This is easy to deal with for the Gibbs sampler: when generating the normal with mean $a(i)$ and variance $b(i)$, just condition on the normal lying between 0 and 1. Allison Gibbs [41] showed that this Markov chain was rapidly mixing. In [41] the Wasserstein metric was used to assess convergence because it was easier to work with; in [54] the method was improved to handle total variation distance. Therefore the details of the Wasserstein metric are omitted here.

The truncation complicates matters: the distributions at each node are no longer shifts from the same family of distributions. For example, a normal with mean 0.5 and variance 1 conditioned to lie in $[0,1]$ is not a shifted version of a normal of mean 0.7 and variance 1 conditioned to lie in $[0,1]$. So the multishift coupling used in the previous section no longer works here. On the other hand, using the inverse transform method to generate the two normals does give a monotonic update function.

A solution lies in the Metropolis-Hastings approach from Section 1.8.2. In this approach a state is proposed to be moved to, an acceptance ratio formed, and if a $\mathsf{Unif}([0,1])$ falls below the acceptance ratio, the state is moved to the proposed state. Otherwise it stays where it is.

Suppose our proposed state is found by taking each node in turn, then independently adding U_i, which is a uniform $[-\varepsilon,\varepsilon]$ to the value at that node, where $\varepsilon > 0$ is a small value. When ε is very small, it is likely that such a proposed move will be accepted, regardless of the initial state. Let $U = (U_1,\ldots,U_{\#V})$. Then

$$\frac{f_X(x+U)}{f_X(x)} \geq \exp\left(-\frac{\#V(2\varepsilon+\varepsilon^2)}{2\sigma^2} - \frac{\gamma^2}{2}\#E(2\varepsilon+\varepsilon^2)\right).$$

Making $\varepsilon = \Theta(\min\{\sigma^2/\#V, 1/[\gamma^2\#E]\})$ gives $f_X(x+Y)/f_X(x) \geq 1/2$, and so there is a $1/2$ chance that the Metropolis-Hastings move will be accepted starting from any state.

Since a uniform is being added to each node value, multishift coupling could be used to couple the proposed state for all initial states. However, there is a another way of coupling that is simpler to implement.

This method works as follows. Suppose the initial value x_i is in the interval $[a,b]$, and the goal is to couple $x_i + U_i$ together where $U_i \sim \text{Unif}([-\varepsilon,\varepsilon])$. Then from a, the goal is to generate from $[a-\varepsilon,a+\varepsilon]$, from b, the goal is to generate from $[b-\varepsilon,b+\varepsilon]$, and from $x_i \in (a,b)$, the goal is to generate from $[x_i-\varepsilon,x_i+\varepsilon]$.

Note that all of these intervals lie in the interval $[a-\varepsilon,b+\varepsilon]$. So AR can be used to sample as follows: generate $A_i \sim \text{Unif}([a-\varepsilon,b+\varepsilon])$. If $A_i \in [x_i-\varepsilon,x_i+\varepsilon]$ then $\phi(x_i,U) = A_i$. Otherwise start over. Repeat until all $x_i \in [a,b]$ have been assigned.

Here is why this is good: if the first A_i falls in $[b-\varepsilon,a+\varepsilon]$, then for all $x_i \in [a,b]$, $A_i \in [x_i-\varepsilon,x_i+\varepsilon]$. Hence the proposed move is the same for all states.

Inside the Gibbs sampler, it is necessary to generate a normal random variable with mean $a(i)$ and variance $b(i)$ truncated to lie in $[0,1]$. One way to generate a $\mathsf{N}(a(i),b(i))$ is to take a standard normal, multiply by $\sqrt{b(i)}$, and add $a(i)$. In order for the result to lie in $[0,1]$, the initial standard normal must have been in $[-a(i)/\sqrt{b(i)},(1-a(i))/\sqrt{b(i)}]$. To obtain a standard normal in $[a',b']$, use the inverse transform method.

If Φ is the cdf of a standard normal, and $U \sim \text{Unif}([0,1])$, then $\Phi^{-1}(U)$ will be a standard normal. Furthermore, $(\Phi(b')-\Phi(a'))U+\Phi(a')$ will be uniform over $[\Phi(a'),\Phi(b')]$, so then $\Phi^{-1}((\Phi(b')-\Phi(a'))U+\Phi(a'))$ is a standard normal conditioned to lie in $[a',b']$.

This gives the following Gibbs sampler update. As earlier, let $\deg(i)$ denote the number of neighbors of node i in the graph.

`Truncated_autonormal_Gibbs_update`	*Input:* x, U, v,	*Output:* $x(i)$
1)	$b(i) \leftarrow (\sigma^{-2}+\deg(i)\gamma^2)^{-1}$, $a(i) \leftarrow \left[\sigma^{-2}y(i)+\gamma^2\sum_{j:\{i,j\}\in E} x(j)\right] b(i)$	
2)	$a' \leftarrow -a(i)/\sqrt{b(i)}$, $b' \leftarrow (1-a(i))/\sqrt{b(i)}$	
3)	$Z \leftarrow \Phi^{-1}((\Phi(b')-\Phi(a'))U+\Phi(a'))$	
4)	$x(i) \leftarrow a(i)+\sqrt{b(i)}Z$	

Next is the Metropolis-Hastings update.

`Truncated_autonormal_MH_update`	*Input:* $x_{\min}, x_{\max}, x$,	*Output:* y, `cflag`
1)	`cflag`$\leftarrow$TRUE	
2)	For all $i \in V$	
3)	Draw $U_i \leftarrow \text{Unif}([x_{\min}-\varepsilon,x_{\max}+\varepsilon])$	
4)	If $U_i \notin [x_{\max}-\varepsilon,x_{\min}+\varepsilon]$ then `cflag`$\leftarrow$FALSE	
5)	If $U_i \notin [x-\varepsilon,x+\varepsilon]$	
6)	$U_i \leftarrow \text{Unif}([x-\varepsilon,x+\varepsilon])$	
7)	$U \leftarrow (U_1,\ldots,U_{\#V})$	
8)	$r \leftarrow f_X(x+U)/f_X(x)$	
9)	Draw U_r uniformly from $[0,1]$	
10)	$y \leftarrow x\mathbb{1}(U_r \le r)+(x+U)\mathbb{1}(U_r > r)$	
11)	If $U_r > 1/2$ then `cflag`$\leftarrow$FALSE	

Putting these two together gives a block update that either coalesces, or does not.

```
Truncated_autonormal_block   Input: x, t   Output: x, cflag

x_min ← (0,0,...,0), x_max ← (1,1,...,1)
For t' from 1 to t
   Draw v uniformly from V, U uniformly from [0,1]
   x_min(v) ← Truncated_autonormal_Gibbs_update(x_min, U, v)
   x_max(v) ← Truncated_autonormal_Gibbs_update(x_max, U, v)
   x(v) ← Truncated_autonormal_Gibbs_update(x, U, v)
   (x, cflag) ← Truncated_autonormal_MH_update(x_min, x_max, x)
```

Now that the block structure is in place, ROCFTP can be run.

```
Truncated_autonormal_ROCFTP   Input: k, t   Output: y

i ← 0, x ← (0,0,...,0)
Repeat
   (x, cflag) ← Truncation_autonormal_block(x, t)
Until cflag
For i from 1 to k
   Repeat
      y_i ← x
      (x, cflag) ← Truncation_autonormal_block(x, t)
   Until cflag
```

6.6 Discrete auxiliary variables

Breyer and Roberts [14] noted that even when a sample space is continuous, often there are discrete auxiliary variables that can be added to the space. With the auxiliary variables, using a Gibbs sampler that bounces back and forth between the continuous and discrete space forces the infinitely many states of the continuous part to move to a finite number of states in the discrete part.

To illustrate this idea, consider a problem of sampling from the posterior of a mixture problem. The description follows that of [14]. Suppose that data $\eta_1, \ldots, \eta_n$ is drawn from a mixture of r different probability densities $f_1, \ldots, f_r$. That is, v_i has density $f = \sum_{i=1}^r w_i f_r$, where $(w_1, \ldots, w_r)$ form a probability vector.

Suppose that initially $(w_1, \ldots, w_r)$ is given a Dirichlet distribution where all the parameters are 1.

Definition 6.1. *The* probability simplex *of dimension r consists of vectors $(w_1, \ldots, w_r)$ where $w_i \geq 0$ for all i and $w_1 + \cdots + w_r = 1$.*

Definition 6.2. *A* Dirichlet distribution *with r parameters $(\alpha_1, \ldots, \alpha_r)$ is the r-dimensional generalization of the beta distribution. It has density $B(\alpha)\prod_{i=1}^r x_i^{\alpha_i - 1}$ with respect to uniform measure over the probability simplex. Here the normalizing constant $B(\alpha) = \prod_{i=1}^r \Gamma(\alpha_i)/\Gamma(\sum_{i=1}^r \alpha_i)$.*

The following helpful fact gives an easy way to simulate from a Dirichlet distribution.

Fact 6.3. *Suppose $\{A_{ij}\}$ are an iid collection of* $\mathsf{Exp}(1)$ *random variables. Let* $w_i = \sum_{j=1}^{k_i} A_{ij}$ *and* $S = \sum_{i=1}^{r} w_i$ *Then* $(w_1/S,\ldots,w_r/S) \sim Dir(k_1,k_2,\ldots,k_r)$

The data can be modeled as coming from the following three-step experiment. First, choose $(w_1,\ldots,w_n) \sim \mathsf{Dir}(1,1,\ldots,1)$. Second, for each $i \in \{1,\ldots,n\}$, choose Z_i from $\{1,\ldots,r\}$ using $\mathbb{P}(Z_i = j) = w_j$. Finally, choose η_i from density f_{Z_i}.

The goal is to sample from the posterior distribution on $(w_1,\ldots,w_r)$ given the data $\eta_1,\ldots,\eta_n$. One way is to use a Gibbs sampler which, given the w values, generates new Z_i conditioned on η, and then given the Z_i, generates new w conditioned on η.

Given η_i and w, the conditional probability that $Z_i = j$ is proportional to $w_j f_j(\eta_i)$. The Dirichlet prior is conjugate, conditioned on the values $Z_1,\ldots,Z_n$, but the distribution of the w is still Dirichlet. Let $N_i(Z) = \sum_{j=1}^{n} \mathbb{1}(Z_j = i)$. Then

$$[w|Z] \sim \mathsf{Dir}(N_1(Z)+1,\ldots,N_r(Z)+1). \tag{6.10}$$

So this update function operates as follows.

`Mixture_Gibbs`	*Input:* current w	*Output:* next w
1)	Draw $w'_1,\ldots,w'_r$ iid $\mathsf{Exp}(1)$	
2)	For i from 1 to n	
3)	Draw Z_i using $\mathbb{P}(Z_i = j) = w_j f_j(\eta_i)/\sum_k w_k f_k(\eta_i)$	
4)	Draw $a \leftarrow \mathsf{Exp}(1)$	
5)	$w'(Z_i) \leftarrow w'(Z_i) + a$	
6)	$w \leftarrow w'/\sum_i w'_i$	

To determine coalescence for this chain, a minimum state $w_{\min}$ and maximum state $w_{\max}$ will be kept. Then it is necessary to sample the Z_i from w' when all that is known is that $w_{\min} \leq w' \leq w_{\max}$.

In [14] they used what they called a basin system, but here an approach more similar to [51] will be utilized.

The idea is best illustrated with an example. Suppose that my goal is to sample simultaneously from the two unnormalized vectors $v_1 = (2,2,1)$ and $v_2 = (1,1,2)$. Both of these vectors are less than or equal to $(2,2,2)$. So first I draw a sample using (unnormalized) probability vector $v_3 = (2,2,2)$, call it I. Suppose that the draw is $I = 1$. Then $v_1(1) = 2$ so I accept I as a draw from v_1 with probability $2/2 = 1$. On the other hand, $v_2(1) = 1$, so I only accept I as a draw from v_2 with probability $1/2$. Suppose I reject I as a draw from v_2. Then I must draw again from $(2,2,2)$, and accept or reject with the appropriate probability. This continues until I have an acceptance for both v_1 and v_2.

Now consider the general problem of drawing simultaneously from all vectors w' satisfying $w_{\min} \leq w' \leq w_{\max}$. Then $w_{\max}$ takes the role of v_3 in the example above, since it upper bounds all the w'. Given I a draw from $w_{\max}$, it should be accepted as a draw from w' with probability $w'(I)/w_{\max}(I)$. This acceptance probability is at least $w_{\min}(I)/w_{\max}(I)$.

Draw U uniform over $[0,1]$. If $U \leq w_{\min}/w_{\max}$, then I is certainly accepted for all w', and the process can stop. If U is greater, however, then I might not be accepted for all w'. In this case, add the state I to the set of possible values for the component, and start over.

Of course, once I is part of the possible state values, it is not necessary to see if I is accepted again, so after n steps, the procedure will terminate.

That completes the first step: from w_{min} and w_{max} find bounding states for the Z_i. Now turn it around: given a bounding state for the $Z_i's$, how can this be used to get a new w_{min} and w_{max}?

Here is one way to think about this that will be helpful. Each w_i is assigned one exponential with mean 1 initially. For every decision to put $Z_i = j$, w_j is granted one more exponential. Therefore if $Z_i = \{a_1, a_2, \ldots, a_k\}$ where $k > 2$, then every $w_{max}(a_\ell)$ is given the exponential. When $k = 1$, then both w_{min} and w_{max} are given the exponential.

The resulting bounding chain update looks like this.

`Mixture_bounding_chain`   *Input:* current w_{min}, w_{max}, Z *Output:* w'_{min}, w'_{max}, Z

1)   Draw $w'_1, \ldots, w'_r$ iid $\mathsf{Exp}(1)$, $w'_{max} \leftarrow w'$, $w'_{min} \leftarrow w'$
2)   For i from 1 to n
3) $Z_i \leftarrow \{\}, w \leftarrow w_{max}$
4)   Repeat
5)   Draw I from unnormalized density w, draw $U \leftarrow \mathsf{Unif}([0,1])$
6) $w(I) \leftarrow 0, Z_i \leftarrow Z_i \cup \{I\}$
7)   Until $w = (0,0,\ldots,0)$ or $U \le w_{min}(I)/w_{max}(I)$
8)   Draw $a \leftarrow \mathsf{Exp}(1)$
9)   For every $j \in Z_i$
10) $w'_{max}(j) \leftarrow w'_{max}(j) + a$
11) $w'_{min}(j) \leftarrow w'_{min}(j) + a \cdot \mathbb{1}(\#Z_i = 1)$

Initialize with $w_{min} = (0,0,\ldots,0)$ and $w_{max} = (\infty, \infty, \ldots, \infty)$ and now ROCFTP can be used to obtain samples.

6.7   Dominated coupling from the past

Dominated coupling from the past (also known as coupling into and from the past) can be used to deal with state spaces that are unbounded (as continuous state spaces often are.) This method was created by Kendall [81] for spatial point processes, and later developed in more generality by Kendall and Møller [82]. In Chapter 7 the spatial point process applications are discussed.

The idea is to build two stochastic processes, an underlying process $\{X_t\}_{t \in \{\ldots, -2, -1, 0, 1, 2, \ldots\}}$ and a dominating process $\{D_t\}_{t \in \{\ldots, -2, -1, 0, 1, 2, \ldots\}}$ that controls the value of the X_t for all t.

Typically the D_t process is a simpler process than that driving the X_t process. It is necessary that the stationary distribution of the D_t process is easy to simulate from. In fact, the following ingredients are needed to use dominated coupling from the past.

1. A dominating process whose stationary distribution is easy to sample from.

2. A way to run the dominating process backwards in time.
3. A way to impute the forward updating uniforms from the backward dominating process.
4. A way to determine if the original process coalesces.

Put these ingredients together to get dominated CFTP.

`Dominated_cftp`

---

1) $t \leftarrow -1, t' \leftarrow 0$
2) Draw D_0 from the stationary distribution of the dominating process
3) Repeat
4)   Draw $D_{t'-1}, D_{t'-2}, \ldots, D_t$ from the reverse dominating process
5)   Run X_t forward to 0 from all X_t allowed by the dominating process D_{-t}, conditioned on the values of $D_{t'-1}, \ldots, D_0$
6) $t' \leftarrow t, t \leftarrow 2t$
7) Until coalescence occurs

Section 5.2.1 discussed how to run a Markov chain backwards in time, and then how to run it forward again conditioned on some information about the forward chain. Now consider an example of this where the state space in continuous.

6.7.1 *Example: Vervaat perpetuities*

Consider the Vervaat perpetuity problem.

Definition 6.3. *Let $W_1, W_2, \ldots$ be an iid sequence of random variables. Then a* perpetuity *is the random variable*

$$Y = W_1 + W_1 W_2 + W_1 W_2 W_3 + \cdots. \tag{6.11}$$

(Sometimes the term perpetuity is also used to refer to the distribution of Y.)

Say $W \sim W_i$. Assume that $W \geq 0$ with probability 1, and that $\mathbb{E}[W] < 1$. Then Y exists and is finite almost surely. It also satisfies the distributional fixed-point equation $Y \sim W(1+Y)$.

From an approximate sampling standpoint, generating Y is easy. Simply generate $W_1, W_2, \ldots, W_t$ and truncate the infinite series. But perfect sampling is more difficult.

This problem arises in several contexts, among them an asymptotic approximation to the running time of the `QuickSelect` algorithm of Hoare [49]. In this application, $W \sim \text{Unif}([0,1])$.

Definition 6.4. *A perpetuity is* Vervaat *if there is $\beta \in (0, \infty)$ such that $W \sim U^{1/\beta}$, where $U \sim \text{Unif}([0,1])$.*

The simplest known representation of the density of Y is as an infinite series. The first perfect simulation algorithm for this problem used the `Bounded_mean_Bernoulli` method of Section 2.6. In [27], Devroye found a simple unnormalized density that served as as upper bound on the density of Y from which it was possible to sample, and then (relatively complicated) upper and lower bounding functions for the unnormalized density to use with `Bounded_mean_Bernoulli` in the AR step. However, all that was shown in [27] was that the algorithm ran in finite time, and that the expected value was not computed (and is not known to be finite).

A Vervaat perpetuity has a natural Markov chain associated with it: the update function $\phi(x,u) = u^{1/\beta}(1+x)$ is a recurrent aperiodic Harris chain whose stationary distribution is the Vervaat perpetuity distribution.

In [36], one method for using dominated CFTP for the Vervaat perpetuity chain was given. The state space $[0,\infty)$ is unbounded, and so it is necessary to use dominated CFTP.

The dominating chain for this problem will always satisfy $0 \leq X_t \leq D_t$. For simplicity, only the $\beta = 1$ case will be considered here; see [36] for more details for the general case. Then $X_{t+1} \sim (1+X_t)U$, where $U \sim \text{Unif}([0,1])$. When $X_t \geq 5$, then $1+X_t \geq 6$, and there is at least a two-thirds chance that $(1+X_t)U \leq X_t - 1$.

So the dominating chain D_t operates according to a simple asymmetric random walk with partially reflecting boundary. The update function for the chain is $\phi_D(d,u) = d + \mathbb{1}(u > 2/3) - \mathbb{1}(u \leq 2/3, x \geq 5)$.

Fact 6.4. *The stationary distribution of the chain with update function* ϕ_D *is* $\pi(i) = (1/2)^{i-4}\mathbb{1}(i \in \{5,6,\ldots\})$.

This dominating process is reversible. In fact, any Markov chain on the integers where the state changes by at most one at each step is reversible.

Definition 6.5. *A chain on* $\{0,1,2,\ldots\}$ *is a* birth-death chain *if* $\mathbb{P}(|X_{t+1} - X_t| \leq 1) = 1$.

Fact 6.5. *All birth-death chains with a stationary distribution are reversible with respect to that distribution.*

Proof. Suppose that π is stationary for a birth-death chain. Prove the result by induction. The base case is when $i = 0$.

$$\pi(\{0\}) = p(0,0)\pi(\{0\}) + p(1,0)\pi(\{1\}),$$
$$\pi(\{0\})(1 - p(0,0)) = p(1,0)\pi(\{1\}).$$

Since $1 - p(0,0) = p(0,1)$, the base case holds.

For the induction hypothesis, suppose $p(i,i+1)\pi(\{i\}) = p(i+1,i)\pi(\{i+1\})$. Then for $i+1$,

$$\pi(\{i+1\}) = p(i,i+1)\pi(\{i\}) + p(i+1,i+1)\pi(\{i+1\}) + p(i+2,i+1)\pi(\{i+2\}).$$

Using reversibility on the first term of the right-hand side gives

$$\pi(\{i+1\}) = p(i+1,i)\pi(\{i+1\}) + p(i+1,i+1)\pi(\{i+1\}) + p(i+2,i+1)\pi(\{i+2\})$$

which gives

$$\pi(\{i+1\})(1 - p(i+1,i) - p(i+1,i+1)) = p(i+2,i+1)\pi(\{i+2\}).$$

Using $1 - p(i+1,i) - p(i+1,i+1) = p(i+1,i+2)$ completes the induction and the proof. □

As noted earlier, the reason such a chain is called reversible is because it looks the same whether it is run in the forward or reverse direction.

Lemma 6.3. *Suppose $X_a \sim \pi$ for $\{X_t\}$ a reversible Markov chain. Then for any $b > a$,*

$$(X_a, X_{a+1}, \ldots, X_b) \sim (X_b, X_{b-1}, \ldots, X_a). \tag{6.12}$$

Proof. Just use induction on $b - a$. The statement that $(X_a, X_{a+1}) \sim (X_{a+1}, X_a)$ is just reversibility. Applying reversibility $b - a$ times gives the induction step. □

Because the chain is reversible, it is easy to simulate in the reverse direction; the updates are done in exactly the same way as forward updates. So for each t, set $D_{t-1} = \phi(D_t, U_{t-1})$.

Now consider how to simulate uniforms for the forward chain conditioned on the values of the dominating chain. Suppose $D_i = j$ and $D_{i+1} = j - 1$. Then the forward move had to be down, and U_i must lie in $[0, 2/3]$. So draw U_i as a uniform conditioned to lie in $[0, 2/3]$. On the other hand, if $D_i = j$ and $D_{i+1} = j + 1$, then the forward move had to be up, so U_i is uniform over $(2/3, 1]$.

Next, how can the original Vervaat perpetuity process be updated in order to coalesce the continuity of possible states down to only a few states? Recall $X_{t+1} = U_{t+1}(1 + X_t)$. Another way to state this is that $[X_{t+1} | X_t] \sim \mathsf{Unif}([0, 1 + X_t])$.

This can be broken down into two pieces. If $U \leq 1/(1 + X_t)$, then $U(1 + X_t) \sim \mathsf{Unif}([0,1])$. So generate a second uniform $U' \sim \mathsf{Unif}([0,1])$, and let X_{t+1} be U'. Otherwise, let X_{t+1} be $U(1 + X_t)$ as before. This is encoded in the following update function,

$$\phi(x, u, u') = \mathbb{1}(u \leq 1/(1+x))u' + \mathbb{1}(u > 1/(1+x))u(1+x). \tag{6.13}$$

Some things to note about this update function: For any values of u and u' in $[0,1]$, this function is monotonic in x. Second, if $X_{t+1} = \phi(X_t, U_{t+1}, U'_{t+1})$, and $U_{t+1} < 1/(1 + D_t)$, then $X_{t+1} = U'_{t+1}$ and is completely independent of the value of X_t. Coalescence has occurred! The result is the following algorithm:

`Vervaat_perpetuity_dominated_cftp`

$t \leftarrow -1$, $t' \leftarrow 0$, draw D_0 from $\mathbb{P}(D_0 = j) = (1/2)^{j-4}\mathbb{1}(j \in \{5, 6, \ldots\})$
`cflag` $\leftarrow$ FALSE
Repeat
    For k from $t' - 1$ to t
        Draw $C \leftarrow \mathsf{Bern}(1/3)$
        $D_k \leftarrow D_{k+1} + C - (1 - C)\mathbb{1}(D_{k+1} > 5)$
        If $D_{k+1} > D_k$ then draw $U_{k+1} \leftarrow \mathsf{Unif}((2/3, 1])$
        If $D_{k+1} \leq D_k$ then draw $U_{k+1} \leftarrow \mathsf{Unif}([0, 2/3])$
        $U'_{k+1} \leftarrow \mathsf{Unif}([0,1])$
    $X_t \leftarrow D_t$
    For k from t to -1
        $X_{k+1} \leftarrow \mathbb{1}(U_{k+1} \leq 1/(1 + X_k))U'_{k+1} + \mathbb{1}(U_{k+1} > 1/(1 + X_k))(1 + X_k)U_{k+1}$
        If $X_{k+1} \leq 1$, then `cflag` $\leftarrow$ TRUE

14) $t' \leftarrow t, t \leftarrow 2t$
15) Until `cflag` is true

6.8 Using continuous state spaces for permutations

Sometimes discrete spaces can be used to make continuous spaces coalesce faster. In the other direction, sometimes use of a continuous space can simplify perfect simulation for a discrete chain.

As an example of this, consider the set of permutations on n objects. In Sections 1.6.2 and 4.5, examples of distributions on the set of permutations were given.

In particular, for the linear extensions of a partial order (Section 4.4), bounding chains were used to do CFTP because the discrete chain does not admit a monotonic update function.

However, by embedding the set of permutations in a continuous space, it is possible to create a Markov chain with a monotonic update function!

For the set S_n of permutations on n, objects, let $\mathscr{C}_n$ be those vectors in $[0,1]^n$ whose coordinates take on unique values. That is, $\mathscr{C}_n = \{x \in [0,1]^n : (\forall i \neq j)(x(i) \neq x(j))\}$. Then each vector in $\mathscr{C}_n$ encodes a permutation in S_n.

Definition 6.6. *Let $f_{perm} : \mathscr{C}_n \to S_n$ be the function $f_{perm}(x) = \sigma$, where σ is the unique permutation such that*

$$x(\sigma(1)) < x(\sigma(2)) < x(\sigma(3)) < \cdots < x(\sigma(n)). \tag{6.14}$$

For instance, $f_{\text{perm}}((0.23, 0.14, 0.75, 0.60)) = (2,1,4,3)$. In fact, f_{perm} maps into the permutation $(2,1,4,3)$ if and only if $x(2) < x(1) < x(4) < x(3)$. An easy way to read off the permutation from the vector is that $\sigma(i) = j$ means that there are exactly j components of x that are at most $x(i)$.

Note that if a point X is chosen uniformly from $\mathscr{C}_n$, then from a symmetry argument it follows that for all σ, $\mathbb{P}(f_{\text{perm}} = \sigma) = 1/n!$.

When σ is a linear extension of a poset, then $\sigma(i) \prec \sigma(j)$ implies that $i < j$. In the continuous space, this is the same as saying that if $a \prec b$, then $x(a) < x(a)$.

Therefore, let

$$\mathscr{C}_{\text{LE}} = \mathscr{C}_n \cap \{x : (a \prec b \to x(a) < x(b))\}. \tag{6.15}$$

For σ a linear extension of a poset $\preceq$, the same symmetry argument as before gives that for a draw $Y \sim \textsf{Unif}(\mathscr{C}_{\text{LE}})$, $\mathbb{P}(f_{\text{perm}}(Y) = \sigma) = 1/Z$, where Z is the number of linear extensions of the poset.

Note that it is easy to construct either a Gibbs or Metropolis-Hastings style Harris chain whose stationary distribution is uniform over $\mathscr{C}_{\text{LE}}$. For convenience of notation, create the dummy components $x(0) = 0$ and $x(n+1) = 1$, and assume $0 \prec i \prec n+1$ for all $i \in \{1, \ldots, n+1\}$.

Then, for instance, you can choose I uniformly from $\{1, 2, \ldots, n\}$, and update $x(I)$ by choosing uniformly from the set of values such that $x(I)$ remains in $\mathscr{C}_{\text{LE}}$. To be precise, choose $x(I)$ uniformly from the interval $[a,b]$, where $a = \max_{j \prec i}\{x(j)\}\}$, and $b = \min_{i \prec j}\{x(j)\}$.

Consider the componentwise partial order on elements of $\mathscr{C}_{LE}$ where $(x(1),\ldots,x(n)) \preceq (y(1),\ldots,y(n))$ if and only if $x(i) \leq y(i)$ for all i. By placing this partial order on elements of $\mathscr{C}_{LE}$ (not to be confused with the partial order that defines the set of linear extensions), a monotonic update function can be created. Note that in the fact below, the update function is defined for all vectors in $[0,1]^n$, not just those that correspond to permutations.

Fact 6.6. *Say that $x_1 \preceq x_2$ for x_1 and x_2 in $[0,1]^n$ if for all $j \in \{1,2,\ldots,n\}$, $x_1(j) \leq x_2(j)$. For $x \in [0,1]^n$, and given $i \in \{1,2,\ldots,n\}$ and $u \in [0,1]$, let $a(x) = \max_{j \prec i}\{x(j)\}$ and $b(x) = \min i \prec j\{x(j)\}$. Consider the update function $y = \phi(x,i,u)$ defined as $y(j) = x(j)$ for $j \neq i$, and $y(i) = (b(x) - a(x))u + a(x)$. Then ϕ is a monotonic update function.*

Proof. Note for $x_1 \preceq x_2$ in $[0,1]^n$,

$$a(x_1) = \max_{j \prec i}\{x_1(j)\} \leq \max_{j \prec i}\{x_2(j)\} = a(x_2). \tag{6.16}$$

Similarly $b(x_1) \leq b(x_2)$, and that makes $y(i)$ smaller coming from x_1 than from x_2. Since for $j \neq i$ $x_1(j) \leq x_2(j)$, the result immediately follows. □

Therefore in principle, monotonic CFTP could be used on this chain starting from the minimum state $(0,0,\ldots,0)$ and maximal state $(1,1,\ldots,1)$. Note that neither the minimum nor the maximal state actually correspond to permutations, but that is not important! The only properties that are important is that they do bound any state that does correspond to a permutation, and the chain being run is monotonic even with these extra states added in to the state space.

As with other Gibbs sampler chains discussed earlier, this chain started from different states will never actually coalesce down to a single state. There are two ways to solve this problem. The first is to use the multishift coupling for uniforms from earlier. With this approach, the update function will bring the continuous state space down to a single state.

But in this case, there is another approach. It is not necessary for this problem for CFTP to return a single state from $\mathscr{C}_{\text{LE}}$ as long as every state returned maps into the same permutation. This uses the idea of interwoven vectors that was introduced in [57].

Definition 6.7. *Two vectors x_1 and x_2 in $[0,1]^n$ are* interwoven *if for the permutation $\sigma = f_{perm}(x_1)$,*

$$x_1(\sigma(1)) < x_2(\sigma(1)) \leq x_1(\sigma(2)) < x_2(\sigma(2)) \leq \cdots \leq x_1(\sigma(n)) < x_2(\sigma(n)). \tag{6.17}$$

Figure 6.4 (which was figure 1 of [57]) shows two interwoven vectors. The purpose of interweaving is as follows: any vector trapped between two interwoven vectors maps to the same permutation as the two original vectors.

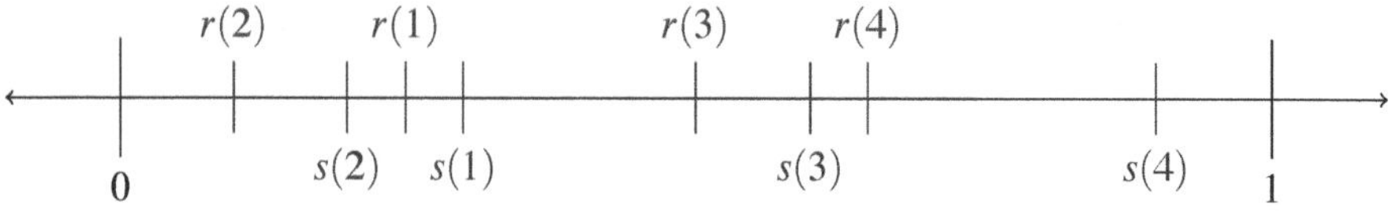

Figure 6.4 *Interwoven vectors:* $r = (0.25, 0.1, 0.6, 0.75)$, $s = (0.3, 0.2, 0.7, 0.9)$, *and* $\sigma = (2,1,3,4)$.

Lemma 6.4. *Suppose* $x_1 \preceq x_2$ *are elements of* $\mathscr{C}_{LE}$. *Then* $(\forall x \in \mathscr{C}_{LE} : x_1 \leq x \leq x_2)(f_{perm}(x_1) = f_{perm}(x) = f_{perm}(x_2))$ *if and only if* x_1 *and* x_2 *are interwoven.*

Proof. Suppose $x_1 < x_2$ are interwoven. Let x be a vector with distinct coordinates satisfying $x_1 \preceq x \preceq x_2$, and let $\sigma = f(x_1)$. Then

$$x_1(\sigma(1)) \leq x(\sigma(1)) \leq x_2(\sigma(1)) \leq x_1(\sigma(2)) \leq \cdots \leq x_1(\sigma(n)) \leq x(\sigma(n)) \leq x_2(\sigma(n)).$$

Removing the x_1 and x_2 vectors from the above inequality and using the fact that $x \in \mathscr{C}_{\text{LE}}$ gives

$$x(\sigma(1)) < x(\sigma(2)) < \cdots < x(\sigma(n))$$

which implies that $f_{\text{perm}}(x) = \sigma$.

Now for the other direction. Suppose $x_1 < x_2$ are not interwoven, and again let $\sigma = f_{\text{perm}}(x_1)$. Then for all i it is true that $x_1(\sigma(i)) < x_2(\sigma(i))$. Since x_1 and x_2 are not interwoven, there must be an i such that $x_2(\sigma(i)) > x_1(\sigma(i+1))$. Let $c_1 = (2/3)x_1(\sigma(i+1)) + (1/3)x_2(\sigma(i))$ and $c_2 = (1/3)x_1(\sigma(i+1)) + (2/3)x_2(\sigma(i))$ so that $x_1(\sigma(i+1)) < c_1 < c_2 < x_2(\sigma(i))$.

Now create two vectors x and y as follows. Let $y(\sigma(j)) = x(\sigma(j)) = x_2(\sigma(j))$ for all $j \notin \{i, i+1\}$. Let $x(\sigma(i)) = c_1, x(\sigma(i+1)) = c_2$ and $y(\sigma(i)) = c_2$ and $y(\sigma(i+1)) = c_1$. Then both x and y are at least x_1 and at most x_2, but $f_{\text{perm}}(x) \neq f_{\text{perm}}(y)$ as these permutations differ by a single transposition. □

To use this with CFTP, consider a step that takes the state in $\mathscr{C}_{\text{LE}}$, generates a permutation from that state, and then uses the permutation to generate a state in $\mathscr{C}_{\text{LE}}$.

This last part is straightforward. Given a permutation σ, generate uniformly from the set of $\mathscr{C}_{\text{LE}}$ that maps to σ by first generating n uniform points $U_1, \ldots, U_n$ in $[0,1]$, and find the order statistics $U_{(1)} < U_{(2)} < \cdots < U_{(n)}$ (they will be distinct with probability 1.) The state

$$(U_{(\sigma(1))}, U_{(\sigma(2))}, \ldots, U_{(\sigma(n))}) \tag{6.18}$$

will be a uniform draw from the states in $\mathscr{C}_{\text{perm}}$ that map to σ.

The pseudocode, then, for generating a partial order according to this method is as follows. First consider the step inside the continuous space.

`Linear_extension_Gibbs_step`
Input: $x \in \mathscr{C}_{\text{LE}}, u \in [0,1], i \in \{1,\ldots,n\}$, *Output:* x

---

1) Let a be 0 if no elements precede i, or $\max_{j:j \prec i} x(j)$ otherwise
2) Let b be 1 if i precedes no other elements, or $\min_{j:i \prec j} x(j)$ otherwise
3) $x(i) \leftarrow a + (b-a)u$

Next make a step that moves from the continuous state to a permutation, then back to a continuous state.

`Permutation_step`
Input: $x \in \mathscr{C}_{\text{LE}}, (u_1,\ldots,u_n) \in [0,1]^n$ *Output:* x

---

1) For $i \in \{1,\ldots,n\}$, let $\sigma(i) \leftarrow \#\{j : x(j) \leq x(i)\}$
2) Let $(u_{(1)},\ldots,u_{(n)})$ be the order statistics of $u_1,\ldots,u_n$
3) For $i \in \{1,\ldots,n\}$, let $x(i) \leftarrow u_{(\sigma(i))}$

Finally, use CFTP to generate a uniform draw from $\mathscr{C}_{\text{LE}}$.

`Linear_extension_cftp` *Input:* t *Output:* X

---

1) Draw $U_1, U_2, \ldots, U_t$ iid $\text{Unif}([0,1])$
2) Draw $I_1, \ldots, I_t$ iid $\text{Unif}(\{1,2,\ldots,n\})$
3) Draw $W_1, W_2, \ldots, W_n$ iid $\text{Unif}([0,1])$
4) $x_{\min} \leftarrow (0,0,\ldots,0)$, $x_{\max} \leftarrow (1,1,\ldots,1)$
5) For t' from 1 to t
6) $\quad$ $x_{\max} \leftarrow$ `Linear_extension_Gibbs_step`$(x_{\max}, U_{t'}, I_{t'})$
7) $\quad$ $x_{\min} \leftarrow$ `Linear_extension_Gibbs_step`$(x_{\min}, U_{t'}, I_{t'})$
8) If $x_{\max}$ and $x_{\min}$ are interwoven
9) $\quad$ $x \leftarrow x_{\max}$
10) Else
11) $\quad$ $x \leftarrow$ `Linear_extension_cftp`$(2t)$
12) $\quad$ For t' from 1 to t
13) $\quad\quad$ $x \leftarrow$ `Linear_extension_Gibbs_step`$(x, U_{t'}, I_{t'})$
14) $x \leftarrow$ `Permutation_step`$(x_{\max}, (W_1,\ldots,W_n))$

So why go to all this trouble, when the bounding chain approach of Section 4.4 is guaranteed to generate a sample in $\Theta(n^3 \ln(n))$ time? Recall that method requires $\Theta(n^3 \ln(n))$ now matter how many precedence equations there are in the poset. Even if there are no elements i and j with $i \prec j$, it will still take on the order of $n^3 \ln(n)$ steps. Call this example the empty poset.

Using the continuous approach on the empty poset, the above procedure will only take $\Theta(n \ln(n))$ steps to finish, as the state will coalesce once every dimension has been chosen, so the number of steps needed is exactly the coupon collector problem.

6.8.1 Height-2 posets

Another situation where the continuous version provably helps is when dealing with posets of height 2. Caracciolo et. al. referred to this as the *corrugated surface* problem [20].

Definition 6.8. *The* height *of a poset is the largest integer h such that there exist distinct elements $a_1,\ldots,a_h$ of the poset with $a_1 \preceq a_2 \preceq \cdots \preceq a_h$.*

Consider the graph that has nodes corresponding to elements of the state space, and an edge between i and j if $i \prec j$ or $j \prec i$. Then let Δ be the maximum degree of this graph. So $\Delta = \max_i \{\#j : i \prec j \text{ or } j \prec i\}$.

First show that the chain is likely to move together if the number of steps is large.

Lemma 6.5. *Consider lines 5 through 7 of* `Linear_extension_cftp`*. If $t \geq 2n((2\Delta-1)\ln(8n^2)+\ln(4n))$, then*

$$\mathbb{P}\left(\max_i |x_{\min}(i) - x_{\max}(i)| > 1/(8n^2)\right) \leq 1/4. \tag{6.19}$$

Proof. Let d_k be the vector $x_{\max} - x_{\min}$ after executing lines 6 and 7 exactly k times. Then $d_0 = (1,1,\ldots,1)$. The proof uses the following potential function on nonnegative vectors:

$$\phi(d) = \sum_{i=1}^{n} d(i)^{2\Delta-1}. \tag{6.20}$$

The goal is to show that the expected value of $\mathbb{E}[d_k]$ decreases exponentially as k increases. Now

$$\mathbb{E}[\phi(d_{k+1})|d_k] = \mathbb{E}\left[\sum_{i=1}^{n} d_{k+1}(i)^{2\Delta-1}\middle| d_k\right]. \tag{6.21}$$

At the $k+1$st step, only coordinate I_{k+1} could change. Hence

$$\mathbb{E}[\phi(d_{k+1})|d_k] = \phi(d_k) + \mathbb{E}[d_{k+1}(I_{k+1})^{2\Delta-1} - d_k(I_{k+1})^{2\Delta-1}|d_k]. \tag{6.22}$$

Suppose that I_{k+1} is not preceded by any other element. Then

$$a_{\min} = a_{\max} = 0, \quad b_{\min} = \min\{1, \inf_{j:I_{k+1}\prec j} x_{\min j}(j)\}, b_{\max} = \min\{1, \inf_{j:I_{k+1}\prec j} x_{\max j}(j)\} \tag{6.23}$$

which makes $d_{k+1}(I_{k+1}) = U_{k+1}(b_{\max} - b_{\min})$.

If there is no element j such that $I_{k+1} \prec j$ then $b_{\max} - b_{\min} = 0$. Otherwise there exists j^* where $x_{\min}(j^*) = b_{\min}$. By definition $b_{\max} \leq x_{\max}(j^*)$, which means $b_{\max} - b_{\min} \leq x_{\max}(j^*) - x_{\min}(j^*) = d_k(j^*)$. Using $d_k(j^*)^{2\Delta-1} \leq \sum_{j:I_{k+1}\prec j} d_k(j)^{2\Delta-1}$ gives

$$[U_{k+1}(b_{\max} - b_{\min})]^{2\Delta-1} \leq U_{k+1}^{2\Delta-1} \sum_{j:I_{k+1}\prec j} d_j^{2\Delta-1}. \tag{6.24}$$

Since $U_{k+1} \sim \textsf{Unif}([0,1])$, taking the expectation of both sides gives

$$\mathbb{E}[d_{k+1}(I_{k+1})^{2\Delta-1}|d_k, I_{k+1}] = \mathbb{E}[U_{k+1}(b_{\max} - b_{\min})] \leq \frac{1}{2\Delta} \sum_{j:I_{k+1}\prec j} d_k(j)^{2\Delta-1}. \tag{6.25}$$

A similar result holds if I_{k+1} is an element that precedes no other element. Taking the expectation over I_{k+1} to remove the conditioning on I_{k+1} gives

$$\begin{aligned}\mathbb{E}[d_{k+1}(I_{k+1})^{2\Delta-1}|d_k] &\leq \frac{1}{n}\sum_{i=1}^{n}\frac{1}{2\Delta}\sum_{j:j\prec i \text{ or } i\prec j} d_k(j)^{2\Delta-1} \\ &= \frac{1}{2\Delta n}\sum_{j}\underbrace{\sum_{i:i\prec j \text{ or } j\prec i}}_{\text{at most } \Delta \text{ terms}} d_k(j)^{2\Delta-1} \leq \frac{1}{2n}\phi(d_k).\end{aligned}$$

Next note that

$$-\mathbb{E}[d_k(I_{k+1})|d_k] = -\mathbb{E}[\mathbb{E}[d_k(I_{k+1})|d_k, I_{k+1}]] = -\sum_i \frac{1}{n} d_k(i)^{2\Delta-1} = -\frac{1}{n}\phi(d_k), \quad (6.26)$$

which gives the exponential decrease (on average) that we were looking for.

$$\mathbb{E}[\phi(d_{k+1})|d_k] \leq \phi(d_k) + \frac{1}{2n}\phi(d_k) - \frac{1}{n}\phi(d_k) = \phi(d_k)\left(1 - \frac{1}{2n}\right). \quad (6.27)$$

A simple induction then gives $\mathbb{E}[\phi(d_k)] \leq n(1-1/(2n))^k$. Using $1-x \leq \exp(-x)$, after $k = 2n((2\Delta-1)\ln(8n^2)+\ln(4n))$ steps, $\mathbb{E}[\phi(d_k)] \leq (1/4)(1/(8n^2))^{2\Delta-1}$, so by Markov's inequality, $\mathbb{P}(\phi(d_k) > (1/(8n^2))^{2\Delta-1}) \leq 1/4$. But for any i, $d_k(i)^{2\Delta-1} \leq \phi(d_k)$, which gives the result. □

Lemma 6.6. *Consider lines 5 through 7 of* `Linear_extension_cftp`*. If* $t \geq 2n((2\Delta-1)\ln(8n^2)+\ln(4n))$*, then the probability that* $x_{\max}$ *and* $x_{\min}$ *are interwoven at the next line is at least* $1/2$*.*

Proof. Start by supposing that x is uniform over $\mathscr{C}_{\text{LE}}$. Then $x_{\min} \leq x \leq x_{\max}$, and by the previous lemma after $t \geq 2n((2\Delta-1)\ln(8n^2)+\ln(4n))$ steps the chance that any coordinate of $x_{\min}$ is farther than $1/(8n^2)$ from a coordinate of $x_{\max}$ is at most 1/4.

After this fixed number of steps, x is still uniform over $\mathscr{C}_{\text{LE}}$. Let p be the chance that any two coordinates of x are closer together than $1/(4n^2)$.

From our discussion of moving from the continuous state to the permutation state back to the continuous state from earlier, we know the coordinates of x can be viewed as the result of n independent uniforms drawn from $[0,1]$. For $i \neq j$, the chance that $|U_i - U_j| \leq 1/(4n^2) \leq 1/(2n^2)$. There are $\binom{n}{2}$ such i and j, so by the union bound, $p \leq n(n-1)/(4n^2) \leq 1/4$.

Suppose for all j, the distance from $x_{\min}(j)$ to $x_{\max}(j)$ is at most $1/(8n^2)$ and $x_{\min}(j) \leq x(j) \leq x_{\max}(j)$. Suppose also that $|x(j) - x(j')| > 1/(4n^2)$. Then $x_{\min}$ and $x_{\max}$ must be interwoven. See Figure 6.5 for an illustration.

To see this analytically, let $i \in \{1,\ldots,n-1\}$. Then

$$x_{\min}(\sigma(i+1)) - x_{\max}(\sigma(i)) = a - b - c, \text{ where}$$

$$a = x_{\min}(\sigma(i+1)) - x(\sigma(i)), \quad b = x(\sigma(i+1) - x_{\max}(\sigma(i)), \quad c = x_{\max}(\sigma(i)) - x(\sigma(i)).$$

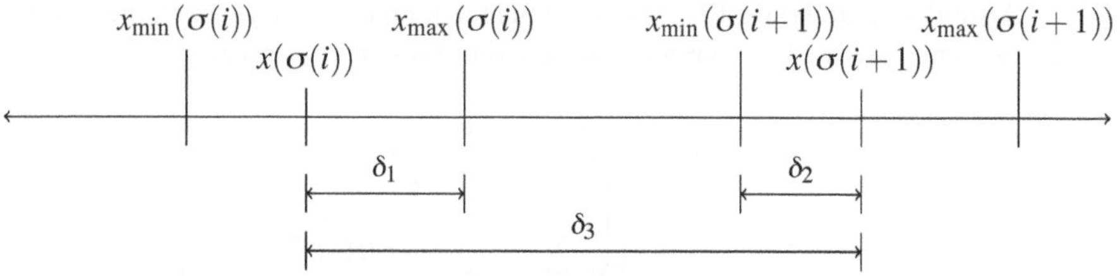

Figure 6.5 *If* δ_1 *and* δ_2 *are at most* $1/(8n^2)$, *and* $\delta_3 > 1/(4n^2)$, *then* $x_{\min}(\sigma(i+1)) > x_{\max}(\sigma(i))$.

If $a > 1/(4n^2)$, $b \leq 1/(8n^2)$ and $c \leq 1/(8n^2)$ for all i, then $x_{\min}(\sigma(i+1)) > x_{\max}(\sigma(i))$ for all i, which makes the two vectors interwoven. By the union bound, the probability that these inequalities hold for a, b, c for all i is at least $1/2$.

□

Since each step in this chain takes $O(\Delta)$ time, this gives an algorithm for generating perfect samples from height 2 posets in $O(n\Delta^2 \ln(n))$ time. In the worst case, $\Delta = \Theta(n)$, and this is no better than the bounding chain. However, for sparse graphs (like lattices), Δ can be small or even fixed, which gives an $O(n \ln(n))$ time algorithm.

Chapter 7

Spatial Point Processes

With dancing, you have to know spatial movement with somebody. It is steps. It's literally steps and knowing how close to be or how far away. You have to have the beat in the right place with the camera.

Channing Tatum

Hundred-year oaks in a forest. Typos in a book. Raindrops on a parking lot. These are examples of data sets that can be modeled using spatial point processes. In a spatial point process, there is a state space S, and some subset of points in the space comprise the process. Usually the points mark an occurrence of some event. Town locations on a map, a list of homes containing someone with a disease, any time the data is locational in time or space, it can be modeled with a spatial point process.

The state space does not have to be purely geometric: if $S = \mathbb{R}^2 \times (-\infty, 0]$, the points could represent the location of the town and the time (with 0 denoting the present) that the town was founded.

Among spatial point processes, the most common model is the Poisson point process (PPP) and its variants. A PPP can be defined mathematically in two different ways. The first approach utilizes the idea of a random measure. The second approach comes at the process from a simulation point of view. Naturally this second method is of more value here!

Our basic presentation here follows notation from Streit [120]; see Chapter 1 of that text for more detail. Throughout this chapter, S will denote the space in which the points of the PPP occur. This space is often a subset of m-dimensional Euclidean space $S \subseteq \mathbb{R}^m$. This space covers the examples treated in this chapter.

An element of a PPP is an unordered set of points in space. (A more general definition of PPP allows repeated points in the space, but making the points distinct allows for great simplification in the description and still covers most applications.) This configuration could contain no points, in which case it is empty, and written as $\emptyset$. As usual, let $\#P$ denote the number of points in the configuration P. Let $S^{[n]}$ denote the set of size n subsets of S (note $S^{[0]} = \{\emptyset\}$.) Then the complete PPP is an element of $S^{[0]} \cup S^{[1]} \cup S^{[2]} \cup \cdots$, which is called the *exponential space*. (See [21] for a more formal definition.)

Definition 7.1. *For $S \subset \mathbb{R}^m$, the* intensity *is a density with respect to Lebesgue measure over S so that for any $A \subseteq S$ of bounded size, $\int_{x \in A} \lambda(x)\,dx < \infty$. If $\int_{x \in A} \lambda(x)\,dx = 0$ for all A of measure 0, call the intensity* continuous.

This of course does not imply that $\int_S \lambda(x)\,dx$ is finite, only that every bounded window is. Since the finiteness is only guaranteed for a bounded window, that is what will be simulated from. Not the complete Poisson point process, but the finite number of points that fall into some bounded window.

A PPP can be described by a simulation procedure that tells how to generate a PPP over a set $S' \subset S$ where $\int_{x \in S'} \lambda(x)\,dx$ is finite.

Definition 7.2. *A random set of points P in the exponential space of S is a* Poisson point process of continuous intensity λ *if given $P \cap A$, the distribution of $P \cap B$ is given by the following simulation technique:*

1. *Draw $N \sim \text{Pois}(\int_{x \in B \cap A^C} \lambda(x)\,dx)$.*
2. *Draw $P_1, \ldots, P_N$ independently from unnormalized measure λ over BA^C.*

When $\lambda = 1$ everywhere, call this the standard PPP *over the space.*

Note that if $\lambda \neq 1$, the PPP drawn can be viewed as coming from a distribution that has a density with respect to a different PPP.

Lemma 7.1. *Let P_1 denote a PPP with intensity measure ν_1, and let P_2 be a draw from a PPP with intensity measure ν_2, which has density f_2 with respect to ν_1. Then the density of the distribution of P_2 with respect to P_1 is*

$$f(p) = \exp(\nu_1(S) - \nu_2(S)) \prod_{v \in p} f_2(v). \tag{7.1}$$

(See Proposition 3.8 of [105].) In particular, if ν_1 is just Lebesgue measure, this expression gives the density of the PPP with respect to a standard PPP.

7.1 Acceptance/rejection

The simplest protocol for perfect simulation of spatial point processes is once again acceptance/rejection. An AR-style algorithm operates in the same fashion as for discrete problems. The ingredients consist of a target density f (possibly unnormalized), density g (possibly unnormalized) from which it is possible to sample directly, and a constant c such that for all configurations x, $f(x)/[cg(x)] \leq 1$.

Then AR first draws a sample X from g. Accept the sample as a draw from f with probability $f(X)/[cg(X)]$. Otherwise, reject and start over.

To illustrate these ideas, consider the Strauss process. This is an example of a repulsive process, which penalizes point processes every time a pair of points in the process is closer than R to each other.

Definition 7.3. *The* Strauss process *has a density with respect to a standard PPP. There are three parameters of the model, $\beta > 0$, $\gamma \in [0,1]$, and $R > 0$. The density is then*

$$f(x) = \beta^{\#x} \gamma^{p_R(x)}, \tag{7.2}$$

where $p_R(x)$ is the number of pairs of points in x such that the distance between them is at most R.

Acceptance/rejection for drawing from the Strauss process over a bounded window first begins by drawing a process with density $\beta^{\#x}$ against a standard PPP. This part is easy: just generate a PPP (call it P) with constant intensity $\lambda(x) = \beta$ throughout using Definition 7.2. Then $\beta^{\#x}\gamma^{p_R(x)}/\beta^{\#x} \leq 1$, so set $c = 1$. So accept P as a draw from the Strauss process with probability $\gamma^{p_R(P)}$.

This works well if γ is close to 1. However, if β and S are large, and γ is small, this will give an exponentially small chance of acceptance.

7.1.1 Covering points

Consider the following model with parameters λ and R greater than 0, and a special point $v \in S$. The goal is to draw a spatial point process of rate λ such that there exists at least one point in the process that lies within distance R of v. Let $B(v,R)$ denote the set of points within distance R of v (also known as the ball of radius R around v). Then the density of a configuration x is $f_v(x) = \mathbb{1}(B(v,R) \cap x \neq \emptyset)$.

Again basic AR can be used for this process. Repeatedly draw a PPP of rate λ until the draw contains a point within distance R of v. Of course, if λ is small, this might take a large number of draws in order to achieve the goal. The goal is to create a method that works well even when λ is very small.

To understand the technique that solves this problem, consider the following two-step procedure. Suppose that we have an algorithm to generate from f_v. First draw $P \sim f_v$, then uniformly select one of the points of $P \cap B(v,R)$ to receive a mark. Let m denote the marked point. The probability that m is the marked point given P is $1/\#(x \cap B(v,R))$. Therefore, the density of the marked process is just $f_{v,\text{mark}}(x,m) = [f_v(x)/\#(x \cap B(v,R))]\mathbb{1}(m \in x \cap B(v,R))$.

At this point, we do not have a method for obtaining draws from f_v, so we cannot create a draw from $f_{v,\text{mark}}$ by first drawing P and then uniformly marking a point. On the other hand, suppose that we could draw from $f_{v,\text{mark}}$. Then, to draw from f_v, first draw from $f_{v,\text{mark}}$ and then just throw away the mark at m leaving x. Since there are $\#(x \cap B(v,R))$ possible points that could have been marked, each with density $f_v(x)/\#(x \cap B(v,R))$, summing gives that the density of the result of this procedure is f_v. So the problem of simulating from f_v has been reduced to drawing from $f_{v,\text{mark}}$.

Drawing from $f_{v,\text{mark}}$ can be accomplished using AR. First draw from $g_{\text{mark}}(x,m) = \mathbb{1}(m \in x \cap B(v,R))$. This can be done by drawing the marked point m first uniformly from $B(v,R)$. Conditioned on this point being in the point process P, the rest of P can be found by simply drawing a point process from S and taking the union with m.

This gives a draw P from $g_{\text{mark}}(x,m)$. Accept P as a draw from $f_{v,\text{mark}}$ with probability $f_{v,\text{mark}}(x)/g_{v\text{mark}} = 1/\#(x \cap B(v,R))$. When λ is small, $\#(x \cap B(v,R))$ is likely to be 1 and so the algorithm will take only a small number of draws.

So AR for f_v works well when λ is large, and $f_{v,\text{mark}}$ works well when λ is small. The interruptibility of AR can be used to combine the two to get an algorithm that works well whether λ is large or small.

`Covering_PPP`
Input: S, v, R *Output:* Poisson point process $\{X_1, \ldots, X_N\} \sim f_v$

1) `cflag` $\leftarrow$ FALSE
2) Let μ be the product of λ and the Lebesgue measure of S
3) Repeat
4) Draw N as a Poisson random variable with mean μ
5) Draw $X_1, \ldots, X_N$ iid uniformly from S
6) If $B(v,R) \cap \{X_1, \ldots, X_N\} \neq \emptyset$, then `cflag` $\leftarrow$ TRUE
7) Else
8) Draw N as a Poisson random variable with mean μ
9) Draw $X_1, \ldots, X_N$ iid uniformly from S
10) Draw $X_{N+1} \leftarrow \text{Unif}(B(v,R))$, $N \leftarrow N+1$
11) Draw $U \leftarrow \text{Unif}([0,1])$
12) `cflag` $\leftarrow U \leq 1/\#(B(v,r) \cap \{X_1, \ldots, X_N\})$
13) Until `cflag`

Lemma 7.2. *The expected number of random variables generated by* `Covering_PPP` *is at most* $(2+\mu)(1+2/\sqrt{3})$.

Proof. Let P be a Poisson process drawn from S. Then to generate P requires generating N with mean μ, and then generating N points. So on average this takes $1+\mu$ points. Lines 8 through 12 require one more draw (for X_{N+1}), so take $2+\mu$ points on average.

Let p be the probability that such a draw P has $B(v,R) \cap P \neq \emptyset$. Note that lines 8 through 12 only happen with probability $1-p$, and then only fail with probability at most p. Hence, if T is the number of random variables drawn by the algorithm

$$\mathbb{E}[T] \leq 1 + \mu + (1-p)[2 + \mu + p\mathbb{E}[T]], \tag{7.3}$$

so $\mathbb{E}[T] \leq (2+\mu)(2-p)/(1-p+p^2) \leq (2+\mu)(1+2/\sqrt{3})$. □

7.2 Thinning

The simulation view of a PPP was used in the previous section. A different view concentrates on the expected number of points that fall into any bounded region.

Lemma 7.3. *For any bounded set $A \subseteq S$, and P a Poisson point process of rate λ over S, $\mathbb{E}[\#(A \cap P)] = \int_A \lambda(x)\, dx$.*

Proof. This follows immediately from the simulation description. □

Lemma 7.4. *Let A and B be any two nonoverlapping measurable subsets of S. Then for P a PPP of rate λ over S, $P \cap A$ and $P \cap B$ are independent.*

Proof. Again this follows directly from the simulation definition. □

What is perhaps surprising is that these properties actually characterize a Poisson point process (a fact that will be used here without proof). They also lead directly to the notion of thinning.

Suppose that P is a Poisson point process of rate λ. For each point x in P, independently flip a coin with probability $p(x)$ of heads. Let P' be the point process consisting of points of P that received a head.

Then if $P \cap A$ and $P \cap B$ are independent, so are $P' \cap A$ and $P' \cap B$. Informally, for a measurable set A, because each point in $P \cap A$ is only retained for P' with probability $p(x)$, the expected number of points in P' is $\int_A \lambda(x)p(x)\,dx$. That means that P' satisfies the two conditions of being a Poisson point process over S of rate λp. These ideas can be summarized as follows.

Definition 7.4. *Suppose for a PPP P, each point $X \in P$ has an independent Bernoulli B_X with mean $\alpha(X)$. Then $Q = \{X \in P : B_X = 1\}$ forms the* thinned *point process.*

Theorem 7.1. *Let P be a PPP of intensity λ, and Q is thinned from P using function α. Then Q is also a PPP of intensity $\alpha\lambda$.*

See [105, p. 23] for a formal proof.

A thinned process can be used to build a perfect simulation algorithm when the intensity λ is given by a density f_λ with respect to Lebesgue measure that is uniformly bounded above by M over the space S.

Bounded_PPP
Input: S, f_λ, M *Output:* Poisson point process P

1) Let μ be M times the Lebesgue measure of S
2) Draw N as a Poisson random variable with mean μ
3) Draw $X_1, \ldots, X_N$ iid uniformly from S
4) For each $x \in \{X_1, \ldots, X_N\}$
5) Draw $U \leftarrow \text{Unif}([0,1])$
6) If $U > f_\lambda(x)/M$, then $P \leftarrow P \setminus \{x\}$

The expected running time of the algorithm is proportional to the Lebesgue measure of S times M, and so it is important to get as tight a bound on f_λ as possible.

7.2.1 Example: a normal-intensity PPP

For example, suppose the goal is to generate a Poisson point process on the one-dimensional interval from 0 to 1, subject to intensity $\lambda(x) = (2\pi)^{-1}\exp(-x^2/2)$. This intensity has maximum value $(2\pi)^{-1}$ at $x = 0$. So first generate a Poisson point process of intensity $(2\pi)^{-1}$ over the interval $[0,1]$. Then, for each point in the process, accept the point as coming from the desired process with probability $\exp(-x^2/2)$. The result will have the target density.

7.2.2 Using thinning to make the window easier

Thinning can also be used to simplify the window over which sampling occurs. Suppose that the goal is to generate a Poisson process of rate λ over an unusual bounded region S. Suppose that S' is also bounded, and $S \subset S'$, but it is much easier to sample from λ over S' than from S.

One way to describe sampling from a Poisson point process over S is as sampling from a point process over S' with density $f(x) = \mathbb{1}(x \cap S^C = \emptyset)$. This density can then be used with the thinning approach. Points that fall into S^C are retained with probability 0, and points that fall into S are retained with probability 1. Put more simply, generate a Poisson point process from over S', and then only retain those points that fall into S. The result will be a Poisson point process over S.

7.3 Jump processes

While AR is usually the easiest method to use, the running time for basic AR does tend to grow exponentially with the size of S. Therefore a more sophisticated method is needed for generating samples from large windows. What will be used is a special type of Markovian process called a jump process. Preston [111] gave a method for creating a jump process whose stationary distribution is a PPP with density f.

Definition 7.5. *A process* $\{X_t\}_{t\in\mathbb{R}}$ *is a* jump process *if given* $X_t = x$*, the amount of time that the process stays at state* x *has an exponential distribution with mean* $1/\alpha(x)$*. The state then jumps to a new state using kernel* K *independent of the past history of the chain. (This is also known as a* continuous-time Markov chain.*)*

Preston's framework had two types of moves in the chain.

1. The *birth rate* $b(x,v)$ is the rate at which a point v is "born" and added to the configuration.
2. The *death rate* $d(x,v)$ is the rate at which a point v "dies" and is removed from the configuration.

The simplest form of a birth-death chain has $d(x,v) = 1$ for all $v \in x$. That means that every point $v \in x$ is given an exponential random variable of rate 1. At the time indicated by that exponential random variable, the point v "dies" and is removed from the configuration.

At the same time, the space S has a birth rate that controls how quickly points are born. The overall rate is $b = \int_{v\in S} b(x,v)\ dx$. When a "birth" occurs, the location of the birth is chosen according to the unnormalized density $b(x,\cdot)$.

Use the notation $x+v$ to indicate the PPP that has all of the points in x plus the point v. To achieve a stationary density f, it is necessary to balance the rate of births and deaths.

Theorem 7.2. *Suppose*

$$f(x)b(x,v) = f(x+v)d(x+v,v). \tag{7.4}$$

Then f is a stationary density for the jump process.

[For a proof, see [111].]

When $d(x,v) = 1$ for all $v \in x$, the goal is to have $f(x)b(x,v) = f(x+v)$. Moreover, the birth rate can be viewed in a Metropolis way, or in a thinning way. That is, if the birth rate is $b'(x,v)$, and the birth of point v is accepted with probability $r(x,v)$, then the final birth rate is effectively $b'(x,v)r(x,v)$.

As in Chapter 1, the end goal is to have the limiting distribution of the chain be the Harris distribution.

Definition 7.6. *A jump process* $\{X_t\}$ *over* Ω *is a* continuous-time Harris chain *if there exists* $A, B \subseteq \Omega$ *and* $\varepsilon > 0$ *for* $x \in A, y \in B$, *and a probability measure* ρ *with* $\rho(B) = 1$ *where*

1. *For* $T_A = \inf\{t \geq 0, X_t \in A\}$, $(\forall z \in \Omega)(\mathbb{P}(T_A < \infty | X_0 = z) > 0)$.
2. *If* $x \in A$, *then* $\alpha(x) \geq \varepsilon$. *Moreover, the amount of time that the state stays in* x *is* $\min(T_1, T_2)$, *where* $T_1 \sim \exp(\varepsilon)$, $T_2 \sim \exp(\alpha(x) - \varepsilon)$. *If the state jumps after* T_1 *time, then the new state is drawn from measure* ρ, *and does not depend on* x.

Since a continuous-time Harris chain only jumps after a random amount of time, such chains can never be periodic. The notion of recurrent is precisely the same for these new chains as given in Definition 1.30. The analogue of Theorem 1.3 is nearly identical except aperiodicity is not required.

Theorem 7.3 (Ergodic Theorem for continuous-time Harris chains). *Let* X_n *be a recurrent continuous-time Harris chain with stationary distribution* π. *If* $\mathbb{P}(R < \infty | X_0 = x) = 1$ *for all* x, *then as* $t \to \infty$, *for all measurable sets* C *and starting points* x:

$$|\mathbb{P}(X_t \in C | X_0 = x) - \pi(C)| \to 0. \tag{7.5}$$

So to show that the limiting distribution of a Preston birth-death chain is the stationary distribution, it suffices to show that the chain is recurrent.

7.3.1 *Example: Strauss process*

Consider how the Preston chain works for the Strauss process. The rate of deaths is always 1. The initial rate of births is β. Then when a point v attempts to be born, it is accepted and added to the configuration x with probability $\gamma^{p_R(x+v)-p_R(x)}$. This makes the effective birth rate $b(x,v) = \beta\gamma^{p_R(x+v)-p_R(x)}$, and hence (7.4) is satisfied.

This is implemented in the following pseudocode. Here m is Lebesgue measure for S a bounded subset of $\mathbb{R}^n$.

`Strauss_update`	*Input:* current state x,	*Output:* next state x
1)	Draw $T_B \leftarrow \text{Exp}(\beta\nu(S))$	
2)	Draw $T_D \leftarrow \text{Exp}(\#x)$	
3)	If $T_D < T_B$ (then there is a death)	
4)	Draw v uniformly from x, $x \leftarrow x - v$	
5)	Else	
6)	Draw v over $m(S)$	
7)	Draw $U \leftarrow \text{Unif}([0,1])$	
8)	If $U < \gamma^{\#\{w \in x: \text{dist}(v,w) \leq R\}}$	
9)	$x \leftarrow x + v$	

7.3.2 Locally stable

The Strauss process density has the property of being locally stable.

Definition 7.7. *A PPP density is* locally stable *if there exists a constant M such that for all configurations x and points v, $f(x+v) \leq Mf(x)$.*

That is, addition of a point does not change the value of the density by more than a constant factor. For the Strauss process, $M = \beta$.

In general, for a locally stable process over a state space S with Lebesgue measure $m(S)$, the update procedure is as follows.

`Locally_stable_update` *Input:* current state x, *Output:* next state x

```
Draw T_B ← Exp(M · m(S))
Draw T_D ← Exp(#x)
If T_D < T_B (then there is a death)
    Draw v uniformly from x, x ← x − v
Else
    Draw v uniformly over S
    Draw U ← Unif([0,1])
    If U < f(x+v)/[Mf(x)]
        x ← x + v
```

Exponentials have the nice property that the probability that one exponential is smaller than an independent exponential is proportional to the rates.

Fact 7.1. *Suppose $A_1 \sim \textsf{Exp}(\lambda_1)$ and $A_2 \sim \textsf{Exp}(\lambda_2)$ are independent exponentially distributed random variables. Then $\mathbb{P}(A_1 < A_2) = \lambda_1/(\lambda_1 + \lambda_2)$.*

So if one just keeps track of the number of points in the PPP, then if this number is i, there is an $i/[i + Mm(S)]$ chance of a death that reduces i by 1, and a $Mm(S)/[1 + Mm(S)]$ chance of a birth that increases i by 1.

Fact 7.2. *Consider a Markov chain on $\{0, 1, 2, \ldots\}$ where there is a constant C such that $\mathbb{P}(N_{t+1} = i+1 | N_t = i) = C/[C+i]$ and $\mathbb{P}(N_{t+1} = i-1 | N_t = i) = i/[C+i]$. Such a chain is aperiodic and recurrent, returning to the state $i = 0$ infinitely often.*

The idea is that since C is fixed, the chance of moving upwards goes to 0 as the state increases. Hence the state never becomes too large. See for instance [84] for a formal proof.

What this means is that the dominating process will return to the empty set infinitely often. That implies the following result.

Fact 7.3. *This locally stable update gives an aperiodic, recurrent Harris chain with stationary density f, so the continuous time analogue of Theorem 1.3 guarantees that the distribution of the state of the process as t goes to infinity will approach the stationary distribution.*

7.4 Dominated coupling from the past for point processes

Kendall and Møller [82] noted that implicit in the `Locally_stable_update` is a dominating process that allows the use of dominated CFTP. For an evolving PPP X_t, the dominating process Y_t contains every point in X_t and possibly some extra ones because it allows every birth into the state.

`Locally_stable_dominating_update` *Input:* x, y *Output:* x, y

Draw $T_B \leftarrow \text{Exp}(M \cdot m(S))$
Draw $T_D \leftarrow \text{Exp}(\#y)$
If $T_D < T_B$ (then there is a death)
    Draw v uniformly from y, $y \leftarrow y - v$, $x \leftarrow x - v$
Else
    Draw v uniformly over S
    Draw $U \leftarrow \text{Unif}([0,1])$
    $y \leftarrow y + v$
    If $U < f(x+v)/[Mf(x)]$
        $x \leftarrow x + v$

This update function is monotonic. To see why, suppose that the input satisfies $x \subseteq y$. When $T_D < T_B$, that means that a point in y is about to die. Since x cannot have a birth without a birth in y, the death of the point in y must also come before any birth in x. If this point in y that is about to die is also in x, then it also dies in x. (Note that if $v \notin x$ then $x - v = x$). Either way, after the death event it still holds that $x \subseteq y$.

On the other hand, if $T_B \leq T_D$, then there is a point v that is added to y. It might or might not (depending on the value of U) be added to x, but it is always added to y. Hence, at the end of a birth event, $x \subseteq y$.

So Y_t is a dominating process for X_t. Moreover, Y_t is a very simple birth-death process. The stationary distribution for the Y_t process is a PPP with rate $\lambda(x) = M$ for all $x \in S$, so it is easy to generate a sample from Y_t.

Formally Y_t is a *marked process*, because for each point in Y_t, there is a choice of uniform U_t called a *mark*. The marks are sufficient to determine the X_t, so it is the marked Y_t and X_t processes that are coupled together.

Now in order to used dominated CFTP, it is also necessary to be able to run the dominating process Y_t backwards in time. This is straightforward from a computational point of view, since the dominating process is reversible. However, it can get a bit confusing, as when time is running backward, a "birth" in the reverse process is a "death" in the forward process. And a "death" in the backward process is a "birth" in the forward process.

The dominating process can be thought of as maintaining a list of birth and death events that happen backwards in time. Call this list EL for event list. Let $D(0)$ denote the configuration of the dominating process at time 0. Then sometime before time 0, a birth or death event occurred. Let $D(-1)$ denote the state of the dominating process immediately before this event occurred.

Similarly, let $D(-i)$ denote the state of the dominating process just before time t_i, where exactly i birth or death events occurred in interval $[t_i, 0)$. Suppose that the current event list holds the first n events that happened before time 0. This is necessary to be able to extend this list backwards in time to $n' > n$ events. This is done by the following pseudocode:

`Generating_event_list`
Input: event list EL of length n, $D(-n)$, $n' > n$
Output: event list EL of length n', $D(-n')$

For i from n to $n'-1$ do
    Draw $B \leftarrow \textsf{Exp}(\#D(-i))$, Draw $D \leftarrow \textsf{Exp}(M\nu(S))$
    If $B < D$ (then there is a birth)
        Draw v uniformly from x, $U \leftarrow \textsf{Unif}([0,1])$
        Add [birth v U] to EL and set $D(-i-1) \leftarrow D(-i) - v$
    Else
        Draw v over $\nu(S)$
        Add [death v 0] to EL and set $D(-i-1) \leftarrow D(-i) + v$

Once the birth event list is generated, then run the chain forward using the dominated process. If the chain couples from all starting states, then the resulting state must be stationary. The pseudocode is as follows.

`Dominated_cftp_for_spatial_point_processes` *Output:* $X \sim f$

Start with EL empty
Draw $N \leftarrow \textsf{Pois}(M\nu(S))$
Draw $D_1, \ldots, D_N$ independently from $\nu(\cdot)$ over S
Let $D(0) \leftarrow \{D_1, \ldots, D_N\}$
$n \leftarrow 1$, $n' \leftarrow 0$
Repeat
    $(\text{EL}, D(-n)) \leftarrow$ `Generating_event_list`$(\text{EL}, D(-n'))$
    For every $X(-n) \subseteq D(-n)$
        Run $X(-n)$ forward to $X(0)$ using the events in EL
Until all the resulting $X(0)$ values are the same
Output $X(0)$

Theorem 7.4 (Kendall and Møller [82]). *The above procedure generates $X \sim f$ with probability 1.*

Proof. Set $Y_n(-n) = \emptyset$ for all n, and use the top n events in the EL list to move Y_n forward to $Y_n(0)$. Then as the number of events grows to infinity, the amount of time in the chain for the Y_n process is going to infinity, so in the limit as n goes to infinity, the distribution of $Y_n(0)$ is going to the stationary density f from Theorem 7.3.

Now let $T = \min\{n : D(-n) = \emptyset\}$. Then for $n > 0$, $Y_n(0)$ is always the same, and equal to X, the output of dominated CFTP. But the only way that a sequence of equal random variables can converge (in measure) to density f, is if every single one has density f. □

As with regular CFTP, it is usually not necessary to actually run the $X(-n)$ process forward from every single starting state. Instead, a bounding chain that is coupled with the dominating chain can be used to test coalescence. The bounding chain consists of a pair of point processes $A \subseteq B \subseteq D$ such that the state X always satisfies $A \subseteq X \subseteq B$.

Points in A are definitely in the state X, but points in BA^C might be in X, or they might not be.

The idea of coupling behind dominating CFTP can also be used to better estimates of ergodic averages, see [99, 100]. In addition, it can also be used for thinning and transformation of spatial processes, which in turn can be useful for model assessment [104, 97].

7.4.1 *Example: dominated CFTP for the Strauss process*

For Strauss, and $A \subseteq X \subseteq B$, the lowest chance of a point v being added is when $X = B$, and the highest chance is when $X = A$. So the bounding states can be updated as follows, given an event list.

`Strauss_bounding_update`
Input: event list EL with n events, $D(-n)$ *Output:* A, `cflag`

$A \leftarrow \emptyset$, $B \leftarrow D(-n)$
For i from n to 1 do
  Let (`type` v U) be the nth event in EL
  If `type` is death then
    $A \leftarrow A - v$, $B \leftarrow B - v$
  Else
    $p_{R,A}(v) \leftarrow \#\{w \in A : \text{dist}(v,w) \leq R\}$
    $p_{R,B}(v) \leftarrow \#\{w \in B : \text{dist}(v,w) \leq R\}$
    If $U < \gamma^{p_{R,B}}$ then $A \leftarrow A + v$
    If $U < \gamma^{p_{R,A}}$ then $B \leftarrow B + v$
If $A = B$ then `cflag` is true

Further examples of dominating CFTP for spatial point processes are discussed in [6, 5].

7.4.2 *Running time*

To understand how long this process must be run before coalescence occurs, it helps to understand how exponential random variables work over small time steps.

Suppose that $A \sim \exp(\lambda)$. Then conditioned on $A > t$, $\mathbb{P}(A \in (t, t+h] | A > t) = \lambda h + o(h)$. Hence for each of the points in A_t or B_t (or both), the chance that the point dies in interval $[t, t+h]$ is $\lambda h + o(h)$.

Points that are in $B_t \setminus A_t$ that die help coalescence occur, since this brings the sets A_t and B_t closer together. For B_t and A_t to move farther apart, a point must be born within distance R of a point in $B_t \setminus A_t$, and $U > \gamma$. Hence the chance that $B_t \setminus A_t$ grows by a point is bounded above by $\beta \nu(B_R)(1-\gamma)h + o(h)$. Here $\nu(B_R)$ is the maximum size of a ball of radius R for a point in the space.

Using this gives the following lemma.

Lemma 7.5. *Consider an interval of time $[t,0]$, where $A_t = \emptyset$ and $B_t = D_t \cap S$, where D_t is the dominating process. Then the chance that $A_0 = B_0$ is at most*

$$\beta\nu(S)\exp(-[1-(1-\gamma)\beta\nu(B_R)]t). \tag{7.6}$$

Proof. Let $N_t = \#(B_t \setminus A_t)$. Then if a point in $B_t \setminus A_t$ dies, N_t is reduced by one; if a point is born within distance R of $B_t \setminus A_t$ and $U > \gamma$, then N_t grows by one.

So for $t \in [t',0]$ and $h > 0$:

$$\mathbb{E}[N_{t+h}|\mathscr{F}_t] \leq N_t - hN_t + h(1-\gamma)\beta\nu(B_R)N_t + o(h). \tag{7.7}$$

Letting $n_t = \mathbb{E}[N_t]$ and taking the expectation of both sides gives

$$n_{t+h} \leq n_t - hn_t + h(1-\gamma)\beta\nu(B_r)n_t + o(h). \tag{7.8}$$

Subtracting n_t, dividing by h, and taking the limit as h goes to 0 gives $n_t' \leq [\beta(1-\gamma)\nu(B_r) - 1]n_t$.

This differential inequality gives the upper bound $n_t \leq n_{t'}\exp([\beta(1-\gamma)\nu(B_R) - 1](t-t'))$. The number of points in $D_{t'}$ is Poisson with mean $\beta\nu(S)$, which completes the proof. □

Therefore, as long as the parameters are in the regime where $(1-\gamma)\beta\nu(B_R) < 1$, efficient generation from the Strauss process is possible.

7.5 Shift moves

The running time results indicate that when β is small, the procedure will be efficient. This result uses the birth and death moves introduced by Preston [111]. By adding a third type of move, a shift move [55], it is possible to improve the analysis further.

A shift move can be viewed as when one point is added while another is destroyed. This is equivalent to shifting the point from one location to another. The governing equation for shifts is similar to that for births and deaths.

Theorem 7.5 ([55]). *Let $b(x,v)$ be the rate at which a point v is born to configuration x, let $d(x+v,v)$ be the rate at which point v in configuration $x+v$ is dying, let $s(x+v,v,w)$ be the rate at which point v in configuration $x+v$ is being shifted to point w, and let $s(x+w,w,v)$ be the rate at which point w in configuration $x+w$ is being shifted to point v. Suppose*

$$f(x)b(x,v) = f(x+v)d(x+v,v)$$
$$f(x+v)s(x+v,v,w) = f(x+w)s(x+w,w,v).$$

Then f is a stationary density for the jump process.

Recall Preston's approach, which keeps deaths occurring at a constant rate of 1, and then just modifies the birth rate. In order to keep things simple, the shifts can be just a subset of the births.

Here is the idea. In a process like the Strauss process, there is a proposed birth

v. There are a number of neighbors $w_1, \ldots, w_k$ where each has a γ chance (independently) of not affecting the birth of v in the slightest.

However, it could be that exactly one neighbor is blocking the draw of v. In this case, flip a p_s-coin, and if it comes up heads, keep the birth and kill the neighbor. In other words, shift the neighbor w to v.

This move maintains both the number of points in the configuration and the number of pairs within distance R of each other. Also, the rate of shifting from $x+v-w$ to $x+w-v$ are exactly equal, so the conditions of the theorem are satisfied.

This gives rise to the following pseudocode.

Strauss_shift_update *Input:* current x, *Output:* next x

Draw $T_B \leftarrow \text{Exp}(\beta \nu(S))$
Draw $T_D \leftarrow \text{Exp}(\#x)$
If $T_D < T_B$ (then there is a death)
    Draw v uniformly from x, $x \leftarrow x - v$
Else
    Draw v over $\nu(S)$
    $N_v \leftarrow \{w : \text{dist}(v,w) \leq R\}$
    For each $w \in N_v$
        Draw $C \leftarrow \text{Bern}(\gamma)$, If $C = 1$ then $N_v \leftarrow N_v - w$
    If $\#N_v = 0$ or ($\#N_v = 1$ and $U < p_s$)
        $x \leftarrow x \setminus N_v + v$

Now examine how to update the bounding chain. Recall that the bounds are $A \subseteq x \subseteq B$, so that points in A are definitely in the configuration, while points in $B \setminus A$ might be in the configuration, or they might not be. So both possibilities have to be considered when deciding what to do at the next step.

Let $N_{A,v} = \{w \in A : \text{dist}(v,w) \leq R\}$, and $N_{B,v} = \{w \in B : \text{dist}(v,w) \leq R\}$. Then each point in $N_{B,v}$ is retained independently with probability $1-\gamma$. If it is removed from $N_{B,v}$, it is also removed from $N_{A,v}$. Now there are three cases to consider.

Case 1: $\#N_{A,v} = \#N_{B,v} = 0$. This is the easiest case, since here v will be born regardless of whether or not a shift is attempted.

Case 2: $\#N_{A,v} = 0$, $\#N_{B,v} = 1$. This is an interesting case. Suppose a shift is attempted. Then the fact that there is a $w \in N_{B,v} \setminus N_{A,v}$ means that this point w might or might not be part of the configuration. If it is not part of the configuration, then the point v is born normally. If it is part of the configuration, then w is shifted to v. Either way, w is now removed from the configuration, and v is added!

Case 3: $\#N_{A,v} = 1$, $\#N_{B,v} > 1$. When there is no shift move then the point in $N_{A,v}$ prevents v from being born. When there is a shift move, bad things happen: v might or might not be born, and the point $w \in N_{A,v}$ might or might not be shifted away. The result is that v and w are both removed from A and added to B.

Case 4: $\#N_{A,v} = 0$, $\#N_{B,v} > 2$. In this case, whether or not there is a shift move the fate of v is uncertain, and so v is added to B but not to A.

This is summarized in the following pseudocode.

Strauss_shift_bounding_update *Input:* x, A, B *Output:* x, A, B

```
Draw T_B ← Exp(βν(S))
Draw T_D ← Exp(#x)
If T_D < T_B (then there is a death)
    Draw v uniformly from x, x ← x − v
Else
    Draw v over ν(S)
    N_{A,v} ← {w ∈ A : dist(v,w) ≤ R}, N_{B,v} ← {w ∈ B : dist(v,w) ≤ R}
    For each w ∈ N_B
        Draw C ← Bern(γ), If C = 1 then N_{B,v} ← N_{B,v} − w and N_{A,v} ← N_{A,v} − w
    If #N_{A,v} = #N_{B,v} = 0
        A ← A + v, B ← A + v
    Else if #N_{A,v} = 0, #N_{B,v} = 1
        If U < p_s then A ← A + v, B ← A + v \ N_{B,v}
        Else B ← A + v
    Else if #N_{A,v} = 1, #N_{B,v} ≥ 1
        If U < p_s then A ← A + v \ N_{A,v}, B ← A + v
    Else if #N_{A,v} = 0, #N_{B,v} ≥ 2
        A ← A, B ← A + v
```

Lemma 7.6. *Consider an interval of time $[t,0]$, where $A_t = \emptyset$ and $B_t = D_t \cap S$, where D_t is the dominating process. Then the chance that $A_0 = B_0$ in the bounding update with shifts with $p_S = 1/4$ is at most*

$$\beta\nu(S)\exp(-[1-(1/2)(1-\gamma)\beta\nu(B_R)]t). \tag{7.9}$$

Proof. As earlier, let $N_t = \#(B_t \setminus A_t)$, $n_t = \mathbb{E}[N_t]$, and consider what happens in a tiny time interval of length h.

There could be a death of a point in $B_t \setminus A_t$, reducing N_t by 1. Then there are the births to consider. Given the chosen birth point v, divide the space S into four subregions corresponding to the four cases.

$$\begin{aligned}
A_0 &= \{s : \#N_{A,v} = 0, \#N_{B,v} = 0\} \\
A_1 &= \{s : \#N_{A,v} = 0, \#N_{B,v} = 1\} \\
A_2 &= \{s : \#N_{A,v} = 1, \#N_{B,v} \geq 1\} \\
A_3 &= \{s : \#N_{A,v} = 0, \#N_{B,v} \geq 2\}
\end{aligned}$$

Points born into A_0 do not change N_t. A point born into A_1 decreases N_t if $U < p_s$, otherwise it increases it by 1. Points born into A_2 can increase N_t by 2 if $U < p_s$, otherwise N_t is unchanged. Last, points born into A_3 always increase N_t by 1.

Putting this together, the overall rate of change in n_t is

$$\begin{aligned}
n_{t+h} &\leq n_t - hn_t + h\mathbb{E}[\nu(A_1)(-p_s + (1-p_s))] + h\mathbb{E}[\nu(A_2)][2p_s] + h\mathbb{E}[\nu(A_3)] + o(h) \\
&= n_t + h[-n_t + (1-2p_s)\mathbb{E}[\nu(A_1)] + 2p_s\mathbb{E}[\nu(A_2)] + \mathbb{E}[\nu(A_3)]] \\
&= n_t + n_t h[-1 + (1/2)\mathbb{E}[\nu(A_1) + \nu(A_2) + 2\nu(A_3)]]
\end{aligned}$$

where the last line follows from setting $p_s = 1/4$.

Note that any point in A_3 neighbors at least two points in $B \setminus A$, while points in A_2 and A_1 neighbor at least one. The expected number of points to survive the γ-coin weeding process is $(1-\gamma)n_t$. Since each covers an area of measure $\nu(B_R)$, and the intensity is β, that means that

$$\mathbb{E}[\nu(A_1) + \nu(A_2) + 2\nu(A_3)] \leq (1-\gamma)n_t\beta\nu(B_R). \tag{7.10}$$

Hence

$$n_{t+h} \leq n_t + n_t h[-1 + (1/2)(1-\gamma)\beta] + o(h).$$

Solving the differential inequality gives the result. □

Notice that the inclusion of the shift move has improved the analysis to the point where twice as large values of β are now shown to run efficiently!

7.6 Cluster processes

Another common model for spatial data is a *cluster process*. Here the set of parent points follows a PPP, and then each parent point (possibly) gives rise to more daughter points. The collection of daughter points forms the cluster process. These types of point processes arise naturally in situations such as disease points or plant growth.

Consider such a point process that lives in the plane. Suppose that the goal is to simulate such a process perfectly over a finite observation window. Then one approach would be to simulate the parent process in the window, and then the daughter points.

This, however, will give a simulation with too few points: it could be that a parent point outside of the observation window could give rise to a daughter point that falls back inside the observation window.

Brix and Kendall [15] noted that this problem could be solved by simulating exactly those parent points on the infinite plane that gave rise to daughter points that lie inside the observation window.

Let W denote the observation window. In the general case, allow each parent point x to have a random mark mark M, so that each point is of the form $[x;M]$. This mark should have finite expectation ($\mathbb{E}[M] < \infty$).

Furthermore there is an intensity $k(x,\cdot)$ for daughter points that depends on the location of the parent point x. Then each parent point has a set of daughter points that is a PPP with intensity $Mk(x,\cdot)$.

So for a parent point x, what is the chance that it has a daughter point that falls in W? Well, the daughter points inside W from x form a PPP, and the chance that a PPP contains at least one point is $1-\exp(-\mu)$, where μ is the mean number of points that fall inside the window.

To find the mean number of points, first let $K(x,W) = \int_{y \in W} k(x,\cdot)\, d\mathbb{R}^2$. That would be the mean number of daughter points to fall in W if the mark M was always

1. Since M is in fact random, it is necessary to integrate over all values of the mark to get:

$$p(x) = 1 - \int_{m \geq 0} \exp(-mK(x,W))\, dM(m) = 1 - \mathscr{L}_M(K(x,W)), \qquad (7.11)$$

where $\mathscr{L}_M$ is the Laplace transform of the mark distribution M.

We only care about a parent point if it has a daughter point that falls into W. Otherwise, the parent point might as well be deleted. So in a sense, all the points in the parent process are being thinned based on their location. The chance that the point is retained is $p(x)$. Given that a parent point is retained, then choose a number of daughter points to fall in W conditional on at least one point in the daughter process falling in W.

So how can such a process be realized? The steps are as follows.

1. First create an upper bounding density $\lambda_u(x) \geq p(x)$ and $\int_S \lambda_u(x)\, d\mathbb{R}^2 < \infty$.
2. Generate a PPP X from S with density λ.
3. For each point X, retain X with probability $p(x)/\lambda(x)$.

7.6.1 *Example: Gaussian daughters*

As an example of how this method can be carried out in practice, consider a window W, which is $[-1,1]^2$. The parent process has constant density $\lambda(\cdot) = \lambda$. Each point in the parent process has mark value $M = 1$. Also, each point has a daughter process whose intensity is a standard bivariate Gaussian, so that

$$k(x,w) = \frac{1}{2\pi} \exp\left(-\frac{1}{2}\|w - x\|^2\right).$$

In order to use the method from Section 7.6, it is necessary to give an upper bound on $k(x,w)$ that integrates to a finite value over the plane. Note that for $w \in W$, $\|w\| \leq \sqrt{2}$. So from the triangle inequality, $\|w - x\| \geq \|x\| - \sqrt{2}$. If $\|x\| \geq 2\sqrt{2}$, then $\|w - x\| \geq \|x\|/2$ and $\|w - x\|^2 \geq \|x\|/4$. If $\|x\| < 2$, then $\|w - x\|^2 \geq 0$.

Hence

$$k(x,w) \leq \frac{1}{2\pi}\left[\exp\left(-\frac{1}{2}\frac{\|x\|^2}{4}\right)\mathbb{1}(\|x\| \geq 2\sqrt{2}) + \mathbb{1}(\|x\| < 2\sqrt{2})\right].$$

Using $\mathbb{1}(\|x\| < 2\sqrt{2})/(2\pi) \leq \exp(-1)$ gives

$$k(x,w) \leq \exp\left(-\frac{1}{2}\frac{\|x\|^2}{4}\right),$$

and so setting $\lambda(x) = \exp(-\|x\|^2/8)$ gives us the upper bound on the density.

Since $\lambda(x)$ integrates to 8π over $\mathbb{R}^2$, on average 8π points will be drawn in the upper bounding process. Once the location of each point x is drawn, it is easy to compute $\mathscr{L}k(x,W)$ and so $p(x)$. Accept each point as being a daughter spawning point with probability $p(x)/\lambda(x)$. Finally, for each accepted point, draw a Poisson random variable with mean $\mathscr{L}k(x,W)$, and draw this many daughter points for the window according to $k(x,\cdot)$.

The resulting union of daughters of these thinned parent points is the draw from the cluster process for the observed window.

Møller and Rasmussen [102] later extended this method to Hawkes processes where each daughter point can itself be the parent for a new point, which can in turn be a parent to daughter points, and so on indefinitely.

7.7 Continuous-time Markov chains

Recall that jump processes are also known as *continuous-time Markov chains*. These types of Markovian processes arise in many models, especially in queues.

7.7.1 Monotonic CFTP for Jackson queuing networks

Definition 7.8. *A* queue *consists of a set of customers waiting in line to be served. When a customer arrives, the length of the queue increases by 1. When a customer is served (which can only happen when the queue length is positive), the queue length decreases by 1.*

Definition 7.9. *When the customers arrive to the queue according to a Poisson process with fixed rate λ, there is a single server who serves customers one at a time, and the service time is exponentially distributed with rate μ, then it is an* M/M/1 *queue. (The first M means that the arrivals are memoryless, the second M means the services are memoryless, and the 1 indicates that there is a single server.)*

This basic notion of queues allows for the length of queue to be arbitrarily large, but most queues in practice have a limit on their size.

Definition 7.10. *A* capacitated queue *has a capacity C for the queue. If there is an arrival to the queue when the length is already C, the arrival is turned away and disappears. An M/M/1 queue with capacity C is denoted as M/M/1/C. If the queue has infinite capacity it is denoted M/M/1/∞.*

M/M/1/C queues form a finite-state continuous-time Markov chain, and have been well studied. The stationary distribution is simple and easy to simulate from. Things become more interesting when once a customer leaves the queue, that customer might enter into another queue.

Definition 7.11. *A* Jackson network *consists of a set of nodes V where once a customer is served at node i, it joins the queue at node j with probability p_{ij}. With probability $1-\sum_{j\neq i} p_{ij}$, it leaves the system. The service rate λ_i is constant for each node i. The network is* open *if there is an arrival process of rate λ, where each arrival goes to queue i with probability p_{0i}, and $\sum_i p_{0i} = 1$.*

If each queue in the Jackson network has arrivals and services that are exponentially distributed, then the Jackson network will be a continuous-time Markov chain. When the capacity of all the queues is infinite, the stationary distribution has a well-known product form, but when the capacities are finite this form becomes difficult to compute.

Monotone CFTP was first applied to this problem by Mattson [90]. Events consist of one of two types: arrival to a queue, and service from a queue. Since the arrival and service processes are Poisson, each can be considered separately. Recall the following well-known fact about exponential random variables.

Lemma 7.7. *Let $X_1,\ldots,X_n$ be independent exponentially distributed random variables, with $X_i \sim \mathsf{Exp}(\lambda_i)$. Let $Y \sim \mathsf{Exp}(\sum_i \lambda_i)$. Then $Y \sim \min\{X_1,\ldots,X_n\}$ and $\mathbb{P}(X_i = \min\{X_1,\ldots,X_n\}) = \lambda_i/(\sum_j \lambda_j)$.*

The number of events that occurs in an interval $[t_1,t_2]$ will be the length of the interval, $t_2 - t_1$, times the rate at which events occur. In a Jackson network, the total rate of events is $\lambda + \sum_i \lambda_i$. This allows the building of a set of service and arrival times over an interval. In the following pseudocode, a positive integer value at $a(t)$ indicates that there was an arrival at node $a(t)$, while an ordered pair (i,j) in the list indicates that there was an attempted service at node i, and after the service (if it took place) the customer moved to the queue at node j. If the pair is of the form $(i,0)$ then the customer left the queue after service.

```
Jackson_network_event_list    Input: t1 < t2    Output: event list a

Draw N ← Pois((t2 − t1)[λ + λi])
Let a be an empty list
For t from 1 to N do
   Draw X ← Bern(λ/[λ + Σi λi])
   If (X = 1) then (there is an arrival event)
      Draw I so that P(I = i) = p0i
      Append I to a
   Else (there is a service)
      Draw I so that P(I = i) = λi/[Σi λj]
      Draw J so that P(J = j) = pij, P(J = 0) = 1 − Σk pik
      Append (I,J) to list a
```

For example, if for interval $[-5,0]$ the output was $a = (3,(2,1),2)$, that means that during the time from -5 up to 0, there was an arrival to node 3 followed by a service at node 2 that then moved to queue 1, and an arrival at node 2.

To actually update the state of the Jackson network, use the events provided to change the values at each queue. Let $(C_1,\ldots,C_m)$ be the capacities of the m queues in the network.

```
Jackson_network_update    Input: x, a    Output: x

For t from 1 to the number of events in a do
   If a(t) = i for some node i
      x(i) ← x(i) + 1(x(i) < Ci)
   Else a(t) = (i, j)
      x(j) ← x(j) + 1(x(i) > 0)1(x(j) < Cj)
      x(i) ← x(i) − 1(x(i) > 0)
```

It is straightforward to check that for any pair of vectors $x \leq y$ and any event list a, it holds that $\texttt{Jackson\_network\_update}(x,a) \leq \texttt{Jackson\_network\_update}(y,a)$. Therefore it can be used to run monotone CFTP using the minimal state of all 0's and the maximal state of $(C_1,\ldots,C_m)$.

`Jackson_network_mcftp` *Input: t* *Output: x*	
1)	$a \leftarrow$ `Jackson_network_event_list`(t)
2)	$x_{\text{min}} \leftarrow (0,0,\ldots,0)$, $x_{\text{max}} \leftarrow (C_1,\ldots,C_m)$
3)	$x_{\text{min}} \leftarrow$ `Jackson_network_update`(x_{min},a)
4)	$x_{\text{max}} \leftarrow$ `Jackson_network_update`(x_{max},a)
5)	If $x_{\text{min}} = x_{\text{max}}$
6)	$x \leftarrow x_{\text{min}}$
7)	Else $x \leftarrow$ `Jackson_network_mcftp`$(2t)$
8)	$x \leftarrow$ `Jackson_network_update`(x,a)

Chapter 8

The Randomness Recycler

Of course there are many ways we can reuse something. We can dye it. We can cut it. We can change the buttons. Those are other ways to make it alive. But this is a new step to use anything—hats, socks, shirts. It's the first step in the process.

Issey Miyake

The last four chapters all utilized coalescence of Markov chains in order to achieve perfect samples. With these techniques you could either be read once or interruptible, but not both. The techniques of this and the next two chapters rely on different ideas that move beyond Markov chains. This enables the construction of linear time, read-once, and interruptible algorithms for sampling from high-noise Markov random fields, along with algorithms for distributions on permutations where coalescence is difficult to achieve.

8.1 Strong stationary stopping time

Before CFTP was introduced, a perfect simulation method using Markov chains was introduced by Aldous and Diaconis [3], called strong stationary stopping times (SSST). In general it is more difficult to find an SSST than a CFTP algorithm for a Markov chain, but an SSST does give a read-once interruptible algorithm. Moreover, the idea leads to the far more general notation of the randomness recycler which is applicable in most instances that CFTP is.

Recall that a stopping time T with respect to a sequence $X_1, X_2, \ldots$ is a random variable T such that the event $\{T \leq n\}$ is measurable with respect to the information contained in $X_0, X_1, \ldots, X_n$.

Definition 8.1. A stationary stopping time *is a stopping time T such that $X_T \sim \pi$, where π is the stationary distribution for the process.* A strong stationary stopping time *is a stationary stopping time where T and X_T are independent even conditioned on the value of X_0.*

Once an SSST exists, the perfect simulation algorithm is simple: run the process $X_1, X_2, \ldots$ forward until T occurs, then report X_T. This gives a read-once, interruptible (because T and X_T are independent) perfect simulation algorithm.

For example, consider simple random walk with partially reflecting boundaries on $\{1,2,3\}$. In this chain the transition probabilities $p(a,b) = \mathbb{P}(X_{t+1} = b|X_t = a)$ are $p(1,1) = p(1,2) = p(2,1) = p(2,3) = p(3,2) = p(3,3) = 1/2$.

The stationary distribution for this Markov chain is uniform over $\{1,2,3\}$. Therefore, a simple way of generating a stationary stopping time is to let $Y \sim \mathsf{Unif}(\{1,2,3\})$ and then let $T_1 = \inf\{t : X_t = Y\}$. Then $X_T = Y$ and so has the correct distribution, but T_1 and X_{T_1} are not independent conditioned on X_0. For instance, when $X_0 = 1$, $\mathbb{P}(T_1 > 2|X_t = 3) = 1$ but $\mathbb{P}(T_1 > 2) < 1$. So this is an example of a stationary time, but not a strong stationary time.

Here is an SSST for this Markov chain. Say that the chain tries to move right when $X_{t+1} > X_t$, or when $X_{t+1} = X_t = 3$. Denote this type of step with an r. Similarly, the chain tries to move left when $X_{t+1} < X_t$ or $X_{t+1} = X_t = 1$. Denote this with an ℓ. So if the chain started at $X_0 = 2$ and the first four steps where $r\ell\ell r$, then $X_4 = 2$ as well.

The SSST looks at the chain in groups of two moves. First, wait until $X_{3k} = 1$ for some integer k. Next, look at the next two moves of the chain. If they are either rr, $\ell\ell$, or ℓr, then since each of these are equally likely to occur, X_{3k+2} is equally likely to be in $\{1,2,3\}$.

That is, for

$$T_2 = \inf\{3k+2 : X_{3k} = 1, \text{ and the next two moves are either } rr,\ \ell\ell, \text{ or } \ell r\}, \quad (8.1)$$

then $X_{T_2} \sim \mathsf{Unif}(\{1,2,3\})$. Moreover, knowing the value of T_2 does not change the distribution of X_{T_2}, even when conditioned on X_0.

8.1.1 *Example: SSST for simple symmetric random walk on* $\{1,2,\ldots,n\}$

Now extend the walk to $\{1,2,\ldots,n\}$ so that $p(a,a-1) = p(a,a+1) = 1/2$ for all $a \in \{2,\ldots,n-1\}$, and $p(1,1) = p(n,n) = 1/2$.

To extend the SSST to this more complicated chain, it is easiest to consider how the distribution of X_t evolves over time. Suppose $X_0 = 1$. This can be represented by the probability vector that places 100% of the probability on state 1, namely, $(1,0,0,\ldots,0)$. After one step in the chain, the probability vector will be $(1/2,1/2,0,0,\ldots,0)$ indicating that $\mathbb{P}(X_1 = 1|X_0 = 1) = \mathbb{P}(X_1 = 2|X_0 = 1) = 1/2$, so now $X_1 \sim \mathsf{Unif}(\{1,2\})$.

Now take one more step in the chain. The probability vector is now $(1/2,1/4,1/4,0,0,\ldots,0)$. The state X_2 is no longer uniform over a set of states. However, the distribution of X_2 is the equal mixture of two uniforms, one uniform over $\{1,2,3\}$, and one uniform over $\{1\}$. See Figure 8.1.

Given $X_2 = x$, draw $Y|X_2 = x$ as uniform over $[0,\mathbb{P}(X_2 = x)]$. Then $(X_2,Y) \sim \mathsf{Unif}(\{(x,y) : 0 \le y \le \mathbb{P}(X_2 = x)\})$.

Note that $[X_2|Y \le 1/2] \sim \mathsf{Unif}(\{1,2,3\})$. Similarly, $[X_2|Y > 1/2] \sim \mathsf{Unif}(\{1\})$.

In general, suppose that $X_t \sim \mathsf{Unif}(\{1,\ldots,i\})$ for $i \in \{2,\ldots,n-1\}$. Then X_{t+1} will be a mixture of the uniform distribution over $\{1,\ldots,i-1\}$ and the uniform distribution over $\{1,\ldots,i+1\}$). When $X_{t+1} > i-1$, then X_{t+1} must have been a draw from the the latter distribution. When $X_{t+1} \le i-1$, there is a 50% chance that X_{t+1} is uniform over $\{1,\ldots,i-1\}$, and a 50% chance that it is uniform over $\{1,\ldots,i+1\}$.

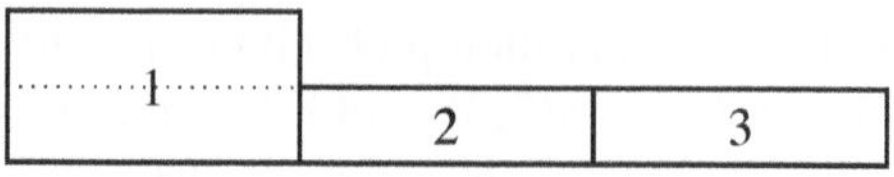

Figure 8.1 *The density of X_2 for simple symmetric random walk on $\{1,2,\ldots,n\}$ conditioned on $X_0 = 1$. The dotted line shows that this density is actually a mixture of the uniform distribution over $\{1\}$ with the uniform distribution over $\{1,2,3\}$.*

This gives the following SSST-based algorithm for generating from the stationary distribution of simple symmetric random walk.

```
SSST_for_ssrw

X ← 1, X* ← 1
Repeat
   Draw U_1 ~ Unif([0,1])
   X ← X − 1(U_1 ≤ 1/2, X > 1) + 1(U_1 > 1/2, X < n)
   Draw U_2 ~ Unif([0,1])
   If X ≤ X* − 1 and U_2 ≤ 1/2
      X* ← X* − 1
   Else
      X* ← X* + 1
Until X* = n
```

Lemma 8.1. *Let (X_t, X_t^*) denote the value of (X, X^*) after t steps through the repeat loop. Then for any $x^* \in \{1,\ldots,n\}^n$ such that $\mathbb{P}((X_0^*, X_1^*, \ldots, X_t^*) = x^*) > 0$,*

$$[X_t | (X_0^*, X_1^*, \ldots, X_t^*) = x^*] \sim \textsf{Unif}(\{1,\ldots,x_t^*\}). \tag{8.2}$$

Proof. The proof is by induction on t. The base case when $t = 0$ has $(X_0, X_0^*) = (1,1)$, so the result holds trivially.

Now assume that (8.2) holds for t, and consider what happens at time $t+1$. Then

$$\mathbb{P}(X_{t+1} = i | X_{t+1}^* = x_{t+1}^*, \ldots, X_0^* = x_0^*) = \frac{\mathbb{P}(X_{t+1} = i, X_{t+1}^* = x_{t+1}^* | X_t^* = x_t^*, \ldots, X_0^* = x_0^*)}{\mathbb{P}(X_{t+1}^* = x_{t+1}^* | X_t^* = x_t^*, \ldots, X_0^* = x_0^*)}.$$

At this point, there are several cases to consider.

For instance, suppose that $x_{t+1}^* = x_t^* + 1$ where $x_t^* \in \{2,\ldots,n-1\}$.

If $1 < i < x_t^*$ (which equals $x_{t+1}^* - 1$ is this case) then X_t is either $i-1$ or $i+1$, each with probability $1/x_t^*$ given $X_t^* = x_t^*$ by the induction hypothesis. Therefore

$$\mathbb{P}(X_{t+1} = i, X_{t+1}^* = x_t^* + 1 | X_t^* = x_t^*, \ldots, X_0^* = x_0^*) = \frac{1}{x_t^*} \cdot \frac{1}{2} \cdot \frac{1}{2} + \frac{1}{x_t^*} \cdot \frac{1}{2} \cdot \frac{1}{2} = \frac{1}{x_t^*} \cdot \frac{1}{2}.$$

If $i = x_t^*$ or $i = x_t^* + 1$, then X_t must be $i-1$, and

$$\mathbb{P}(X_{t+1} = i, X_{t+1}^* = x_t^* + 1 | X_t^* = x_t^*, \ldots, X_0^* = x_0^*) = \frac{1}{x_t^*} \cdot \frac{1}{2}.$$

Therefore

$$\mathbb{P}(X_{t+1}^* = x_t^* + 1 | X_t^* = x_t^*, \ldots, X_0^* = x_0^*) = \sum_{i=1}^{x_{t+1}^*} \frac{1}{x_t^*} \cdot \frac{1}{2} = \frac{x_{t+1}^*}{2x_t^*},$$

and for all $i \in \{1, \ldots, x_t^* + 1\}$,

$$\mathbb{P}(X_{t+1} = i | X_{t+1}^* = x_t^* + 1, \ldots, X_0^* = x_0^*) = (2x_t^*)^{-1} / [x_{t+1}^* / (2x_t^*)] = 1/x_{t+1}^*,$$

which completes the induction in this case.

The other cases are shown in a similar fashion. □

So the value of X^* either goes up or down at each step. On average, it goes up slightly more often than it goes down. Consider the following well-known lemma that follows from the Optional Sampling Theorem. (For a proof, see for instance, [31].)

Lemma 8.2. *Suppose $\{Y_t\}_{t=0,1,\ldots}$ is a stochastic process on $\{0,1,2,\ldots\}$ with $\mathbb{E}[Y_{t+1}|Y_t] \leq Y_t - \delta$ for some constant $\delta > 0$, and $|Y_{t+1} - Y_t| \leq M$ for all t. Let $T = \inf\{t : Y_t = 0\}$. If $\mathbb{E}[Y_0]$ is finite then $\mathbb{E}[T] \leq \mathbb{E}[Y_0]/\delta$.*

This fact can be transformed into many different results. For instance, if Y_t is on a state space of the form $\{1,2,\ldots,N\}$, $|Y_t - Y_{t+1}| \leq 1$ for all t and $\mathbb{E}[Y_{t+1}|Y_t] \geq Y_t + f(i)$ for all $Y_t \leq i$, then $\mathbb{E}[\inf\{t : Y_t = i+1 | Y_0 = i\}] \leq 1/f(i)$. Using this fact, it is possible to bound the mean of the SSST for this chain.

Lemma 8.3. *Starting with $X_0^* = 1$, let $T = \inf\{t : X_t^* = n\}$. Then $\mathbb{E}[T] < n(n-1)/2$.*

Proof. How much does X_t^* increase on average as each step? Well, X_t^* increases by 1 in `SSST_for_ssrw` when $X_{t+1} \in \{1, \ldots, X_t^* - 1\}$ and $U_2 > 1/2$ or when $X_{t+1} \in \{X_t^*, X_{t+1}^*\}$.

The chance of this occurring is

$$\begin{aligned}\mathbb{P}(X_{t+1}^* = X_t^* + 1) &= \frac{2}{X_t^*} \cdot \mathbb{P}(U_1 > 1/2) + \frac{2}{X_t^*} \mathbb{P}(U_1 \leq 1/2)\mathbb{P}(U_2 > 1/2) + \frac{X_t^* - 2}{X_t^*} \mathbb{P}(U_2 > 1/2) \\ &= \frac{1}{2} + \frac{1}{2X_t^*}.\end{aligned}$$

That means $\mathbb{P}(X_{t+1}^* = X_t^* - 1) = [1/2] - [1/(2X_t^*)]$, and

$$\mathbb{E}[X_{t+1}^* | X_t^*] = 1/X_t^*. \tag{8.3}$$

So that means when $X_t^* = 1$, then on average one step is needed to move it to an X^* value of 2. When $X_t^* = i$, on average at most $1/(1/i) = i$ steps are needed to move the process to an X^* value of $i+1$. To move from 1 up to n then, on average

$$1 + 2 + \cdots + (n-1) = n(n-1)/2$$

steps are needed. □

At each step, the goal is to make the X_t^* state larger, since when $X_t^* = n$, it is known that X_t is uniform over $\{1,2,\ldots,n\}$. In the good case, $X_{t+1}^* = X_t^* + 1$. But when $X_{t+1} < X_t^*$, this good case only happens with probability $1/2$.

When the good case is rejected, the state has lost some of its randomness. It is known that the state is no longer uniform from $\{1,\ldots,X_t^*\}$. However, all is not lost: conditioned on rejection, X_{t+1} is uniform over $\{1,\ldots,X_t^* - 1\}$. Reclaiming at least part of the distribution of X_{t+1} after rejection is called *recycling*. Using this recycling idea to keep the rejected sample as random as possible forms the basis of the randomness recycler (RR).

1. There is an updating process $\{X_t\}$.
2. There is an indexing process $\{X_t^*\}$ so that $[X_t|X_t^*]$ has a known distribution.
3. There is a value of X_t^* such that $[X_t|X_t^* = x_\pi^*] \sim \pi$.
4. At each step, the configuration X_t can be updated.
5. This allows the distribution indexed by X_t^* to be updated.
6. Usually, when the distribution is closer to the target distribution, there is a chance of rejecting the configuration as coming from the new distribution.
7. There is a sense in which the indexed distribution is either closer or farther away from the target distribution.

In this example X_t is a Markov chain, in which case $T = \inf\{t : X_t^* = x_\pi^*\}$ gives a SSST for the Markov chain. In general, for the randomness recycler there is no requirement that X_t must be a Markov chain. This gives flexibility in updating X_t that allows for more targeted application of random moves. That in turn allows for easier construction of linear time perfect simulation algorithms for MRF's.

8.2 Example: RR for the Ising model

Before giving the general format of the randomness recycler, consider a specific example of the Ising model on a graph $G = (V,E)$. Recall that here a configuration labels each node 1 or -1, and the distribution π has density that is proportional to $\exp(-\beta H(x))$, where $-H(x)$ is the number of edges whose endpoints receive the same label.

Begin by considering the graph that is a 4-cycle. In this graph $V = \{a,b,c,d\}$, and $E = \{\{a,b\},\{b,c\},\{c,d\},\{d,a\}\}$. The goal is to draw $X \sim \pi$, but start with an easier goal. Can we draw $X \sim \pi$ conditioned on all the values of X being 1? Of course, just make $X(a) = X(b) = X(c) = X(d) = 1$. Say that nodes a,b,c, and d are frozen at the value 1.

Now consider taking a step towards the final sample by "unfreezing" node a. The goal is to make $X_2 \sim \pi(\cdot|X(b) = X(c) = X(d) = 1)$. This is also easy. Since two of the neighbors of a are still frozen at 1, $X_2(a) = 1$ with probability proportional to $\exp(2\beta)$, and $X_2(a) = -1$ with probability proportional to $\exp(0\beta) = 1$. The normalizing constant to make these into probabilities is $1+\exp(2\beta)$. Therefore,

$$\mathbb{P}(X(a)=1|X(b)=X(c)=X(d)=1)=\frac{\exp(2\beta)}{1+\exp(2\beta)},$$
$$\mathbb{P}(X(a)=-1|X(b)=X(c)=X(d)=1)=\frac{1}{1+\exp(2\beta)}.$$

Where things get interesting is the next step. What should the probability be that $X(b)=1$, or $X(b)=-1$? Well, the probability of unfreezing node b to label 1 or -1 should depend on what happened at the choice of a. That is, this can be thought of as a two-stage experiment: first unfreeze a, and then try to unfreeze b. First note that conditioned on $X(c)=X(d)=1$, the probabilities for, say, $(X(a),X(b))=(1,-1)$ only depend on the edges leaving the unfrozen nodes. Let A be the event that $X(c)=X(d)=1$ and that the label for a was chosen using $X(b)=1$. Then the goal is to have

$$\mathbb{P}((X(a),X(b))=(1,1)|A)\propto\exp(3\beta) \qquad \mathbb{P}((X(a),X(b))=(1,-1)|A)\propto\exp(\beta)$$
$$\mathbb{P}((X(a),X(b))=(-1,1)|A)\propto\exp(\beta) \qquad \mathbb{P}((X(a),X(b))=(-1,-1)|A)\propto\exp(\beta)$$

because this would be a draw from $\pi(\cdot|A)$.

Break down the first expression:

$$\begin{aligned}\mathbb{P}((X(a),X(b))=(1,1)|A)&=\mathbb{P}(X(a)=1|A,X(b)=1)\mathbb{P}(X(b)=1|A,X(a)=1)\\&=\frac{\exp(2\beta)}{Z_1}\cdot\mathbb{P}(X(b)=1|A,X(a)=1).\end{aligned}$$

Here Z_1 is the normalizing constant for the choice of $X(a)$ (note $Z_1=\exp(2\beta)+1$ in this example). For the right-hand side to be proportional to $\exp(3\beta)$, it must be true that $\mathbb{P}(X(b)=1|A,X(a)=1)\propto\exp(\beta)$.

Repeating this calculation for the other three cases gives the following four expressions.

$$\begin{aligned}\mathbb{P}(X(b)=1|A,X(a)=1)&\propto\exp(\beta)\\\mathbb{P}(X(b)=-1|A,X(a)=1)&\propto\exp(-\beta)\\\mathbb{P}(X(b)=1|A,X(a)=-1)&\propto\exp(\beta)\\\mathbb{P}(X(b)=-1|A,X(a)=-1)&\propto\exp(\beta)\end{aligned}$$

Here is the subtle point: to ensure that $(X(a),X(b))\sim\pi(\cdot|A)$, it is necessary that the same normalizing constant be applied to $\mathbb{P}(X(b)=i|X(a)=j)$ for all i and j. In this case, set $Z_2=\exp(\beta)+\exp(\beta)=2\exp(\beta)$ so, for instance, $\mathbb{P}(X(b)=-1|A,X(a)=1)=\exp(-\beta)/[2\exp(\beta)]$ and

$$\mathbb{P}((X(a),X(b)=(1,-1)|A)=\frac{\exp(2\beta)}{Z_1}\cdot\frac{\exp(-\beta)}{Z_2}=\frac{\exp(\beta)}{Z_1Z_2},$$

and the same overall normalizing constant of Z_1Z_2 holds for the other three cases.

Now, using Z_2 when conditioning on both $X(a)=1$ and $X(a)=-1$ means that $\mathbb{P}(X(b) = 1|A, X(a) = -1) + \mathbb{P}(X(b) = -1|A, X(a) = -1) = 1$, but $\mathbb{P}(X(b) = 1|A, X(a) = -1) + \mathbb{P}(X(b) = -1|A, X(a) = -1) < 1$ for $\beta > 0$. What should be done with the rest of the probability?

In this case, there is no valid unfrozen choice of $X(b)$ to make, so the node b should remain frozen at 1. But this comes at a cost: the only way that b could have remained frozen is if $X(a) = 1$. So by making this choice of keeping b frozen, a must be refrozen at the value of 1.

So just as in the SSST for the random walk from the previous section, the RR algorithm for the Ising model attempts to advance by unfreezing a node, but sometimes must retreat by refreezing neighboring nodes.

This refreezing of neighboring nodes is bad, but it is not as bad as the general acceptance/rejection approach, which throws away entirely the old sample whenever rejection occurs. Here, some remnant of the sample can be kept, as only the neighbors of the node are frozen. This process of keeping some of the sample is called *recycling*, and this is the key to making this approach linear time for nontrivial values of β.

This approach can be broken down into the following seven pieces.

1. There is the set of configurations from which we are attempting to sample.
2. There is a set of frozen nodes whose label is fixed.
3. Conditioned on the labels of the frozen nodes, there exists a distribution on the unfrozen nodes.
4. There is a target distribution over the set of configurations.
5. When all the nodes are unfrozen, the distribution on unfrozen nodes matches the target distribution.
6. When all the nodes are frozen, it is easy to sample.
7. There is a method of moving from the current state and current set of frozen nodes to a new state and set of frozen nodes such that the correct distribution on unfrozen nodes is maintained.

Let x^* denote the set of frozen nodes in a graph, and consider the general problem of how a frozen node $v \in x^*$ should be attempted to be made unfrozen. For $c \in \{-1, 1\}$, let f_c denote the number of frozen neighbors of v that have color c, and n_c the number of unfrozen neighbors of v that have color c.

Suppose for concreteness that v is frozen at label 1. Then the factor for the weight from edges adjacent to v whose other endpoint is not frozen is $\exp(n_1\beta)$. When v is unfrozen, if v is unfrozen at 1, then the weight should gain an extra factor of $\exp(f_1\beta)$, since all of those f_1 edges would now move from an unfrozen node to a frozen node.

If v is unfrozen at -1, then the weight should gain an extra factor of $\exp([f_{-1} + n_{-1}]\beta)$, but also a factor of $\exp(-n_1\beta)$, since those n_1 edges no longer contribute to the weight. Hence

$$\mathbb{P}(X(v) = 1|X(x^*), n_1, n_{-1}, f_1, f_{-1}) \propto \exp(f_1\beta)$$
$$\mathbb{P}(X(v) = -1|X(x^*), n_1, n_{-1}, f_1, f_{-1}) \propto \exp((f_{-1} + n_{-1} - n_1)\beta).$$

To find the constant of proportionality, it is necessary to maximize $\exp(f_1\beta) +$

$\exp((f_{-1}+n_{-1}-n_1)\beta)$ over choices of n_{-1} and n_1. For $\beta > 0$, this is maximized when $n_1 = 0$. Therefore, for node v having $\deg(v)$ neighbors, the proposal kernel is

$$\mathbb{P}(X(v) = 1|X(x^*), n_1, n_{-1}, f_1, f_{-1}) = \frac{\exp(f_1\beta)}{\exp(f_1\beta) + \exp((\deg(v) - f_1)\beta)}$$

$$\mathbb{P}(X(v) = -1|X(x^*), n_1, n_{-1}, f_1, f_{-1}) = \frac{\exp((f_{-1}+n_{-1}-n_1)\beta)}{\exp(f_1\beta) + \exp((\deg(v) - f_1)\beta)}.$$

If the node v is being unfrozen from -1, a similar calculation gives a proposal kernel of

$$\mathbb{P}(X(v) = -1|X(x^*), n_1, n_{-1}, f_1, f_{-1}) = \frac{\exp(f_{-1}\beta)}{\exp(f_{-1}\beta) + \exp((\deg(v) - f_{-1})\beta)}$$

$$\mathbb{P}(X(v) = 1|X(x^*), n_1, n_{-1}, f_1, f_{-1}) = \frac{\exp((f_1+n_1-n_{-1})\beta)}{\exp(f_{-1}\beta) + \exp((\deg(v) - f_{-1})\beta)}.$$

This gives the following overall algorithm.

RR_for_Ising

$x \leftarrow (1,1,\ldots,1)$, $x^* \leftarrow V$
Repeat
  Let v be any node in x^*, $a \leftarrow x(v)$
  For $c \in \{1,-1\}$
    $f_c \leftarrow \#\{w \in x^* : \{v,w\} \in E, x(w) = c\}$
    $n_c \leftarrow \#\{w \notin x^* : \{v,w\} \in E, x(w) = c\}$
  $Z \leftarrow \exp(f_a\beta) + \exp((\deg(v) - f_a)\beta)$
  Draw C from $\{-1,1,r\}$ where $\mathbb{P}(C = a) = \exp(f_a\beta)Z^{-1}$,
    $\mathbb{P}(C = -a) = \exp((f_{-a}+n_{-a}-n_a)\beta)Z^{-1}$, $\mathbb{P}(C = r) = 1 - \mathbb{P}(C \in \{-1,1\})$
  If $C \in \{-1,1\}$
    $x(v) \leftarrow C$, $x^* \leftarrow x^* \setminus \{v\}$
  Else
    $x^* \leftarrow x^* \cup \{w : \{v,w\} \in E\}$
Until $x^* = \emptyset$

The proof of the algorithm's correctness is deferred to the next section when the RR approach will be shown to be generally valid.

Note that when β is small, $\mathbb{P}(C = 1)$ and $\mathbb{P}(C = -1)$ will both be close to $1/2$ and so the probability of recycling ($\mathbb{P}(C = r)$) will be very small. That makes it likely that a node will be unfrozen, and that the algorithm will progress quickly towards the final state.

With this in hand, it is possible for small β to show that the running time of RR for Ising is linear in the size of the graph.

Lemma 8.4. *Let*

$$\delta = 1 - (\Delta+1)\frac{\exp(\beta\Delta) - \exp(-\beta\Delta)}{\exp(\beta\Delta) + 1}.$$

If $\delta > 0$, then the expected number of steps used by RR_for_Ising *is bounded above by $\#V/\delta$.*

Proof. Let Y_t denote the number of nodes frozen at each step. When $Y_t = 0$, the algorithm terminates. Each step of RR can be viewed as beginning by reducing the number of frozen nodes by 1 by unfreezing node v.

Consider

$$p = \mathbb{P}(C = r) = \frac{\exp((f_{-a} + n_{-a} + n_a)\beta) - \exp((f_{-a} + n_{-a} - n_a)\beta)}{\exp(f_a\beta) + \exp((f_{-a} + n_{-a} + n_a)\beta)}. \tag{8.4}$$

Fix the value of f_a and f_{-a}, and let $k = n_a + n_{-a}$ be fixed as well. Then

$$p = \frac{\exp((f_{-a} + k)\beta) - \exp((f_{-a} + k - 2n_a)\beta)}{\exp(f_a\beta) + \exp((f_{-a} + k)\beta)} \tag{8.5}$$

which is increasing in n_a. So to make p as large as possible, set $n_a = k$ and $n_{-a} = 0$. Hence

$$p \leq \frac{\exp((f_{-a} + k)\beta) - \exp((f_{-a} - k)\beta)}{\exp(f_a\beta) + \exp((f_{-a} + k)\beta)}. \tag{8.6}$$

Next, fix $\ell = f_{-a} + k$. Then $p \leq [\exp(\ell\beta) - \exp(\ell - 2k)]/[\exp(f_a\beta) + \exp(\ell\beta)]$. This expression is increasing in k, and so is maximized when $k = \ell$ and $f_{-a} = 0$. Together with $f_a + \ell = \deg(v)$, this gives

$$p \leq \frac{\exp(\ell\beta) - \exp(-\ell\beta)}{\exp((\deg(v) - \ell)\beta) + \exp(\ell\beta)}. \tag{8.7}$$

Differentiating the right-hand side with respect to ℓ and simplifying gives

$$\frac{\beta[2e^{\deg(v)\beta} + 2e^{2\ell\beta}]}{(e^{(\deg(v)-\ell)\beta} + e^{\ell\beta})^2} \tag{8.8}$$

which is always positive, hence maximized when ℓ is as large as possible, namely, $\deg(v)$. Therefore

$$p \leq \frac{\exp(\deg(v)\beta) - \exp(\deg(v)\beta)}{1 + \exp(\deg(v)\beta)}. \tag{8.9}$$

Differentiating again shows this right-hand side is increasing in $\deg(v)$. If Δ is the maximum degree in the graph, then $p \leq [\exp(\Delta\beta) - \exp(\Delta\beta)]/[1 + \exp(\Delta\beta)]$.

Therefore the expected number of nodes that are unfrozen at each step is at least $1 - (\Delta + 1)p \geq 1 - (\Delta + 1)[\exp(\beta\Delta) - \exp(-\Delta\beta)]/[1 + \exp(\beta\Delta)]$. Lemma 8.2 then finishes the proof. □

8.3 The general randomness recycler

The general framework for the randomness recycler employs the following ingredients. These are listed in the same order as the pieces of RR for the Ising model from the previous section.

1. A measurable space $(\Omega, \mathscr{F})$ for configurations called the *(primary) state space*).
2. A measurable space $(\Omega, \mathscr{F}^*)$ that plays an indexing role called the *dual state space.*
3. An index kernel K_I from $(\Omega^*, \mathscr{F}^*)$ to $(\Omega, \mathscr{F})$.
4. A probability measure π on $(\Omega, \mathscr{F})$, which is the target distribution.
5. A special state in the dual space x_π^* such that $K(x_\pi^*, A) = \pi(A)$ for all measurable A.
6. A starting index state x_0^*, such that it is possible to draw variates from $K(x_0^*, \cdot)$.
7. A bivariate kernel K from $\Omega^* \times \Omega$ to itself and a kernel K^* from Ω^* to itself that satisfies the *design property*, defined as follows.

Definition 8.2. *Say that K, K^*, and K_I have the* design property *if*

$$\int_{x \in \Omega} K_I(x^*, dx) K((x^*, x), (dy^*, dy)) = K^*(x^*, dy^*) K_I(y^*, dy). \tag{8.10}$$

In other words, given the index X_t^* and $[X_t | X_t^*] \sim K_I(X_t^*, \cdot)$, after one step to (X_{t+1}^*, X_{t+1}), it should be true that $[X_{t+1} | X_{t+1}^*] \sim K_I(X_{t+1}^*, \cdot)$. The design property is somewhat analogous to the use of reversibility in designing Markov chains. Once the design property holds, the basic RR algorithm falls into place.

Theorem 8.1 (RR Theorem). *Suppose RR is run using kernels that satisfy the design property. Let T be the first time such that $X_t^* = x_\pi^*$. Then T and X_T are independent, and*

$$[X_T | T < \infty] = \pi. \tag{8.11}$$

The following lemma will be useful in proving the theorem.

Lemma 8.5. *After t steps, given the index state x_t^*, the distribution of the current state comes from $K_I(x_t^*, \cdot)$, even when conditioned on the history of the index process. That is to say,*

$$[X_t | X_0^* = x_0^*, \ldots, X_t^* = x_t^*] \sim K_I(x_t^*, \cdot). \tag{8.12}$$

Proof. By induction: the base case is just the definition of K_I. Suppose (8.12) holds after t steps, and consider step $t+1$.

Let A be any measurable set, and the event $H_t = \{X_0^* = x_0^*, \ldots, X_t^* = x_t^*\}$. Then

$$\begin{aligned}
\mathbb{P}(X_{t+1} \in A | H_t, X_{t+1}^* = x_{t+1}^*) &= \frac{\mathbb{P}(X_{t+1} \in A, X_{t+1}^* \in dx_{t+1}^* | H_t)}{\mathbb{P}(X_{t+1}^* \in dx_{t+1}^* | H_t)} \\
&= \frac{\int_{x \in \Omega} \mathbb{P}(X_t = dx, X_{t+1} \in A, X_{t+1}^* \in dx_{t+1}^* | H_t)}{\mathbb{P}(X_{t+1}^* \in dx_{t+1}^* | H_t)} \\
&= \frac{\int_{x \in \Omega} K_I(x_t^*, dx) K((x_t^*, x), (dx_{t+1}^*, A))}{\mathbb{P}(X_{t+1}^* \in dx_{t+1}^* | H_t)} \\
&= \frac{K^*(x_t^*, dx_{t+1}^*) K_I(x_{t+1}^*, A)}{K^*(x_t^*, dx_{t+1}^*)} \\
&= K_I(x_{t+1}^*, A).
\end{aligned}$$

□

Now the proof of the theorem is easy.

Proof RR theorem. Let A be any measurable set. Then

$$\begin{aligned}\mathbb{P}(X_T \in A|T=t) &= \mathbb{P}(X_T \in A|X_0^* \neq x_\pi^*, \ldots, X_{t-1}^* \neq x_\pi^*, x_t^* = x_\pi^*) \\ &= K_I(x_\pi^*, A) \\ &= \pi(A).\end{aligned}$$

□

8.3.1 Simple symmetric random walk

Consider the simple symmetric random walk example. There $\Omega = \Omega^* = \{1,2,\ldots,n\}$, and for $x^* \in \Omega^*$, $K_I(x^*,\cdot) \sim \mathsf{Unif}(\{1,2,\ldots,x^*\})$. The target π is $\mathsf{Unif}(\{1,2,\ldots,n\})$, and $x_\pi^* = n$. It is easy to sample from $x_0^* = 1$.

The bivariate kernel is a two-step procedure. First take one step in the simple random walk to get from state x to y. Then, if $y > x^* - 1$ or an independent $U \sim \mathsf{Unif}([0,1])$ is at most 1/2, let $y^* = x^* + 1$. Otherwise $y^* = x^* - 1$.

Fact 8.1. *The design property holds for* `SSST_for_ssrw`.

Proof. Fix $x^*, y^* \in \{1,\ldots,n\}$ and $y \in \{1,\ldots,n\}$. There are several subcases to consider.

For instance, suppose $y^* = x^* + 1$, $y = x^* + 1$. Then the only way to end at y is for $x = x^*$, and a move to the right occurred. Note $K_I(x^*,\{x\}) = 1/x^*$ in this case, so

$$\int_{x\in\Omega} K_I(x^*,\cdot)K((x^*,x),(dy^*,dy)) = \frac{1}{x^*}\cdot\frac{1}{2}.$$

Now $K_I(y^*,\{y\}) = 1/y^* = 1/(x^*+1)$ in this case, so it remains to compute $K^*(x^*,\{x^*+1\})$.

For each $x \in \{1,2,\ldots,x^*-2\}$, there will be a 50% chance of changing the index from x^* to x^*+1. For $x \in \{x^*-1,x^*\}$, if the move is to the left, there is a 50% chance of increasing the index, but if the move is to the right, then the index always increases. Here there is a 75% chance of increase in these two cases. Therefore

$$K^*(x^*,\{x^*+1\}) = \frac{x^*-2}{x^*}0.5 + \frac{2}{x^*}0.75 = \frac{1}{2}\cdot\frac{x^*+1}{x^*}. \tag{8.13}$$

Combining with $K_I(y^*,\{y\}) = 1/(x^*+1)$ gives

$$K^*(x^*,\{x^*+1\})K_I(y^*,\{y\}) == \frac{1}{2}\cdot\frac{1}{x^*}. \tag{8.14}$$

which matches $\int_{x\in\Omega} K_I(x^*,\cdot)K((x^*,x),(dy^*,dy))$.

There are many other cases to consider, but they are all resolved in a similar fashion. □

8.3.2 The Ising model

In the RR approach for Ising from Section 8.2, ingredient 1 is the set of configurations $\Omega = \{-1,1\}^V$, and ingredient 2 (the dual state space) consists of subsets of nodes (the frozen set) together with labels for each frozen node.

The index kernel (ingredient 3) is just the Ising density conditioned on the values of the labels for the frozen nodes. The target distribution (ingredient 4) is the Ising density for the entire graph, and ingredient 5 is when no nodes are frozen. Ingredient 6 is when all nodes are frozen at label 1, and ingredient 7 is given by the `RR_for_Ising` procedure.

Fact 8.2. *The design property holds for* `RR_for_Ising`.

Proof. Fix x^* and y^* as sets of frozen nodes together with their labels. Then either y^* has one fewer frozen node than x^*, or it has more frozen nodes.

First suppose that v is frozen in x^*, but is unfrozen in y^*. Let y be a state whose frozen nodes in y^* have the correct label.

Then there is exactly one state x such that $K((x^*,x),(y^*,y)) > 0$, namely, the same as y but with the label of v changed to what it was frozen to by x^*.

Let $Z_{x^*} = \sum_{x \text{ consistent with } x^*} \exp(H(x)\beta)$, and let Z, $f_{x(v)}, f_{-x(v)}, n_{x(v)}, n_{-x(v)}$ be as defined in `RR_for_Ising`. Suppose $y(v) = x(v)$. Then

$$K((x^*,x),(y^*,y)) = \exp(H(x)\beta)Z_{x^*}^{-1}\exp(f_{x(v)}\beta)Z^{-1} = \exp(H(y)\beta)Z_{x^*}^{-1}Z^{-1}.$$

Similarly, if $y(v) = -x(v)$, then

$$\begin{aligned} K((x^*,x),(y^*,y)) &= \exp(H(x)\beta)Z_{x^*}^{-1}\exp((f_{-x(v)} + n_{-x(v)} - n_{x(v)})\beta)Z^{-1} \\ &= \exp(H(y)\beta)Z_{x^*}^{-1}Z^{-1}. \end{aligned}$$

Hence the design property holds in this case with $K^*(x^*,y^*) = Z_{x^*}^{-1}Z^{-1}$.

Now suppose that y^* has more frozen nodes than x^*. The only way this could have happened is if rejection occurred. For any y consistent with y^*, $x = y$ is the only state where moving from (x^*,x) to (y^*,y) is possible. In particular:

$$\begin{aligned} K((x^*,x),(y^*,y)) &= \exp(H(x)\beta)Z_{x^*}^{-1}\mathbb{P}(C = r) \\ &= \exp(H(y)\beta)Z_{x^*}^{-1}C(x^*,y^*). \end{aligned}$$

Note that $\exp(H(x)\beta)/\exp(H(y)\beta)$ only consists of factors arising from edges with one endpoint a neighbor of v and the other endpoint a frozen edge in x^*, or an edge both of whose endpoints are neighbors of v that are unfrozen in x^*.

Either way, these factors are recoverable from x^* and y^*. Also $\mathbb{P}(C = r)$ is a function of $f_{x(v)}, f_{-x(v)}, n_{x(v)}, n_{-x(v)}$, all of which is recoverable from x^* and y^*. Therefore all of these factors can be written as $C(x^*,y^*)$.

Since $Z_{x^*}^{-1}C(x^*,y^*)$ is a function of x^* and y^*, the design property holds in this case as well. □

8.3.3 Using Metropolis-Hastings and the design property

One way to achieve the design property is to use a Metropolis-Hastings-like approach. Suppose that the current state is (x^*, x), and the goal is to move to a new index y^*. Then create a proposal state Y for index y^* using kernel $K_{\text{propose}}((x^*, x), \cdot)$, where for all measurable A, $K_I(y^*, A) > 0 \Rightarrow K_{\text{propose}}((x^*, x), A) > 0$.

To get the chance of proposing in state A, integrate over the possible values of X_t:

$$\mathbb{P}(Y \in A | X_t^* = x^*) = \int_{x \in \Omega} K_I(x^*, dx) K_{\text{propose}}((x^*, x), A).$$

The goal is to have $Y \sim K_I(y^*, \cdot)$, so it is important that the right-hand side of the equation above be positive whenever $K_I(y^*, A) > 0$.

Definition 8.3. *Say that measure* ν_1 *is* absolutely continuous *with respect to* ν_2 *if for all measurable* A, $\nu_1(A) > 0 \Rightarrow \nu_2(A) > 0$.

Create a ratio similar to the Metropolis-Hastings ratio:

$$\rho(x^*, y^*, y) = \frac{K_I(y^*, dy)}{\int_{x \in \Omega} K_I(x^*, dx) K_{\text{propose}}((x^*, x), dy)}. \tag{8.15}$$

Because the numerator is absolutely continuous with respect to the denominator, this ratio exists (formally it is called a Radon-Nikodym derivative.)

Those familiar with the Metropolis-Hastings method for Markov chains know that to get reversibility, it is necessary to accept the move with probability equal to the minimum of 1 and the Metropolis-Hastings ratio. The ratio method for RR is not quite so forgiving.

For RR, it is necessary to accept with probability between 0 and 1, and so the ratio ρ must be normalized. Let

$$M(x^*, y^*) = \sup_{y \in \Omega} \rho(x^*, y^*, y)$$

$$p_{\text{accept}}(x^*, y^*, y) = \frac{\rho(x^*, y^*, y)}{M(x^*, y^*)}.$$

Metropolis-Hasting RR then proceeds as follows:

1. Based on the current index x^*, choose a new index y^* to attempt to move to.
2. Propose a new configuration y for the new index using $K_{\text{propose}}((x^*, x), \cdot)$.
3. Accept this state and move to (y^*, y) with probability $p(x^*, y^*, y)$.
4. Otherwise reject, and move to an index z^* so that $K_I(z^*, \cdot)$ is the distribution of x conditioned on rejection occurring.

Lemma 8.6. *Metropolis-Hastings RR satisfies the design property.*

Proof. What is $K^*(x^*, \{y^*\})$ for Metropolis-Hastings RR? In order to make the move, acceptance must occur, and so integrate over all possible values of x given $X_t^* = x^*$, and all possible proposed states y.

$$K^*(x^*,\{y^*\}) = \int_{y\in\Omega}\int_{x\in\Omega} K_I(x^*,dx)K_{\text{propose}}((x^*,x),dy)\frac{\rho(x^*,y^*,y)}{M(x^*,y^*)}$$
$$= \int_{y\in\Omega}\frac{K_I(y^*,dy)}{M(x^*,y^*)} = M(x^*,y^*)^{-1}$$

On the other hand,

$$\int_{x\in\Omega} K_I(x^*,dx)K((x^*,x),(\{y^*\},dy) = \int_{x\in\Omega} K_I(x^*,dx)K_{\text{propose}}((x^*,x),dy)\frac{\rho(x^*,y^*,y)}{M(x^*,y^*)}$$
$$= K_I(y^*,dy)M(x^*,y^*)^{-1} = K_I(y^*,dy)K^*(x^*,y^*),$$

so the design property is satisfied for the move from x^* to z^*.

The proof for the move from x^* to z^* is similar.

□

The difficult part of applying RR is coming up with a good recycling procedure. One such procedure for the Ising model was given previously; the next section gives a very different RR approach to the same problem.

8.4 Edge-based RR

In the previous section the distribution was always the same, the standard Ising model distribution for the graph. A completely different way of using the index set is to let it change the distribution that the current configuration comes from.

For instance, suppose the index is not the set of frozen nodes, but rather a set of edges. If an edge is not in the index, that means that that edge is effectively removed from the graph. This can be viewed as $g_{\{i,j\}}(c_1,c_2) = 1$ for all c_1,c_2, or equivalently, the values of the labels of the endpoints of an edge only affect the density if the edge is part of the index set x^*.

Suppose that $x^* = \emptyset$. Then x^* indicates that none of the edges are contributing to the Ising density, and so it is easy to sample from the correct distribution. Since there is no external magnetic field, the label for each node is independent, and uniformly distributed over $\{-1,1\}$.

Now suppose that an attempt is made to add an edge $\{i,j\}$ to x^* in order to get the next index y^*. Note that the proposed state y is the same as the original state. But $K_I(y^*,\{x\})$ might not equal $K_I(x^*,\{x\})$ depending on the values of $x(i)$ and $x(j)$. If $x(i) = x(j)$, then $K_I(y^*,\{x\})Z(y^*) = \exp(\beta)K_I(x^*,\{x\})Z(x^*)$. If $x(i) = x(j)$, then $K_I(y^*,\{x\})Z(y^*) = K_I(x^*,\{x\})Z(x^*)$. Using the Metropolis-Hastings RR approach from earlier:

$$\rho(x^*,x^*\cup\{i,j\},y) = \exp(\beta\mathbb{1}(x(i) = x(j)))Z(x^*)/Z(y^*).$$

Once again the nuisance factors $Z(x^*)/Z(y^*)$ always appear regardless of the configuration x, and so also appear in $M(x^*, y^*)$. That means these factors cancel out in the acceptance probability p, and

$$p(x^*, x^* \cup \{i, j\}, y) = \exp(-\beta \mathbb{1}(x(i) \neq x(j))).$$

When $x(i) = x(j)$, the acceptance probability is 1 and the new state is accepted. When $x(i) \neq x(j)$, the acceptance probability is $\exp(-\beta)$, and the possibility of rejection exists. Rejection means that $x(i) \neq x(j)$, and so the configuration no longer comes from the target distribution, but the target distribution conditioned on $x(i) \neq x(j)$.

So if (for instance), $x(i) = 1$, then $x(j) = -1$, and there is a connected component of nodes labeled -1 that does not extend back to node i. The idea is to try to isolate these nodes from the rest of the sample by removing edges.

Removing an edge from the index is very similar to adding an edge. The proposed move is to index $y^* = x^* \setminus \{e\}$. If the endpoints of e have the same label in x, then $K_I(x^*, x)\exp(-\beta)Z(x^*) = K_I(y^*, x)Z(y^*)$. If they do not have the same label in x, then $K_I(x^*, x)Z(x^*) = K_I(y^*, x)Z(y^*)$.

Hence $p(x^*, x^* \setminus \{\{i', j'\}\}, x) = \exp(-\beta \mathbb{1}(x(i') = x(j')))$. Now suppose this rejects. The only way this could happen is if $x(j') = x(i')$.

So either the edge is removed, or the state is conditioned on the endpoints of the edges being the same. At this point the edge can be removed, since conditioned on the endpoints having the same value, the edge only changes the unnormalized density by the constant $\exp(\beta)$.

So our index actually consists of three components. The first component is the set of edges included in the graph. The second component is either empty, or consists of a single node of the graph. The third component consists of a set of nodes of the graph.

The configuration is a draw from the Ising model 1) only including edges in the first component in the graph; 2) every label on every node in the third component is the same; and 3) every node in the third component has a different label from the node in the second component.

Now suppose that every edge adjacent to a node in the third component is absent from the first component. Then the nodes of the third component can be relabeled uniformly from $\{-1, 1\}$, and no longer is the configuration conditioned on the third component nodes having the same label (which is different from the second component node.) So at this point the second (and third) components can be changed back to the empty set, and once again work on adding (rather than removing) edges from the first component.

In fact, the edge that we tried to add originally can be added with probability 1: it is the only edge adjacent to node j in $x^*(1)$, so simply draw $x(j)$ as equal to $x(i)$ with probability $\exp(\beta)/[1 + \exp(\beta)]$, otherwise it is the opposite label.

Since every individual step in the chain is following Metropolis-Hastings RR, the final configuration will be a draw from the Ising model. The pseudocode looks like this.

RR_for_Ising_edge

1) $x \leftarrow \mathsf{Unif}(\{-1,1\}^V),\ x^* \leftarrow (\emptyset,\emptyset,\emptyset)$
2) Repeat
3)   Let $e = \{i,j\} \notin x^*(1)$ where $E \setminus x^*(1) \setminus \{e\}$ remains connected
4)   Draw $U \leftarrow \mathsf{Unif}([0,1])$
5)   If $U > \exp(-\beta \mathbb{1}(x(i) \neq x(j)))$
6) $x^*(2) \leftarrow i,\ x^*(3) \leftarrow \{j\}$
7)   While there is an edge $e = \{i',j'\} \in x^*(1)$ where $i' \in x^*(3)$
8) $x^*(1) \leftarrow x^*(1) \setminus \{e\}$, Draw $U \leftarrow \mathsf{Unif}([0,1])$
9)   If $U > \exp(-\beta \mathbb{1}(x(i') = x(j')))$ then $x^*(3) \leftarrow x^*(3) \cup \{j'\}$
10)   Draw $x(x^*(3)) \leftarrow \mathsf{Unif}(\{-1,1\}^{x^*(3)}),\quad x^*(2) \leftarrow \emptyset,\quad x^*(3) \leftarrow \emptyset$
11)   Draw $C \leftarrow \mathsf{Bern}(\exp(\beta)/[1+\exp(\beta)]),\ x(j) \leftarrow x(i)C - x(i)(1-C)$
12) $x^*(1) \leftarrow x^*(1) \cup \{e\}$
13) Until $x^*(1) = E$

Now look at the range of β over which linear expected running time can be shown.

Lemma 8.7. *Let*

$$\delta = 1 - 2(\Delta - 1)(1 - \exp(-\beta))\frac{\exp(\beta\Delta)}{1 + \exp(\beta\Delta)}.$$

If $\delta > 0$, then the expected number of steps used by RR_for_Ising_edge *is bounded above by $\#E/\delta$.*

Proof. In order to show this result, consider

$$Y = \#(x^*(1)) - \#\{\{i',j'\} \in x^*(1) | i' \in x^*(3)\}.$$

The value of Y changes at lines 6, 8, 9, and 12. In order to understand how these affect the value, it is best to consider lines 12 and 6 together, and also 8 and 9 together.

Note that each time line 12 is executed, line 6 might be executed once, or zero times. At the time line 12 is executed, $x^*(3) = \emptyset$, and so Y increases in value by 1.

When line 6 is executed, a node is added to $x^*(3)$ that might be adjacent to as many as $\Delta - 1$ edges in $x^*(1)$. So it could have the effect of decreasing Y by at most $2(\Delta - 1)$. From line 5, line 6 is executed with chance $\mathbb{P}(x(i) \neq x(j), U > \exp(-\beta))$. These are independent, and $\mathbb{P}(x(i) \neq x(j)) \leq 1/2$, so line 6 only occurs with chance at most $(1/2)(1 - \exp(-\beta))$.

Therefore, lines 12 and 6 together contribute an expected change to Y of at least

$$1 - (1/2)(1 - \exp(-\beta))2(\Delta - 1) \geq \delta. \tag{8.16}$$

Now consider lines 8 and 9. Every time line 8 occurs, line 9 occurs, and the expected change in Y can be combined over the two lines. Line 8 removes an edge from both $x^*(1)$ and $\{\{i',j'\} \in x^*(1) | i' \in x^*(3)\}$, therefore line 8 increases Y by 1.

Line 9 has a $\mathbb{P}(x(i) = x(j), U > \exp(-\beta))$ chance of adding at most $2(\Delta - 1)$ to the value of Y. As in the earlier analysis of line 6, $\mathbb{P}(x(i) = x(j), U > \exp(-\beta) =$

Table 8.1 *Comparison of* `RR_for_Ising` *and* `RR_for_Ising_edge` *provable range of* β *where expected running time is linear in the size of the input graph. Note here a.p.t. stands for artificial phase transition. For instance, when* $\beta \in [0, 0.8779)$ *and* $\Delta = 3$, `RR_for_Ising` *is guaranteed to run in linear time. It could still run in linear time for larger values of* β, *but there is no a priori guarantee.*

Δ	β a.p.t. for `RR_for_Ising`	β a.p.t. for `RR_for_Ising_edge`
3	0.09589	0.3951
4	0.05578	0.2566
5	0.03646	0.1902

$\mathbb{P}(x(i) = x(j))(1 - \exp(-\beta)$. Even if all the neighbors of j have the same label as i, the chance that $x(j) = x(i)$ is at most $\exp(\beta\Delta)/[1 + \exp(\beta\Delta)]$.

Hence the expected change in Y from lines 8 and 9 is at most

$$1 - 2(\Delta - 1)(1 - \exp(-\beta))\exp(\beta\Delta)/[1 + \exp(\beta\Delta)] \geq \delta. \tag{8.17}$$

A minor modification of Lemma 8.2 then completes the proof. □

For both `RR_for_Ising` and `RR_for_Ising_edge`, the range of allowable β is $O(1/\Delta)$. However, as Table 8.1 shows, the edge-based RR has a slight edge at the theoretical level, in that it is possible to show that it has an expected running time that is linear over a slightly broader range of β than the node approach.

8.5 Dimension building

8.5.1 *Move ahead one chain*

Now consider a RR approach to the distribution on permutations arising from the Move ahead 1 (MA1) model. See Section 4.5 for the details and motivation. Recall that $x(i)$ denotes the number of the item placed into position i. The goal is to sample from

$$\pi_{\text{MA1}}(x) \propto \prod_{j=1}^{n} p_{x(j)}^{n-j}.$$

Suppose that the items are labeled so that $p_1 \geq p_2 \geq p_3 \geq \cdots \geq p_n$. Then the permutation with greatest weight is $x = (1, 2, \ldots, n)$.

To use RR for this problem, use the frozen positions approach. At the beginning all of the nodes are frozen. Starting with position 1 and working towards position n, attempt to unfreeze each successive position. Given that ℓ is the lowest frozen position, only the first $\ell - 1$ positions are unfrozen, and the distribution on the unfrozen positions is

$$\pi_{x^*}(x) = \prod_{j=1}^{\ell-1} p_{x(j)}^{n-j}. \tag{8.18}$$

When $\ell = 1$ all the nodes are frozen, so it is trivial to sample from, and when $\ell > n$ all the nodes are unfrozen, so this is draw from the target distribution.

As with our other RR algorithms, the key to making this effective is a good proposed move and recycling procedure. Start with the move.

Still using ℓ for the lowest numbered frozen node, suppose $x(\ell) = a$. Suppose that we attempt to unfreeze position ℓ. This can be accomplished by removing item a from the permutation and reinserting it in one of positions 1 through ℓ.

For instance, suppose the current state is 2143 where positions 3 and 4 are frozen (so $\ell = 3$). Then attempt to unfreeze position 3 by taking item 4 and reinserting it in either the first, second, or third position, giving either 4213, 2413, or 2143 as the result.

Note that items after position 3 will remain unchanged no matter what happens. Items after the insertion point but before position ℓ will have one added to their position.

This means that if I is the random insertion point, a good proposal is

$$\mathbb{P}(I = i) \propto p_a^{n-i} \prod_{j=i}^{\ell-1} p_{x(j)}^{-1}. \tag{8.19}$$

Let $a_1 < a_2 < \cdots < a_{\ell-1}$ be the order statistics for $\{x(1), x(2), \ldots, x(\ell-1)\}$. Since the probabilities of the a_i are increasing in i, the expression in (8.19) is maximized when $(x(i), x(i+1), \ldots, x(\ell-1)) = (a_i, a_{i+1}, \ldots, a_{\ell-1})$.

So to draw I, first choose I' by

$$\mathbb{P}(I' = i) = \frac{p_a^{n-i} \prod_{j=i}^{\ell-1} p_{x(j)}^{-1}}{\sum_{i'=1}^{\ell} p_a^{n-i'} \prod_{j=i'}^{\ell-1} p_{x(j)}^{-1}}. \tag{8.20}$$

Then, with probability

$$\frac{p_a^{n-I'} \prod_{j=I'}^{\ell-1} p_{x(j)}^{-1}}{p_a^{n-I'} \prod_{j=I'}^{\ell-1} a_j^{-1}} = \prod_{j=I'}^{\ell-1} \frac{a_j}{p_{x(j)}} \tag{8.21}$$

accept $I = I'$, make the insertion, and increase ℓ by 1.

This makes

$$\mathbb{P}(I = i) = \frac{p_a^{n-i} \prod_{j=i}^{\ell-1} p_{x(j)}^{-1}}{\sum_{i'=1}^{\ell} p_a^{n-i'} \prod_{j=i'}^{\ell-1} a_j^{-1}} \tag{8.22}$$

and since the denominator in a function of the index, the result satisfies the design property.

But what if we reject making I equal to I'? The chance that this happens is

$$1 - \frac{p_a^{n-I'} \prod_{j=I'}^{\ell-1} p_{x(j)}^{-1}}{p_a^{n-I'} \prod_{j=I'}^{\ell-1} a_j^{-1}}, \tag{8.23}$$

which is a function of $x(I'), \ldots, x(\ell)$. Therefore, by freezing nodes from I' onward, the design property is met.

The resulting algorithm looks like this.

RR_for_MA1

$x \leftarrow (1, 2, \ldots, n)$, $\ell \leftarrow 1$
Repeat
  Let $a_1 < a_2 < \ldots < a_{\ell-1}$ be the order statistics of $\{x(1), \ldots, x(\ell-1)\}$
  Draw I so $\mathbb{P}(I' = i) = \dfrac{p_a^{n-i} \prod_{j=i}^{\ell-1} p_{x(j)}^{-1}}{\sum_{i'=1}^{\ell} p_a^{n-i'} \prod_{j=i'}^{\ell-1} p_{x(j)}^{-1}}$
  Draw B as $\text{Bern}(\prod_{j=I'}^{\ell-1} (a_j / p_{x(j)}))$
  If $B = 1$
    $(x(I'), \ldots, x(\ell)) \leftarrow (x(\ell), x(I'), \ldots, x(\ell-1))$, $\ell \leftarrow \ell + 1$
  Else
    $\ell \leftarrow I'$
Until $\ell = n$

8.6 Application: sink-free orientations of a graph

An algorithm of Cohn, Pemantle, and Propp [25] for generating random sink-free orientations of a graph (as in Section 4.6.1) more quickly than using CFTP can also be put in the framework of an RR algorithm.

An orientation assigns to each edge $\{i, j\}$ a direction that is either (i, j) or (j, i). A node i is a *sink* if for all j such that $\{i, j\}$ is an edge, the edge is oriented towards the node i, that is, the direction is (j, i). As the name implies, a sink-free orientation is an orientation with no sinks. The goal is to generate uniformly at random from the set of sink-free orientations of a graph.

A simple AR approach to generating such samples is to choose an orientation for each edge $\{i, j\}$ uniformly from $\{(i, j), (j, i)\}$, and then accept when the resulting orientation is sink-free. This works very well for small graphs, however, when the graph is large, but the degree of nodes is small, the probability that acceptance occurs decreases exponentially in the size of the graph.

While brute force AR does not work well in this case, it does point the way towards an RR algorithm. The index set x^* will be a subset of nodes. The distribution indexed by x^* is the set of orientations of the graph conditioned on the event that every node in x^* is not a sink.

Then when $x^* = \emptyset$ this is simply the set of orientations, and the distribution is easy to sample from (simply orient each edge uniformly at random). When $x^* = V$, all nodes are not sinks, and this is the target distribution. So each step should attempt to add nodes to x^*.

To increase the size of x^*, simply select any node v not in x^*, and look to see if v is not a sink. If v is not a sink, then the orientation (by AR theory) must be a draw uniformly from the orientations where the nodes $x^* \cup \{v\}$ are sink-free.

However, if v is a sink, then the rest of the orientation is no longer uniform over orientations where x^* is sink-free, but is instead conditioned on v being a sink. Recycling this orientation is easy, however: simply reorient uniformly at random any edge adjacent to v. Of course, this might make the neighbors of v into a sink, and so they should be removed from x^*. This gives the following algorithm.

`RR_for_sink-free_orientations`

For each $\{i,j\} \in E$
    Draw $x(\{i,j\})$ uniformly from $\{(i,j),(j,i)\}$
$x^* \leftarrow \emptyset$
Repeat
    Let v be any node in $V \setminus x^*$
    If v is a sink in x
        $x^* \leftarrow x^* \cup \{v\}$
    Else
        Let $N_v = \{w : \{v,w\} \in E\}$
        For each $w \in N_v$, draw $x(\{v,w\})$ uniformly from $\{(v,w),(w,v)\}$
        $x^* \leftarrow x^* \setminus N_v$
Until $x^* = V$

Fact 8.3. `RR_for_sink-free_orientations` *satisfies the design property.*

Proof. There are two cases to consider. Suppose $y^* = x^* \cup \{v\}$. Let y be any orientation where every node in y^* is not a sink. Then there is exactly one x that can move to y^*, so if Z_A counts the number of orientations where every node in A is not a sink,

$$K_I(x^*,x)\mathbb{P}((x^*,x),(y^*,y)) = \frac{1}{Z_{x^*}} = \frac{Z_{y^*}}{Z_{x^*}} \cdot \frac{1}{Z_{y^*}} = \frac{Z_{y^*}}{Z_{x^*}} K_I(y^*,y). \tag{8.24}$$

Since the fraction Z_{y^*}/Z_{x^*} is a function of x^* and y^*, the design property is satisfied.

Now suppose that $y^* \subset x^*$. Let y be any state where the nodes in y^* are not sinks. Then rejection must have occurred because node v was a sink. So there still was exactly one x that would move to y, and since each edge adjacent to y was uniformly oriented, the chance of moving to y was $1/2^{\deg(v)}$. Hence

$$K_I(x^*,x)\mathbb{P}((x^*,x),(y^*,y)) = \frac{1}{Z_{x^*}} \cdot \frac{1}{2^{\deg v}} = \frac{Z_{y^*}}{2^{\deg(v)} Z_{x^*}} \cdot \frac{1}{Z_{y^*}} = \frac{Z_{y^*}}{Z_{x^*}} K_I(y^*,y). \tag{8.25}$$

□

Therefore by Theorem 8.1, `RR_for_sink-free_orientations` is valid. It was further shown in [25] that the running time of this procedure is fast in general compared to the CFTP approach.

Lemma 8.8. *Let T be the random number of nodes that are recycled. Then $\mathbb{E}[T]$ is at most $\binom{n}{2}$.*

See [25] for the proof details.

Chapter 9

Advanced Acceptance/Rejection

Happiness can exist only in acceptance.

George Orwell

Chapter 2 presented the acceptance/rejection (AR) protocol, which runs into trouble when naively applied to permutation problems such as estimating the permanent and high-dimensional Markov random field problems, such as sampling from the Ising model. This is because typically the number of combinatorial objects in a set grows exponentially in the dimension. In this chapter three ideas are introduced to deal with this problem, sequential acceptance/rejection (SAR), partially recursive acceptance/rejection (PRAR), and popping. Together with the ability to preprocess a given problem instance, SAR gave the first perfect simulation algorithms for estimating the permanent that has provably polynomial time over a nontrivial subclass of nonnegative matrices. PRAR allows the acceptance/rejection approach to be applied to Markov random fields such as the Ising or autonormal model without involving Markov chains in any way. Popping allows the pieces of the sample that are blocking acceptance to be pruned away, giving a perfect simulation algorithm for sampling from the rooted trees of a graph.

9.1 Sequential acceptance/rejection

Theoretically, AR can be used on any set where the set has an upper bound. For instance, if A is a finite set with $\#A \leq M$, simply draw $U \sim \text{Unif}([0,M])$ until $U \in (0,\#A]$. If the elements of $\#A$ are numbered, then return element $\lceil U \rceil$.

Unfortunately, numbering the elements of A is not always possible in practice, especially when the upper bound is not given through a complex argument rather than a simple assignment of number to object.

Sequential acceptance/rejection (SAR) can be employed when an upper bound on the partition function can be shown using either induction or strong induction.

For instance, suppose that the goal is to sample from an unnormalized density on the set of permutations on n objects, so $\mathbb{P}(X = \sigma) = w(\sigma)/\sum_{\sigma' \in S_n} w(\sigma')$, where $\sum_{\sigma' \in S_n} w(\sigma')$ is difficult to compute.

When $X(1)$ is assigned, the remaining components $X(2),\ldots,X(n)$ form a permutation of the elements $\{1,2,\ldots,n\} \setminus \{X(1)\}$. A problem where assigning one component leaves a subset of objects of the same type is called *self-reducible*, and this is a key property for using random samples from an unweighted density in order to estimate the partition function (see [72] for details).

For convenience, SAR will be described for discrete sets, but this method applies equally well to continuous problems. For a nonnegative function w, let $Z(A) = \sum_{x \in A} w(x)$, where $x \in C^n$ is a vector in an n-dimensional space whose elements are drawn from C.

Let $i \in \{1, \ldots, n\}$. If for all $x \in A$ with $w(x) > 0$, $x(i) = a_i$, say that the ith component of x is *fixed*, otherwise it is *free*.

Lemma 9.1. *Let $M(A)$ be a function with the following properties.*

- *If $A = \{x\}$, then $M(A) \geq w(x)$.*
- *For any A with free dimension i, and $c \in C$, let $A_c = \{x : x(i) = c\}$. Then*

$$M(A) \geq \sum_{c \in C} M(A \cap A_c). \tag{9.1}$$

Then $M(A) \geq Z(A) = \sum_{x \in A} w(x)$.

Proof. Use induction on the number of free dimensions in A. If A has no free dimensions, then $A = \{x\}$, and $M(A) \geq w(x) = Z(A)$, so the base case is done.

The induction hypothesis is that the result applies when A has k or fewer free dimensions; now consider A with $k+1$ free dimensions. Choose any one of the dimensions. The key observation is that $Z(A) = \sum_{c \in C} Z(A \cap A_c)$. Hence

$$M(A) \geq \sum_{c \in C} M(A \cap A_c) \geq \sum_{c \in C} Z(A \cap A_c) = Z(A). \tag{9.2}$$

[Note it is necessary to use a strong induction hypothesis here and suppose the technique works for k or fewer free dimensions. This is because for certain w functions, fixing one dimension might also effectively fix other dimensions as well!] □

SAR gives samples from unnormalized density $w(x)$ given the ability to compute $M(A)$ with the properties in Lemma 9.1. As before, let $C = \{c_1, \ldots, c_k\}$ denote the set of possible labels for any $x(i)$.

`SAR` *Input: A, Output: X*

$A_0 \leftarrow A$
While $\#A > 1$
    Let i be any free dimension of A
    For all $j \in \{1, \ldots, k\}$, $A_j \leftarrow \{x \in A : x(i) = c_j\}$
    Draw I so $\mathbb{P}(I = j) = M(A_j)/(M(A))$, $\mathbb{P}(I = 0) = 1 - \sum_j M(A_j)/M(A)$
    If $I = 0$ then $X \leftarrow \texttt{SAR}(A_0)$ and return X
    Else $A \leftarrow A_I$
Draw $B \leftarrow \text{Bern}(w(x)/M(\{x\}))$
If $B = 0$ then $X \leftarrow \texttt{SAR}(A_0)$, return X
Else let X be the only element of A

Before showing that the overall algorithm works, it helps to have a preliminary understanding of what happens when lines 2 through 9 complete without recursively calling SAR.

Lemma 9.2. *Let E_x denote the event that* SAR *reaches line 10 with $A = \{x\}$. Then $\mathbb{P}(E_x) = w(x)/M(A_0)$.*

Proof. The proof is by induction on the size of A at line 2. When A has one element, then lines 8 and 9 make the lemma true. Suppose that the lemma holds for $\#A \leq n$, now suppose $\#A = n+1$.

Let $x \in A$. Then $x \in A_j$ for some $j \in \{1,\ldots,k\}$. So $\mathbb{P}(E_x)$ is the probability that set A_j is chosen times the probability that x is chosen uniquely from A_j. This second factor is $w(x)/M(A_j)$ by the strong induction hypothesis. Hence $\mathbb{P}(E_x) = [M(A_j)/M(A)][w(x)/M(A_j)] = w(x)/M(A)$ and the induction is complete. □

Lemma 9.3. *Suppose there exists at least one x with $w(x) > 0$. The output X is a draw from the unnormalized density w. The expected running time is $M(A)/Z(A)$ times the number of free dimensions in A.*

Proof. Conditioned on acceptance occurring (so no recursive calls to SAR are made, $\mathbb{P}(X = x|\text{accept}) \propto w(x)$. Our AR theory then says that $\mathbb{P}(X = x) \propto w(x)$.

The chance of accepting is $\sum_x w(x)/M(A) = Z(A)/M(A)$. The running time result then follows from the basic AR theory. □

As with all AR-type algorithms, if n is the number of steps needed to obtain a single sample, $M(A)/n$ provides an immediate estimate of Z_A or this algorithm can be used as part of a GBAS approximation (see Section 2.5.)

9.2 Application: approximating permanents

In Section 1.6.2 the permanent of a matrix $A = (A(i,j))$ was given as

$$\text{per}(A) = \sum_{\sigma \in S_n} \prod_{i=1}^{n} A(i,\sigma(i)). \tag{9.3}$$

This should not be confused with the determinant of a matrix, defined as

$$\det(A) = \sum_{\sigma \in S_n} (-1)^{\text{sign}(\sigma)} \prod_{i=1}^{n} A(i,\sigma(i)). \tag{9.4}$$

While the problems appear similar on the surface, the determinant of a matrix can be computed using (to the same order) the time needed to perform matrix multiplication, whereas finding the permanent of a matrix is a #P-complete problem. [124].

When A is a nonnegative matrix, the permanent is the normalizing constant of the unnormalized density on permutations

$$w(\sigma) = \prod_{i=1}^{n} A(i,\sigma(i)). \tag{9.5}$$

This problem has been studied extensively. A Markov chain for this distribution was shown to have polynomial running time in 2001 ([71, 73]) but the large constants involved in this result (and subsequent improvements [12]) make it impractical for applications.

The techniques of this section not only generate samples from the distribution given by w, they also immediately give an approximation algorithm for per(A) without any extra work. These methods are provably polynomial when the matrix is dense, with small constants.

9.2.1 Basic AR

Suppose that A' is formed from A by multiplying a row of A by a constant k. Since each term in per(A) will then be multiplied by the same k, $\text{per}(A') = k\,\text{per}(A)$. So if B is formed by dividing each row of A by its largest element,

$$\text{per}(B) = \left[\prod_{i=1}^{n} \max_j A(i,j)\right] \text{per}(A), \tag{9.6}$$

and the weights for B give the same distribution as the weights for A. The difference is that the elements of B now fall into $[0,1]$.

To use the basic AR approach, draw σ uniformly from S_n, the set of all permutations. Then accept as a draw from the weights associated with A with probability $\prod_{i=1}^{n} B(i,\sigma(i))$.

The expected number of samples required is

$$n!/\,\text{per}(B). \tag{9.7}$$

Unfortunately, this can easily be exponential in n. For instance, if A is the identity matrix, so is B, and $\text{per}(B) = 1$, so $n!$ samples are required to generate a single sample using basic AR.

9.2.2 Sequential AR

Sequential acceptance/rejection can be used with better bounds on the permanent to greatly improve upon basic AR for estimating the permanent.

First consider the case when the entries of the matrix (a_{ij}) are either 0 or 1. In this case, $w(\sigma) = 1$ if and only if $a_{i,\sigma(i)} = 1$ for all i. Therefore for each i, there is a set D_i such that $\sigma(i)$ must fall into the set to obtain positive weight. Then the goal is to draw uniformly from the set of permutations such that $\sigma(i) \in D_i$ for all i.

For example, with the matrix

$$A = \begin{pmatrix} 1 & 1 & 0 \\ 0 & 1 & 1 \\ 1 & 0 & 1 \end{pmatrix}, \tag{9.8}$$

$D_1 = \{1,2\}$, $D_2 = \{2,3\}$, and $D_3 = \{1,3\}$. The permanent of this matrix is 2, since $(1,2,3)$ and $(2,3,1)$ are the only two permutations that have $\sigma(i) \in D_i$ for all i.

In the first row, it is possible to have $\sigma(1) = 1$ or $\sigma(1) = 2$. For a matrix A, let $A_{i,j}$ be the matrix where $\sigma(i) = j$, so the entry $A(i, j)$ is unchanged, and the other entries in row i are changed to 0. Then each of these choices gives rise to a subproblem:

$$A_{1,1} = \begin{pmatrix} 1 & 0 & 0 \\ 0 & 1 & 1 \\ 1 & 0 & 1 \end{pmatrix}, \quad A_{1,2} = \begin{pmatrix} 0 & 1 & 0 \\ 0 & 1 & 1 \\ 1 & 0 & 1 \end{pmatrix}, \quad A_{1,3} = \begin{pmatrix} 0 & 0 & 0 \\ 0 & 1 & 1 \\ 1 & 0 & 1 \end{pmatrix}. \tag{9.9}$$

The set of permutations allowed by A consists of the union of the set of allowed permutations of $A_{1,1}$, the set of allowed permutations of $A_{1,2}$, and the set of allowed permutations of $A_{1,3}$. These last three sets are disjoint. So $\text{per}(A) = \text{per}(A_{1,1}) + \text{per}(A_{1,2}) + \text{per}(A_{1,3})$.

In order to use SAR on this problem, it is necessary to have a bound $\text{per}(B) \leq M(B)$ that can be proved inductively, so $M(B) \geq M(B_1) + \cdots + M(B_n)$. The most basic approach uses a very simple bound on the permanent.

Lemma 9.4. *Let $r_i = \#D_i$ be the sum of the elements of row i of the matrix A. Define*

$$M_1(A) = \prod_{i=1}^{n} r_i. \tag{9.10}$$

Then for any row i,

$$M_1(A_{i,1}) + M_1(A_{i,2}) + \cdots + M_1(A_{i,n}) \leq M_1(A), \tag{9.11}$$

and $\text{per}(A) \leq M(A)$.

Proof. Prove (9.11) first. Note that if $A(i, j) = 0$, $M_1(A_{i,j}) = 0$ as well, so (9.11) is really just

$$M_1(A) \geq \sum_{j:A(i,j)=1} M_1(A_{i,j}). \tag{9.12}$$

But for j such that $A(i, j) = 1$, $M_1(A_{i,j}) = M_1(A)/r_i$. Hence

$$\sum_{j:A(i,j)=1} M_1(A_{i,j}) = \sum_{j:A(i,j)=1} \frac{M_1(A)}{r_i} = M_1(A), \tag{9.13}$$

which completes the proof of (9.11).

Now show $\text{per}(A) \leq M(A)$ by induction on the number of rows i with $r_i > 1$. If $r_i = 0$ or $r_i = 1$ for all i then the result is true, so the base case holds.

Let i be any row with $r_i > 1$. Then $Z(A) = \sum_{j:A(i,j)=1} Z(A_{i,j})$, and by induction $Z(A_{i,j}) \leq M_1(A_{i,j})$. So

$$Z(A) \leq \sum_{j:A(i,j)=1} M_1(A_{i,j}) = M_1(A), \tag{9.14}$$

which finishes the proof. □

Therefore the conditions to apply SAR are met. Since the $M_1(A_{i,j})$ are equal and sum to $M_1(A)$ for all j such that $A(i, j) = 1$, at each step choose uniformly from this set of j to be the value for $\sigma(i)$.

Restricted_permutations_basic_bound, *Input: A, Output:* σ, n

```
n ← 0
Repeat
   n ← n+1, i ← 0, σ ← (0,0,...,0)
   Repeat
      i ← i+1
      Draw σ(i) uniformly from {j : A(i,j) = 1}
   Until i = n or σ(i) ∈ {σ(1),...,σ(i−1)}
Until σ ∈ S_n
```

In the matrix of (9.8), $\prod_{i=1}^{3} r_i = 2 \cdot 2 \cdots 2 = 8$, and $\text{per}(A) = 2$. Therefore the chance of accepting is $2/8 = 1/4$ and 4 draws will be necessary on average in order to obtain one sample.

This basic bound is terrible. Even when the matrix is as dense as possible and $r_i = n$ for all i, using this basic bound is a bad idea. Consider the first order Sterling bound on factorials.

Theorem 9.1 (Sterling's formula).

$$\left(\frac{n}{e}\right)^n \sqrt{2\pi n} \leq n! \leq \left(\frac{n}{e}\right)^n e\sqrt{n}. \tag{9.15}$$

If $D_i = n$ for all i, then $\prod \#D_i = n^n$, making the expected number of steps to get a sample at least $e^{n-1}/\sqrt{n}$.

So to get a polynomial time algorithm, it is necessary to find better bounds. This section will present two such bounds, both of which give a polynomial time algorithm for sampling when $\#D_i$ is large (for instance when $\#D_i = n$ for all i) but still tend to be exponential when the $\#D_i$ are small.

9.2.3 Union of D_i bound

This bound takes advantage of the fact that when a 0 in the matrix is changed to a 1, the permanent might stay the same or become larger. So build a new problem where $D_i' = \cup_{i'=1}^{i} D_{i'}$. Algebraically, this means make $A'(i,j) = \max_{i' \leq i} A(i,j)$. Using this on the matrix from (9.8) gives

$$A' = \begin{pmatrix} 1 & 1 & 0 \\ 1 & 1 & 1 \\ 1 & 1 & 1 \end{pmatrix}. \tag{9.16}$$

The reason for doing this is that now $D_i' \subseteq D_{i+1}'$. So when a choice is made for $\sigma(i)$, it reduces the number of possibilities for $\sigma(i+1)$ by 1. In fact, each of the choices $\sigma(1), \ldots, \sigma(i)$ reduces the number of possibilities for $\sigma(i+1)$ by 1. This is the reasoning behind the following result.

Lemma 9.5. *For a* $\{0,1\}$ *matrix* A, *let* $g(i) = \#\{j : \exists i' \leq i \text{ with } A(i',j) = 1\}$.

$$M_2(A) = \prod_{i=1}^{n} \max(0, g(i) - (i-1)). \tag{9.17}$$

If $M_2(A) = 0$ then $\text{per}(A) = 0$. If $M_2(A) > 0$, $g(i'') = 1$ for $i'' < i$, and $g(i) - (i - 1)$ is the number of nonzero entries of A in row i, then

$$M_2(A) = \sum_j M_2(A_{i,j}). \tag{9.18}$$

Before proving the lemma, suppose that M_2 is being used with SAR. Then when a column j is chosen, $A(i, j)$ remains 1, but all other elements in that row and column are removed. So after $i - 1$ steps, the first $i - 1$ rows have exactly $i - 1$ nonzero elements, all in different columns. Moreover, in each of these columns the entry of row i has been zeroed out, so $\#D_i = g(i) - (i - 1)$. So for all $i'' < i$, $g(i'') = 1$ and $\#D_i = g(i) - (i - 1)$, which is why these are the conditions given in the lemma for (9.18).

Proof. If $M_2(A) = 0$, that means there is a row i such that strictly fewer than i columns in rows $1, \ldots, i$ have a 1 anywhere in them. Therefore, no rook placement with i rooks exists for these first i rows. So there is no rook placement for the whole matrix.

Now suppose that $M_2(A) > 0$ and $\prod_{i''=1}^{i-1} \left(\# \left(\cup_{i'=1}^{i} D_{i'} \right) - (i - 1) \right) = 1$. Then for each row i, $\#(\cup_{i'=1}^{i} D_{i'}) - (i - 1) > 0$. Consider how $M_2(A_{i,j})$ and $M_2(A)$ are related. In $A_{i,j}$, the only nonzero entry in row i is in column j.

Let $g^{i,j}(i') = \#\{j : (\exists i'' \leq i')(A_{i,j}(i'', j) = 1\}$. Then

$$M_2(A_{i,j}) = \left[\prod_{i'' < i} g^{i,j}(i'') - (i'' - 1)) \right] (g^{i,j}(i) - (i - 1)) \left[\prod_{i'' > i} g^{i,j}(i'') - (i'' - 1)) \right]$$

$$M_2(A) = \left[\prod_{i'' < i} g(i'') - (i'' - 1)) \right] (g(i) - (i - 1)) \left[\prod_{i'' > i} g(i'') - (i'' - 1)) \right].$$

For $i'' > i$, $\cup_{i' \leq i''} D_{i'} = \cup_{i' \leq i''} D_{i'}^{i,j}$, so the right-most factors are the same in both expressions. The left-most factors of $M_2(A_{i,j})$ and $M_2(A)$ are both 1.

That leaves the center factor. Since $\#(\cup_{i' \leq i} D_i^{i,j} - (i - 1)) = 1$, that means

$$M_2(A_{i,j}) \leq M_2(A)/\#(\cup_{i' \leq i} D_i - (i - 1)). \tag{9.19}$$

But there are exactly $g(i) - (i - 1)$ nonzero entries of A in row i by assumption, so summing over j gives (9.18). $\qquad \square$

Restricted_permutations_union_bound,   *Input: A, Output:* σ, n

---

1) $n \leftarrow 0$
2)   Repeat
3) $n \leftarrow n + 1, i \leftarrow 0$
4)   Repeat
5) $i \leftarrow i + 1$
6)   Draw $\sigma(i)$ uniformly from $\cup_{i'=1}^{i} D_{i'} \setminus \{\sigma(1), \ldots, \sigma(i - 1)\}$
7)   Until $i = n$ or $\sigma(i) \notin D_i$
8)   Until $\sigma \in S_n$

For (9.8), this bound is $(2)(3-1)(3-2)=4$, twice as good a bound as the basic bound. When $\#D_i = n$ for all i, this gives a bound of $n!$, so unlike the simplest bound, this bound is tight for the maximum density matrix.

Of course, it is not necessary to move through the rows from 1 up to n in order. Permuting the rows of a matrix does not change the permanent, but can change this upper bound! Or instead of moving along the rows, one could move across the columns. Again that would not change the permanent, but would change the upper bound being used.

9.2.4 *Generalized Bregman inequality*

Let $r_i = \#D_i$. In 1963, Minc [93] conjectured the following upper bound for the permanent of a matrix with entries in $\{0,1\}$:

$$\text{per}(A) \le \prod_{i=1}^{n} (r_i!)^{1/r_i}. \tag{9.20}$$

This bound is tight when the rows and columns of the matrix can be permuted so that the 1's form into square blocks. For A from (9.8), $(2!)^{1/2} \cdot (2!)^{1/2} \cdot (2!)^{1/2} = 2^{3/2} = 2.828\ldots$, and so knowing that the permanent is an integer actually obtains the correct answer.

Minc's conjecture was proven true by Bregman in 1973 [13]. Unfortunately, the proof did not use self-reducibility, and so does not allow the use of SAR.

To use a Minc-Bregman style bound for SAR, first let $A'_{i,j}$ be the same matrix as A, but with all entries in both row i and column j, except (i,j), zeroed out. So for A from (9.8),

$$A'_{1,1} = \begin{pmatrix} 1 & 0 & 0 \\ 0 & 1 & 1 \\ 0 & 0 & 1 \end{pmatrix}. \tag{9.21}$$

Next the bound needs to be unpacked a bit. Using Stirling's approximation:

$$\begin{aligned}(r!)^{1/r} &\approx \left[\left(\frac{r}{e}\right)^r \sqrt{2\pi r}\right] = \frac{r}{e}(2\pi r)^{1/(2r)} \\ &= \frac{r}{e}\left[\exp\left(\frac{\ln(2\pi)}{2r} + \frac{\ln(r)}{2r}\right)\right] = \frac{r}{e}\left[1 + \frac{\ln(2\pi)}{2r} + \frac{\ln(r)}{2r}\right] + O\left(\frac{\ln(r)^2}{r^2}\right).\end{aligned}$$

So each factor in Minc-Bregman is (to the first order) about $e^{-1}[r + (1/2)\ln(r) + (1/2)\ln(2\pi)]$ The generalized Bregman factor uses this approximation.

Definition 9.1. *The* generalized Bregman factor *is*

$$h_1(r) = [e^{-1} + (1-e^{-1})r]\mathbb{1}(0 < r < 1) + e^{-1}[r + (1/2)\ln(r) + \exp(1) - 1]\mathbb{1}(r \ge 1). \tag{9.22}$$

Table 9.1 illustrates how close h_1 is to the original Bregman factors. Since the original Bregman factor bound was tight, it must be true that $h_1(r) \ge (r!)^{1/r}$. The table shows that in fact the ratio between the two tends to be very small.

In order to show that h_1 can be used for SAR, the following fact about the behavior of h_1 is needed.

Table 9.1 *Comparison of* $h_1(r)$ *to* $(r!)^{1/r}$. *The maximum relative error between the two values is 6.65% at* $r = 3$.

r	1	2	3	4	5	6
$h_1(r)$	1	1.4953	1.9378	2.3586	2.7675	3.1689
$(r!)^{1/r}$	1	1.4142	1.8171	2.2133	2.6051	2.9937

Lemma 9.6. *For all* $a \in [0, \min\{r, 1\}]$,

$$\frac{a}{h_1(r-a)} \leq e \ln\left(\frac{h_1(r)}{h_1(r-a)}\right). \tag{9.23}$$

The proof of this technical lemma, while straightforward, is fairly lengthy. The interested reader can find the entire proof in [60]. Figure 9.1 illustrates this Lemma by plotting the right-hand side of (9.23) minus the left-hand side for various values of r and a.

Instead of partitioning the space of permutations by the choice of $x(1)$, partition by the choice of $x^{-1}(1)$. That is, partition by which nonzero entry in the first column (rather than the first row) is chosen to be part of the permutation. The following lemma shows that this methodology can be used with SAR.

Lemma 9.7. *Let* A *be a matrix whose elements all lie in* $[0, 1]$, $C \subset \{1, 2, \ldots, n\}$, *and* σ *a permutation. Then set* $r_i = \sum_{j \notin \sigma(C)} A(i, j)$ *and*

$$M_3(A, C, \sigma) = \prod_{i \in C} A(i, \sigma(j)) \prod_{i \notin C} h_1(r_i). \tag{9.24}$$

Fix j *such that* $j \notin \sigma(C)$. *Then for* $i \notin C$, *let* σ_i *be a permutation such that* $\sigma_i(k) = \sigma(k)$ *for all* $k \in C$, *and* $\sigma_i(i) = j$. *The key fact about* M_3 *is*

$$M_3(A, C, \sigma) \geq \sum_{i \notin C : A(i,j) > 0} M_3(A, C \cup \{i\}, \sigma_i). \tag{9.25}$$

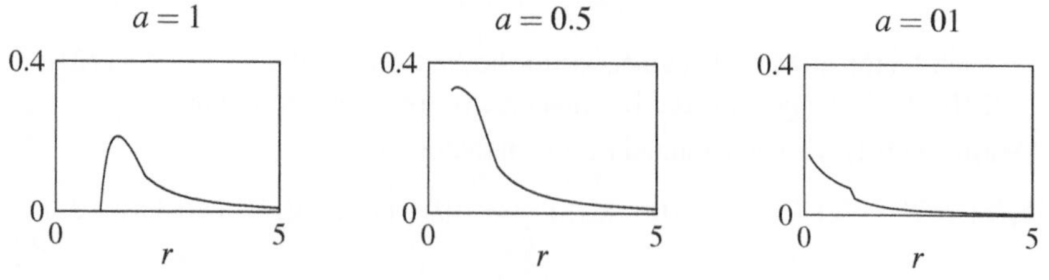

Figure 9.1 *Illustration of Lemma 9.6. The figures plot the right-hand side of Equation* (9.23) *minus the left-hand side. The inequality is actually a strict inequality everywhere except when* $r = a = 1$.

Proof. Fix a column j. Let $C_j = \{i \notin C : A(i,j) > 0\}$. The three cases to consider are when $\#C_j$ is 0, 1, or at least 2.

If $C_j = \emptyset$, then the right-hand side of (9.25) is 0 and the result is trivially true.

The next case is when $C_j = \{i\}$. If $\sigma(i) \neq j$, $M_3(A, C \cup i, \sigma) = 0$ since it includes $A(i, \sigma(i)) = 0$ as a factor. So to show (9.25), it is necessary to show that $M_3(A,C,\sigma) \geq M_3(A, C \cup \{i\}, \sigma_i)$. This holds because $\sum_{j' \notin \sigma_i(C \cup \{i\})} A(i', j') \leq \sum_{j' \notin \sigma_i(C)} A(i', j')$, h_1 is an increasing function, and $A(i,j) \leq h_1(A(i,j)) \leq h_1(r_i)$.

The last case is when $\#C_j \geq 2$. Consider how $M_3(A,C,\sigma)$ and $M_3(A, C \cup \{i\}, \sigma_i)$ are related. First, the row i is zeroed out except at (i,j), changing by a factor of $A(i,j)/h_1(r_i)$. Then all the entries of $C_j \setminus \{i\}$ are zeroed out, changing by a factor of $\prod_{i \in C_j \setminus \{i\}} h_1(r_i - A(i,j))/h_1(r_i)$.

Let $M = M_3(A,C,\sigma)$. Then the above argument gives

$$\begin{aligned}
M_3(A, C \cup \{i\}, \sigma_i) &\leq M \frac{A(i,j)}{h_1(r_i)} \prod_{j \in C_j \setminus \{i\}} \frac{h_1(r_i - A(i,j))}{h_1(r_1)} \\
&= M \frac{A(i,j)}{h_2(r_i - A(i,j))} \prod_{j \in C_j} \frac{h_1(r_i - A(i,j))}{h_1(r_1)} \\
&= M \frac{A(i,j)}{h_1(r_i - A(i,j))} \exp\left(\sum_{i \in C_j} \ln\left(\frac{h_1(r_i - A(i,j))}{h_1(r_1)} \right) \right).
\end{aligned}$$

Summing over i gives

$$\sum_{i \in C_j} M_3(A'_{i,j}) \leq M_3(A) \left[\sum_{i \in C_j} \frac{A(i,j)}{h_2(r_i - A(i,j))} \right] \exp\left(\sum_{i \in C_j} \ln\left(\frac{h_1(r_i - A(i,j))}{h_1(r_1)} \right) \right).$$

Now apply Lemma 9.6, and use $\ln(x) = -\ln(1/x)$ to get

$$\sum_{i \in C_j} M(A, C \cup \{i\}, \sigma_i) \leq M \left[\sum_{i \in C_j} e \ln\left(\frac{h_1(r_i)}{h_1(r_1 - A(i,j))} \right) \right] \exp\left(- \sum_{i \in C_j} \ln\left(\frac{h_1(r_i)}{h_1(r_1 - A(i,j))} \right) \right).$$

The right-hand side is now of the form $eMy\exp(-y)$ where $y \geq 0$. This function has a maximum value of M at $y = 1$, yielding

$$\sum_{i \in C_j} M_3(A, C \cup \{i\}, \sigma_i) \leq M = M_3(A, C, \sigma). \tag{9.26}$$

as desired. □

This result immediately gives a SAR-type algorithm for generating from weighted permutations where the weights all fall in the closed interval from 0 to 1. The expected number of partial permutations generated by the algorithm will be $M(A)/\text{per}(A)$.

`Restricted_permutations_generalized_Bregman_bound`
Input: A, *Output:* σ, n

$n \leftarrow 0$
Repeat
  $n \leftarrow n+1$, $\sigma_{\text{inv}} \leftarrow (0,0,\ldots,0)$, $j \leftarrow 0$
  Repeat
    $j \leftarrow j+1, C_j \leftarrow \{i : A(i,j) > 0\}$
    If $C_j = \{i\}$, then $\sigma(i) \leftarrow j$
    Else if $\exists i \in C_j$ where $A(i,j) = r_i$, then $\sigma_{\text{inv}}(j) \leftarrow i$
      Else
        Calculate $M_3(A)$
        For all $i \in C_j$, calculate $M_3(A'_{i,j})$
        Draw $U \leftarrow [0, M_3(A)]$
        If $U > \sum_{i'=1}^{n} M_3(A'_{i',j})$ then $\sigma_{\text{inv}} \leftarrow$ reject
        Else let $\sigma_{\text{inv}}(j) \leftarrow \min\{i : \sum_{i'=1}^{i} M_3(A'_{i',j}) > U\}$
  Until $\sigma_{\text{inv}} =$ reject or $j = n$
Until $\sigma_{\text{inv}} \neq$ reject
Return σ_{inv}^{-1}

9.2.5 *Lower bounds on permanents*

To give an upper bound on the running time of the algorithm, it is necessary to show that $M(A)/\operatorname{per}(A)$ is small. This requires a lower bound on the permanent.

In the 1920s, Van der Waerden conjectured the following lower bound. This conjecture was independently shown to be true by Falikman [34] and Ergorychev [33].

Theorem 9.2 (Falikman-Egorychev Theorem). *Let A be a nonnegative matrix whose row and column sums are at most 1. Then* $\operatorname{per}(A) \geq n!/n^n$.

A matrix whose row and column sums are 1 is called doubly stochastic. The theorem states that the doubly stochastic matrix with the smallest value of the permanent is the one where every entry in the matrix is $1/n$. That is, the mass of the entries is spread out as far as possible.

In [52], it was considered how to use this theorem to bound the running time of SAR using the Generalized Bregman bound.

Definition 9.2. *Say that a matrix is* regular *if all the row and column sums are the same.*

For an n by n matrix A that is regular with common row sum r, dividing each row by r gives a doubly stochastic matrix. Hence $\operatorname{per}(A) \geq r^n n!/n^n$. This gave rise to the following result.

Lemma 9.8. *Suppose that each row and column sum of a $\{0,1\}$ matrix A equals γn for $\gamma \in [1/n, 1]$. Then* `Restricted_permutations_generalized_Bregman_bound`(A) *takes on average at most $O(n^{1.5+0.5/\gamma})$ steps.*

Proof. From our earlier AR theory, the number of trials will be bounded above by

$\prod_i h_1(r)/[r^n n!/n^n] = (n^n/n!)(h_1(r)/r)^n$. By Sterling's inequality, $n! \geq (n/e)^n\sqrt{2\pi n}$. Hence $n^n/n! \leq e^n(2\pi n)^{-1/2}$. Since $r \geq 1$, $h_1(r) = e^{-1}[r + (1/2)\ln(r) + \exp(1) - 1]$, so

$$\frac{h_1(r)}{r} = e^{-1}\left[1 + \frac{(1/2)\ln(r) + e - 1}{r}\right]$$
$$\leq e^{-1}e^{[(1/2)\ln(r)+e-1]/r}.$$

Therefore,

$$\frac{n^n}{n!}\cdot\left(\frac{h_1(r)}{r}\right)^n \leq \frac{e^n}{\sqrt{2\pi n}}\cdot e^{-n}\exp((1/2)\ln(r) + e - 1)/\gamma$$
$$= (2\pi n)^{-1/2}(\gamma n)^{0.5/\gamma}e^{(e-1)/\gamma}$$
$$= O(n^{-0.5+0.5/\gamma}).$$

Each step requires at most n^2 operations to compute the bounds, making the running time (on average) at most $O(n^{1.5+0.5/\gamma})$ steps. □

9.2.6 Sinkhorn balancing

So how does the lower bound help when the matrix is not regular? Recall that if a row (or column) of the matrix is multiplied by a constant, then the permanent is multiplied by the same constant.

Therefore, by dividing each row by its sum, it is possible to make a matrix whose row entries are all 1, and whose permanent is related in a known way to the original matrix.

However, the column sums might not be 1. So divide the columns by their sums. This might have messed up the rows, however. So next divide the rows by their sum. Continue moving back and forth between rows and columns until the matrix is almost regular. This procedure is known as Sinkhorn balancing [116].

More precisely, if X and Y are diagonal matrices, then XAY is a scaling of A where $\text{per}(A) = \text{per}(XAY)/[\text{per}(X)\,\text{per}(Y)]$. From the definition, the permanent of a diagonal matrix is just the product of its entries. By keeping track of the scaling done to rows and columns at each step, when Sinkhorn balancing is ended the matrices X and Y will be known.

The question then becomes: how many steps must be taken before the resulting matrix is close to balanced? To answer this question, it is necessary to have a notion of how close the matrix is to being doubly stochastic.

Definition 9.3. *Say that diagonal matrices X and Y* scale *A to accuracy α if both the row and columns sums of XAY all fall in $[1-\alpha, 1+\alpha]$.*

In [75], it was shown that the ellipsoid method could be used to obtain such a scaling in $O(n^4\ln(\ln(1/\alpha)n/\alpha))$ time. While a useful theoretical result, the ellipsoid method is very complex to code and can suffer from numerical instability.

In practice, simply employing the method of rescaling the rows, then the columns, then the rows again, and so on until accuracy α is reached, is much faster. It can be shown to require only $O((\alpha^{-1} + \ln(n))\sqrt{n}\log(1/\alpha))$ steps, and each step takes $\Theta(n^2)$ time. This method is called *Sinkhorn balancing.*

`Sinkhorn_balancing`
Input: A, α *Output:* A, s

$s \leftarrow 1$
Repeat
    Let r_i be the sum of row i of A
    Divide row i of A by r_i, multiply s by r_i
    Let c_i be the sum of column i of A
    Divide column i of A by c_i, multiply s by c_i
Until $\max_i\{|r_i - 1|, |c_i - 1|\} \leq \alpha$

Now for matrices B with row and column sums exactly 1, $\text{per}(B) \geq n!/n^n$. When they are only 1 within additive accuracy α, the following holds.

Lemma 9.9. *Let $B = XAY$ be A scaled to accuracy $\alpha \leq 0.385/n$. Then*

$$\text{per}(B) \geq \frac{n!}{n^n}\exp(-4n^2\alpha). \tag{9.27}$$

See [86] for the proof. So for $\alpha = 0.25/n^2$, only a factor of e is lost in the running time of the acceptance/rejection result. Sinkhorn balancing takes $O(n^{4.5}\ln(n))$ steps to accomplish this level of accuracy.

Now that the graph has been made regular, what is the best way to generate samples? Well, Minc's inequality works best when the row sums are as large as possible. To make them larger, simply multiply the row by the inverse of the largest element. That makes the row sums as large as possible while keeping all of the elements of the matrix in the set $[0,1]$.

Next, the SAR method using the generalized Bregman bound can be applied to the regularized matrix to obtain an estimate of p.

`Estimate_permanent`
Input: A, k *Output:* $\hat{a}$

Let n be the number of rows and columns of A
$(A,s) \leftarrow$ `Sinkhorn_balancing`$(A, 0.25/n^2)$
For every row i of A
    $c \leftarrow \max_j A(i,j)$
    Divide row i of A by c, multiply s by c
$R \leftarrow 0$
Repeat k times
    $(\sigma, n) \leftarrow$ `Restricted_permutations_generalized_Bregman_bound`(A)
    Draw $R_1, \ldots, R_n$ iid Exp(1)
    $R \leftarrow R + R_1 + \cdots + R_n$
$\hat{a} \leftarrow M_3(A)s(k-1)/R$

Recall from Section 2.5, the reason for drawing the exponentials at each step is so that $\hat{a}/\operatorname{per}(A)$ is $k-1$ times an inverse gamma distribution with parameters k and 1. Therefore $\mathbb{E}[\hat{a}] = \operatorname{per}(A)$ and the relative error of the estimate does not depend on $\operatorname{per}(A)$ at all.

The balancing at the beginning only needs to be done once, so the running time of the procedure will be the time to do the balancing plus the time to do the acceptance/rejection part. The first step in bounding this running time is to show that the maximum element of each row in the scaled matrix is small when the original matrix A is sufficiently dense.

Lemma 9.10. *Let $\gamma \in (1/2, 1]$, and A be an n-by-n matrix with entries in $[0,1]$ and at least γn entries equal to 1 in every row and column. Let $B = XAY$ with X and Y diagonal matrices where B is doubly stochastic with accuracy α. Then*

$$\max_{i,j} B(i,j) \leq \frac{(1+\alpha)^2}{(2\gamma - 1)n - 3n\alpha - 2}. \tag{9.28}$$

Proof. Fix i and j in $\{1,\ldots,n\}$. Introduce $S_r = \{j' \neq j : A(i,j') = 1\}$ and $S_c = \{i' \neq i : A(i',j) = 1\}$. Given the density of A, $\#S_r$ and $\#S_c$ are at least $\gamma n - 1$. Break A into four submatrices based on S_r and S_c. For subsets S_1 and S_2 of $\{1,\ldots,n\}$, let $A(S_1,S_2)$ be the submatrix of A that contains elements that are in rows of S_1 and columns of S_2. Then

$$A_1 = A(S_r,S_c),\ A_2 = A(S_r,S_c^C),\ A_3 = A(S_r^C,S_c),\ A_4 = A(S_r^C,S_c^C),$$

and B_1,B_2,B_3,B_4 are the corresponding submatrices of B.

Let $s(C)$ denote the sum of the entries of a matrix C. Then since B is doubly stochastic to accuracy $1-\alpha$, each row and column adds up to at least $1-\alpha$. Hence

$$\begin{aligned} s(B_1) + s(B_2) &\geq \#S_c(1-\alpha) \\ s(B_1) + s(B_3) &\geq \#S_r(1-\alpha). \end{aligned}$$

And since each row adds up to at most $1+\alpha$:

$$\begin{aligned} s(B_1) + s(B_2) + s(B_3) + s(B_4) &\leq n(1+\alpha) \\ \#S_c(1-\alpha) + \#S_r(1-\alpha) - s(B_1) + s(B_4) &\leq n(1+\alpha). \end{aligned}$$

Since the submatrix B_4 includes $B(i,i)$, $B(i,i) \leq s(B_4)$ and

$$B(i,i) \leq n(1+\alpha) + s(B_1) - (\#S_c + \#S_r)(1-\alpha). \tag{9.29}$$

From the density assumption, $\#S_c + \#S_r \geq 2\gamma n - 2$. To upper bound $s(B_1)$, first define $x(i) = X(i,i)$ and $y(i) = Y(i,i)$ so that $B(i',j') = x(i')A(i',j')y(j')$. Then

$$s(B_1) \leq \sum_{i' \in S_c} \sum_{j \in S_r} x(i')A(i',j')y(j') = \left[\sum_{i' \in S_c} x(i')\right]\left[\sum_{j' \in S_r} y(j')\right]. \tag{9.30}$$

Given B is doubly stochastic to within accuracy α, $\sum_{i'=1}^{n} x(i')A(i',j)y(j) \leq 1+\alpha$, and $\sum_{j'=1}^{n} x(i)A(i,j')y(j') \leq 1+\alpha$. This gives

$$\sum_{i' \in S_c} x(i') \leq -B(i,j) + \sum_{i'=1}^{n} x(i')A(i',j) \leq -B(i,j) + y(j)^{-1}(1+\alpha)$$

$$\sum_{j' \in S_r} y(j') \leq -B(i,j) + \sum_{j'=1}^{n} A(i,j')y(j') \leq -B(i,j) + x(i)^{-1}(1+\alpha).$$

So $s(B_1) \leq x(i)^{-1}y(j)^{-1}(1+\alpha)$. Now when $A(i,j)=0$, $B(i,j)=x(i)A(i,j)y(j)=0$ as well. When $A(i,j)=1$, $B(i,j)=x(i)y(j)$. So $s(B_1) \leq B(i,j)^{-1}(1+\alpha)^2$.

At this point we have

$$0 \leq B(i,j) \leq n(1+\alpha) + B(i,j)^{-1}(1+\alpha)^2 - (2\gamma n - 2)(1-\alpha). \tag{9.31}$$

Solving gives

$$B(i,j) \leq (1+\alpha)^2/[(2\gamma - 1 - 2\gamma\alpha - \alpha)n - 2]. \tag{9.32}$$

Using $2\gamma + 1 \leq 3$ finishes the proof. □

Lemma 9.11. *Let B be an n-by-n nonnegative matrix with* $\max_{i,j} B(i,j) \leq [(2\gamma - 1)n]^{-1} + o(n^{-1})$ *that is doubly stochastic to accuracy* $0.385/n^2$. *Then for C the matrix formed by dividing each row of $B(i,j)$ by its maximum value,*

$$\frac{M_3(C)}{\text{per}(C)} \leq 5(2\pi n)^{-1/2}(n(2\gamma-1)+o(n))^{(2\gamma-1)/2} e^{(2\gamma-1)(e-1)}. \tag{9.33}$$

Proof. By Lemma 9.9, $\text{per}(B) \geq n!n^{-n}\exp(-4n^2(0.385/n^2)) \geq 0.2n!n^{-n}$. Let m_i denote the maximum element of row i of B. Then $\text{per}(B) = \text{per}(C)\prod m_i^{-1}$. So $\text{per}(C) \geq n!s^{-1}n^{-n}\prod m_i^{-1}$.

For the upper bound, the rows of C add to at least $[1-0.385/n^2]/m_i$. Since for $r \geq 1$, $h_1(r) = e^{-1}[r + (1/2)\ln(r) + e - 1]$, that gives an upper bound $\text{per}(C) \leq e^{-n}\prod_i (m_i^{-1} + (1/2)ln(m_i^{-1}) + e - 1)$.

On the other hand $0.2n!n^{-n} \geq 0.2e^{-n}\sqrt{2\pi n}$. Hence

$$\begin{aligned}
\frac{M_3(C)}{\text{per}(C)} &\geq \frac{e^{-n}\prod_i (m_i^{-1} + (1/2)\ln(m_i^{-1}) + e - 1)}{0.2\sqrt{2\pi n}e^{-n}\prod m_i} \\
&= \prod_i \left(1 + \frac{(1/2)\ln(m_i^{-1}) + e - 1}{m_i^{-1}}\right) \\
&\leq 5(2\pi n)^{-1/2}\exp\left(\sum_i \frac{(1/2)\ln(m_i^{-1}) + e - 1}{m_i^{-1}}\right) \\
&\leq 5(2\pi n)^{-1/2}\exp\left(\sum_i \frac{(1/2)\ln(n(2\gamma-1)+o(n)) + e - 1}{n(2\gamma-1)+o(n)}\right) \\
&\leq 5(2\pi n)^{-1/2}\exp((2\gamma-1)^{-1}[(1/2)\ln(n(2\gamma-1)+o(n)) + e - 1]) \\
&= 5(2\pi n)^{-1/2}(n(2\gamma-1)+o(n))^{(\gamma-1/2)^{-1}} e^{(2\gamma-1)^{-1}(e-1)}.
\end{aligned}$$

□

Each step takes $O(n^2)$ time to execute, giving the following.

Lemma 9.12. *The time needed to estimate* $\text{per}(A)$ *for an n-by-n* $\{0,1\}$ *matrix A with row (or column) sums at least* γn *for some* $\gamma \in (1/2, 1]$ *to within a factor of* $1+\varepsilon$ *with probability at least* $1-\delta$ *is at most*

$$O(n^{4.5}\ln(n) + n^{1.5+(2\gamma-1)/2}\varepsilon^{-2}\ln(\delta^{-1})). \tag{9.34}$$

9.3 Partially recursive acceptance/rejection

Recursive acceptance/rejection (RAR) operates by partitioning the set of dimensions into two (or more) sets, drawing a random sample on each individual set, and then accepting or rejecting based on the samples.

For instance, consider once again the Ising model (with no magnetic field for simplicity) where $w(x) = \prod_{\{i,j\}\in E} \exp(\beta \cdot \mathbb{1}(x(i) = x(j))$. It is easy to restrict this weight to a set of nodes by only considering nodes in the set.

$$w_V(x) = \prod_{\{i,j\}\in E, i\in V, j\in V} \exp(\beta \cdot \mathbb{1}(x(i) = x(j)). \tag{9.35}$$

Now consider a single node $\{v\}$. Then $w_{\{i\}}(1) = w_{\{i\}}(-1) = 1/2$. So this is easy to sample from. Since the weight function $w_{V\setminus\{i\}}$ only considers nodes in $V \setminus \{i\}$, it is a slightly easier problem than the original. Let $X(i)$ be a sample from $w_{\{i\}}$, and $X(V \setminus \{i\})$ be a sample from $w_{V\setminus\{i\}}$. Now consider trying to bring these two samples back together.

Suppose i is adjacent to two edges $\{i, j\}$ and $\{i, j'\}$. Then if $X(i) = X(j)$ then the value of $w_V(X)$ is as large as it can be. But if $X(i) \neq X(j)$ then the sample is missing out on a factor of $\exp(\beta)$. Hence only accept this with probability $\exp(-\beta)$.

Similarly, if $X(i) \neq X(j')$, only accept this sample X as a draw from w_V with probability $\exp(-\beta)$. If rejection occurs, just as in regular AR, start over. This gives the following algorithm.

`Recursive_AR_Ising`, *Input:* V, *Output:* $X \sim w_V$

Repeat
    Let i be any node in V, $T \leftarrow 1$
    $X(i) \leftarrow \mathsf{Unif}(\{-1, 1\})$
    If $V \neq \{i\}$
        Draw $X(V \setminus \{i\}) \leftarrow$ `Recursive_AR_Ising`$(V \setminus \{i\})$
        For all j such that $\{i, j\}$ is an edge
            Draw $U \leftarrow \mathsf{Unif}([0,1])$
            $T \leftarrow T \cdot \mathbb{1}(U \leq \exp(-\beta \mathbb{1}(x(i) \neq x(j)))$
Until $T = 1$

Lemma 9.13. `Recursive_AR_Ising` *returns draws from* $w_V(x)$ *where* w *is the weights for the Ising model with no magnetic field restricted to the nodes in* V.

Proof. Proof by induction on the number of nodes in V. When $\#V = 1$, $X(i) \sim \mathsf{Unif}(\{-1,1\})$, so it is trivially true. Now assume it holds for all V of size $n-1$ where $n > 1$, and consider a set of nodes V of size n.

In line 3 $X(i) \sim w_{\{i\}}$, and in line 5 (by the induction hypothesis) $X(V \setminus \{i\}) \sim w_{V\setminus\{i\}}$. Now $w_V(x)/w_{V\setminus\{i\}}(x) = \prod_{j:\{i,j\}\in E} \exp(\beta \mathbb{1}(x(i) = x(j)))$. Hence $w_v(X)/w_{V\setminus\{i\}}(x) \leq \exp(\beta \cdot \text{deg(i)})$.

Therefore, X should be accepted as a draw from w_V with probability

$$\prod_{j:\{i,j\}\in E} \exp(\beta \mathbb{1}(x(i) = x(j))) \exp(\beta \cdot \deg(v)) = \prod_{j:\{i,j\}\in E} \exp(-\beta \mathbb{1}(x(i) \neq x(j))). \tag{9.36}$$

This is exactly what lines 6 through 9 accomplish. □

So `Recursive_AR_Ising` is a valid AR algorithm. But as with any AR algorithm, the chance of accepting can be quite low, making the algorithm slow. Let $T(n)$ denote the expected number of calls to line 3 when $\#V = n$. Then given a connected graph, the chance of rejection is at least $(1/2)(1-\exp(-\beta))$. Hence $T(n) \geq T(1) + T(n-1) + (1/2)(1-\exp(-\beta))T(n)$. Using $T(1) = 1$ and solving for $T(n)$ gives $T(n) \geq [1+T(n-1)]/[(1/2)(1+\exp(-\beta))]$. So for $\beta > 0$, $T(n) \geq (2/[1+\exp(-\beta)])^n$, and so grows at least exponentially fast in the number of nodes.

9.3.1 *Rearranging the order*

The key to making $T(n)$ polynomial for a wider range of β is to note that in many cases, it is not necessary to execute line 5 before evaluating lines 6 through 9!

In line 8, if $U \leq \exp(-\beta)$, then it is unnecessary to know if $x(i) \neq x(j)$; either way T remains at 1. If the chosen uniform is at most $\exp(-\beta)$ for all j such that $\{i,j\} \in E$ (call this event S_i), then acceptance occurs regardless of the outcome of `Recursive_AR_Ising(V \ {i})`.

So do not execute the call to `Recursive_AR_Ising(V \ {i})` until it is known if S_i occurs or not. When S_i occurs, say that the state on sets $\{i\}$ and $V \setminus \{i\}$ have *separated*, as effectively they are now independent.

On the other hand, what should be done when $U > \exp(-\beta)$ when considering $\{i,j\} \in E$? In this case, it is necessary to learn what the value of node j in the sample for nodes $V \setminus \{i\}$ is. This can be done recursively without having to determine the entire sample.

Call this method partially recursive acceptance/rejection (PRAR) because it does not always use recursion, it just uses enough recursion to decide whether or not to reject at the first step.

The input to `PRAR_Ising` consists of the nodes V where the partial sample is coming from, the set F that is the minimum number of nodes that the sample must determine. The output is $G \subseteq V$ such that $F \subseteq G$ and the state of the sample over these nodes $X(G)$.

`PRAR_Ising`, *Input:* V, F *Output:* $G, X(G) \sim w_G$

1) Repeat
2) Let i be any node in F, $G \leftarrow \emptyset$
3) Repeat
4) Draw $X(i) \leftarrow \mathsf{Unif}(\{-1,1\})$
5) For each node $j \in V$ such that $\{i,j\} \in E$, draw $U_j \leftarrow \mathsf{Unif}([0,1])$
6) $F' \leftarrow \{j : U_j > \exp(-\beta)\}, G \leftarrow \emptyset$
7) If $F' \neq \emptyset$
8) $(G, X(G)) \leftarrow$ `PRAR_Ising`$(V \setminus \{i\}, F')$
9) Until for all $j \in F', X(i) = X(j)$
10) $G \leftarrow G \cup \{i\}, F \leftarrow F \setminus G, V \setminus G$
11) Until $F = \emptyset$

Lemma 9.14. *For Δ the maximum degree of the graph, suppose that*

$$\delta = \exp(-\beta\Delta) - \Delta(1 - \exp(-\beta)) \tag{9.37}$$

is positive. Then on average `PRAR_Ising` *draws a number of random variables bounded above by $\#F(\Delta+1)\delta^{-1}$. In particular, if $0 < \alpha < 1/2$ and $\beta \leq \alpha/\Delta$, then $\delta > 1 - 2\alpha$.*

Proof. At any given time in the algorithm, there might be one or more recursive calls to `PRAR_Ising` active. Let N_t denote the sum of the sizes of the F sets in all of these calls, where t is the number of times line 3 has been executed. At line 9, N can increase by up to Δ, and at line 11, N can go down by at least 1. When β is small, $\mathbb{E}[N_{t+1}|N_t]$ will be smaller than N_t.

That is,

$$\begin{aligned}\mathbb{E}[N_{t+1}|N_t] &\leq N_t - 1 \cdot \mathbb{P}(F' = \emptyset) + \mathbb{E}[F'|F' \neq \emptyset]\mathbb{P}(F' \neq \emptyset)\\ &\leq (N_t - 1)\exp(-\beta\Delta) + \Delta(1 - \exp(-\beta)) = N_t - \delta.\end{aligned}$$

Then Lemma 8.2 shows that the expected number of times line 2 is executed is at most $N_0/\delta \leq \#F/\delta$. Each time line 2 is called, lines 4 and 6 draw at most $\Delta + 1$ random variables, completing the average running time bound. The rest follows from $\exp(-x) \geq 1 - x$. □

9.3.2 Why use PRAR?

Like the randomness recycler, PRAR is a read-once, interruptible algorithm that can often run in linear time. Unlike RR, PRAR only requires knowledge of how each edge and node separately contributes to the distribution.

Consider sampling from C^V with density that is an automodel as in (1.7). Then PRAR works as follows.

`PRAR_Auto_model`, *Input:* V, F *Output:* $G, X(G) \sim w_G$

Repeat
  Let i be any node in F, $G \leftarrow \emptyset$
  Repeat
    Draw $X(i)$ using $\mathbb{P}(X(i) = c) \propto f_i$
    For each node $j \in V$ such that $\{i, j\} \in E$
      Draw $U_j \leftarrow \textsf{Unif}([0,1])$
      $F' \leftarrow \{j : U_j > \max_{c_2} g_{i,j}(X(i), c_2)/\max_{c_1,c_2} g_{i,j}(c_1, c_2)\}$, $G \leftarrow \emptyset$
      If $F' \neq \emptyset$
        $(G, X(G)) \leftarrow$ `PRAR_Ising`$(V \setminus \{i\}, F')$
  Until for all $j \in F'$, $U_j \leq g_{i,j}(X(i), X(j))/\max_{c_1,c_2} g_{i,j}(c_1, c_2)$,
  $G \leftarrow G \cup \{i\}$, $F \leftarrow F \setminus G$, $V \setminus G$
Until $F = \emptyset$

Lemma 9.15. *The output of PRAR is a draw from the target density. For every node i, let*

$$c_i = \sum_{j:\{i,j\} \in E} \mathbb{P}(U_j \leq \max_{c_2} g_{i,j}(X(i), c_2)/\max_{c_1,c_2} g_{i,j}(c_1, c_2). \tag{9.38}$$

Then if $c_i < 1$ for all i, the algorithm terminates in time linear in $\#F$.

Proof. As in the proof of Lemma 9.13, an induction on the nodes of the graph together with basic AR theory shows correctness. The value of c_i is an upper bound on the expected number of nodes in G in a recursive call. When this is bounded above by $\max_{i \in V} c_i < 1$, then the branching process of calls dies out with probability 1, meaning that overall only a finite number of such nodes need be examined. This finite number is independent of the node chosen, so each node in $\#F$ requires an amount of work that is upper bounded by the same amount. □

9.4 Rooted spanning trees by popping

Consider the problem of generating a spanning tree of an undirected graph at random.

Definition 9.4. *A graph $G = (V, E)$ is* connected *if between every pair of nodes $\{i, j\}$ there is at least one path between i and j using edges in E.*

Definition 9.5. *A* spanning tree *of a graph $G = (V, E)$ is a subset of edges T such that for any pair of nodes $\{i, j\}$ in V, there is exactly one path connecting i and j using edges in T.*

Another way to say this is that a spanning tree is a collection of edges that connects the graph, but which does not have any cycles.

Any connected graph has at least one spanning tree, and it can be decided if a graph is connected in linear time. Consider the problem of generating uniformly from the set of spanning trees of a connected graph. Any spanning tree contains $\#E - 1$ edges. Therefore, a simple AR algorithm is to draw a subset of $\#E - 1$ edges uniformly from all edges, and then accept if it happens to be a tree.

This can perform exceedingly poorly. Consider the k-by-k two-dimensional square lattice with $n = k^2$ nodes and $m = 2k(k-1)$ edges. From a result of Chung [24], the number of spanning trees of this graph is at most $n^{-1}4^{n-1}e^{-\alpha(n-16)}$, where $\alpha \approx 0.2200507$. The number of subsets is m choose $(k-1)^2$, which is bounded above by $2(k-1)^2$ choose $(k-1)^2$, a central binomial coefficient. Using $(k-1)^2 = n - 2\sqrt{n}$ together with lower bounds on central binomial coefficients gives that the number of subsets is at least $4^{n-2\sqrt{n}+1}/[2\sqrt{n-2\sqrt{n}+1}]$. Therefore, the probability of acceptance is at most $e^{16\alpha}4^{2\sqrt{n}-1}\exp(-\alpha n)$, which declines exponentially fast in n.

Broder [16] and Aldous [2] independently developed the following perfect simulation algorithm for this problem. Fix any $v_0 \in V$. Run a simple random-walk Markov chain $V_0 = v_0, V_1, V_2, \ldots$ on the nodes of the graph. In such a walk, given the current node v, the next node is a neighbor of v chosen uniformly at random. So $\mathbb{P}(V_{t+1} = w | V_t = v) = 1/\deg(v)$. Then for each node $v \neq v_0$, let $t(v) = \inf\{t : V_t = v\}$.

Since the graph is finite and connected, each $t(v)$ will be finite with probability 1. Then let $T = \cup_{v \neq v_0}\{V_{t(v)-1}, V_{t(v)}\}$. This will be a tree because it consists of those edges taken in the simple random walk that reach a node for the first time. So there cannot be any cycles.

Definition 9.6. *Let $T(v_0) = \inf_{v \neq v_0}\{t(v)\}$ be the number of steps needed for the simple random walk on the graph to reach all the other vertices of the graph. Then $T(v_0)$ is the* cover time *of the graph.*

The Broder method requires a number of steps equal to the cover time of the graph. Wilson [128] managed to improve upon this method, developing a perfect simulation algorithm whose worst-case running time was the cover time of a graph, and whose best-case running time is much faster.

To accomplish this, Wilson created an extension of AR called popping. It is actually easier to describe with a slightly different problem.

Definition 9.7. *A* directed tree *is a tree T where each edge $\{i, j\} \in T$ is oriented either (i, j) or (j, i). If there is a vertex r such that for all $v \neq r$, there is a directed path from v to r, then call r the* root *of the tree.*

Suppose a node r is fixed to be the root. Then each undirected tree corresponds to exactly one directed tree rooted at r. Therefore the problem of sampling uniformly from the spanning trees is equivalent to sampling uniformly from the set of directed trees rooted at r.

Refer to edge (i, j) as outgoing from i. In a rooted directed tree, the root r will have no outgoing edges, and every other node has exactly 1 outgoing edge.

Therefore, the basic AR algorithm is: for each node $i \neq r$, uniformly select from the neighbors of i. If j is selected, add the oriented edge (i, j) to the tree. If the result is a rooted directed tree, accept, otherwise reject and start over. Since the probability of drawing a particular oriented tree in this fashion is $\prod_{i \neq j}\deg(i)^{-1}$, and so independent of the tree drawn, the result is uniform over the set of trees.

Another way to view this basic AR algorithm is to envision each node i having a sequence of iid draws uniformly from the neighbors. Following the notation of [128], call the choices for node i, $S_{i,1}, S_{i,2}, \ldots$, and label this whole sequence S_i. This can be viewed as a stack of random draws, with $S_{i,1}$ on top, followed by $S_{i,2}$, et cetera. The first random choice of directed tree given the stacks is $T = \cup_{i \neq r}\{S_{i,1}\}$.

Given an infinite stack (aka a sequence) $a_1, a_2, \ldots$, let the popping operator be $f_{\text{pop}}(a_1, a_2, \ldots) = (a_2, a_3, \ldots)$. This has "popped" the first value in the sequence off of the stack, and moved up all other items in the stack one position.

In AR, if rejection occurs, each stack is popped once, that is, each S_i is changed to $f_{\text{pop}}(S_i)$. When rejection occurs (so that the $\{S_{i,1}\}$ do not form a tree), there must be a directed cycle $i_1, \ldots, i_k$ in the set of edges $\{S_{i,1}\}$. Instead of popping all the nodes, instead try popping just $i_1, \ldots, i_k$. Stop popping when there is no longer a cycle to pop.

The question with this approach is twofold. First, does the choice of directed cycle change the outcome? Second, once the popping is done, is the set of oriented edges $\{S_{i,1}\}$ independent of what came before? The next lemma answers these questions.

Lemma 9.16 (Theorem 4 of [128]). *The set of choices of which cycle to pop next is irrelevant. Given stacks $\{S_i\}$ for $i \neq r$, either one of two possibilities must occur. Either the algorithm never terminates, or the algorithm terminates at the same place in the stack independent of the set of choices.*

Proof. Fix the set of stacks $\{S_i\}$ for $i \neq r$. Following [128], let C be a cycle that can be popped. In other words, suppose that there exists some k-tuple of cycles to be popped $C_1, C_2, \ldots, C_k = C$ where the last cycle C_k is the same as C.

Suppose that the user decides not to pop C_1 initially, but instead pops $\tilde{C}$. If $\tilde{C}$ shares no vertices with $C_1, \ldots, C_k = C$, then of course these cycles can still all be popped, ending with C. Otherwise, let C_i be the first of these cycles that shares a vertex with $\tilde{C}$.

Suppose C_i and $\tilde{C}$ share vertex w. Then $C_1, \ldots, C_{i-1}$ do not contain w, hence the stack $\{S_w\}$ is the same whether or not $C_1, \ldots, C_{i-1}$ have been popped. Therefore the next vertices after w in C_i and $\tilde{C}$ are actually the same vertex. Continuing this logic, all the vertices of C_i and $\tilde{C}$ must be the same.

Hence popping $\tilde{C} = C_i, C_{i+1}, \ldots, C_k = C$ still allows the popping of C.

This means that if there is an infinite sequence $C_1, C_2, \ldots$ of cycles that can be popped in the stacks, then no matter what choice $\tilde{C}$ is made to pop first, every cycle in the sequence can still be popped. Hence if the algorithm does not terminate, any change in the order of cycle popping will not change that fact.

On the other hand, if the algorithm terminates after making choices $C_1, \ldots, C_n$, then any cycle other than C_1 must equal C_i for some $i > 1$. Hence popping C_i first is merely changing the order in which the cycles are popped.

Once $C_1, \ldots, C_n$ have been popped, the choices of the top of the stack are independent of what came before. The probability of a particular cycle C appearing in the stack is $p(C) = \prod_{i \in C} \deg(i)^{-1}$. Similarly, let $p(T) = \prod_{i \neq r} \deg(i)^{-1}$ be the chance that tree T is at the top of the stacks. Then the chance of getting stacks that after popping cycles $C_1, \ldots, C_n$ ends at tree T is just $[\prod_k p(C_k)]\, p(T)$. Since this factors into the chance of drawing cycles $C_1, \ldots, C_n$ and the tree T, they must be independent of each other. □

This gives rise to the cycle-popping version of the algorithm. Begin with each node (except r) having a uniformly chosen outgoing edge. While there is a cycle in the orientation, pop it by independently choosing new outgoing edges for each node in the cycle.

An equivalent way to view this procedure is as a loop-erased random walk. Consider a random walk that looks like a light cycle from Tron, that records the trail behind it as it moves. (Or for the biologically minded, consider a slug that leaves a slime trail in its wake.) The cycle/slug moves by choosing a new outgoing edge from the node it is currently at, and moving to the next. If it ever returns to a node it previously occupied, that gives a cycle. That cycle can immediately be popped by removing it from the trail. Hence the walk is "loop-erasing."

When the trail reaches the root, or any other node that has a trail to the root, it is unnecessary to continue. The resulting collection of trails will be the loop-oriented random walk.

Generate_random_spanning_tree

---

1) Choose any node r in V to be the root
2) $V_T \leftarrow \{r\}$
3) While $V_T \neq V$ is nonempty
4)   Let v be any node in $V \setminus V_T$
5) $W \leftarrow \{v\}$
6)   While $v \notin V_T$
7)   Draw w uniformly from the neighbors of v
8) $V(w) \leftarrow v, T \leftarrow T \cup \{(v,w)\}$
9)   If $w \in W$
10) $z \leftarrow w$
11)   Repeat
12) $T \leftarrow T \setminus \{V(w), w\}, W \leftarrow W \setminus \{w\}, w \leftarrow V(w)$
13)   Until $w = z$
14) $W \leftarrow w$
15) $V_T \leftarrow V_T \cup W$

While the methods in this section were set up to handle the problem of uniformly generating spanning trees, they can be easily modified to handle sampling from trees where the weight of a tree is the sum of the weights of the edges that comprise the tree. See [128] for details.

Chapter 10

Stochastic Differential Equations

If God has made the world a perfect mechanism, He has at least conceded so much to our imperfect intellect that in order to predict little parts of it, we need not solve innumerable differential equations, but can use dice with fair success.

Max Born

Diffusion models appear throughout applied mathematics, and form a natural stochastic generalization of differential equations (DEs). In a first-order DE such as $dy = f(y,t)\ dt$, the idea is that as time advances a small amount h, the value of y changes by the small amount $f(y,t) \cdot h$. This naturally gives rise to a numerical method: keep updating time by adding h to the current time value at each step, while updating y by adding $f(y,t) \cdot h$ to the current y value at each step. For many DE's, only such numerical solutions are possible.

The above algorithm is known as Euler's method, and is the simplest way to numerically solve a first-order differential equation. Of course, more advanced methods exist (see for instance [68].) Typically, as h goes to 0, the numerical error in the approximation goes to 0 as well. Still, all such methods are always approximate. Except under unusual circumstances, there will always be numerical error.

The idea of a stochastic differential equation (SDE) is to improve the model by adding random movement through the use of a differential Brownian motion, dB_t. An SDE might have the form

$$dX_t = a(X_t)\ dt + b(X_t)\ dB_t. \tag{10.1}$$

Very roughly speaking, dB_t can be thought of as a normal random variable that has mean 0 and variance dt. When Z is a standard normal random variable with mean 0 and variance 1, then cZ has mean 0 and variance c^2. Therefore, an approach to numerically solving an SDE is to use

$$X_{t+h} = X_t + a(X_t)h + b(X_t)\sqrt{h}Z. \tag{10.2}$$

For small h, $\sqrt{h}$ is very much larger than h. Therefore, the normal random variable $\sqrt{h}Z$ with standard deviation $\sqrt{h}$ is spread out much wider than the deterministic part $a(X_t)h$. For h of order $(b(X_t)/a(X_t))^2$, there is a chance (as with the multishift coupling from Chapter 6) that the randomized part of the update is independent of the deterministic part. This is essentially the property that allows perfect simulation to occur.

Suppose $Y_{t+1} = X_t + \sqrt{h}Z$ is put forward as the next candidate for X_{t+1}. Then there should be a reasonable chance of accepting Y_{t+1} as a perfect draw from X_t. Unfortunately, it is unknown how to compute the probability of acceptance for this simple scheme.

Therefore, a more sophisticated approach using densities of the solution of the SDE with respect to Brownian motion will be used. However, these densities are very difficult to compute exactly, and so new Monte Carlo methods were created for generating Bernoulli random variables with the correct mean. To understand these methods, a deeper understanding of Brownian motion is necessary.

10.1 Brownian motion and the Brownian bridge

Definition 10.1. *A stochastic process $\{B_t\}$ is* standard Brownian motion *if*

1. $B_0 = 0$.
2. *The map $t \mapsto B_t$ is continuous with probability 1.*
3. *For $a \leq b \leq c \leq d$, $B_d - B_c$ is independent of $B_b - B_a$.*
4. *For $a \leq b$, $B_b - B_a \sim \mathsf{N}(0, b-a)$.*

A Brownian motion process is defined for all $t \in \mathbb{R}$, therefore, it is not possible to write down (let alone simulate) an entire Brownian motion process at once.

A more achievable goal is to simulate standard Brownian motion at a finite set of times $0 < t_1 < t_2 < t_3 < \cdots < t_n$. Since all the differences $B_{t_i} - B_{t_{i-1}}$ are normally distributed, the joint distribution of $(B_{t_1}, \ldots, B_{t_n})$ will be multivariate Gaussian.

For a standard normal random variable $Z \sim \mathsf{N}(0,1)$ and a constant k, $kZ \sim \mathsf{N}(0,k^2)$. Then since $B_{t_1} - B_0 \sim \mathsf{N}(0,t_1)$, simply generate a standard normal random variable $Z_1 \sim \mathsf{N}(0,1)$, and let B_{t_1} be $B_0 + Z_1\sqrt{t_1}$. Continuing in this fashion, let $B_{t_{i+1}}$ be $B_{t_i} + Z_i\sqrt{t_{i+1} - t_i}$, where $Z_1, \ldots, Z_n$ are iid $\mathsf{N}(0,1)$.

```
Standard_Brownian_Motion
```
Input: $0 < t_1 < t_2 < \ldots < t_n$,   *Output:* $B_{t_1}, B_{t_2}, \ldots, B_{t_n}$

1) $t_0 \leftarrow 0,$, $B_0 \leftarrow 0$
2) For i from 1 to n
3)   Draw Z a standard normal random variable
4) $B_{t_i} \leftarrow B_{t_{i-1}} + Z\sqrt{t_i - t_{i-1}}$

Of course, it is often the case that the Brownian motion needs to start at a location other than 0.

Definition 10.2. *Let $\{B_t\}$ be standard Brownian motion. Then $B_t^x = B_t + x$ is* Brownian motion started at x.

Brownian motion where the values of B_0 and B_T are fixed is called a *Brownian bridge* or *pinned Brownian motion*. (In some texts this terminology is reserved for Brownian motion where $B_0 = B_T = 0$, but here it will be used for any fixed values for B_0 and B_T.) Given the ability to simulate Brownian motion, the following lemma gives a fast way to simulate a Brownian bridge.

Lemma 10.1. *Let $\{B_t\}_{t\in[0,T]}$ be standard Brownian motion. Then*

$$R_t^{x,y} = x + B_t - (t/T)(x + B_T - y) \tag{10.3}$$

has the same distribution as $[\{B_t\}_{t\in[0,T]} | B_0 = x, B_T = y]$.

See [7] for a proof. The corresponding algorithm is as follows.

Standard_Brownian_Bridge *Input:* $0 < t_1 < t_2 < \ldots < t_n, x, y, T$ *Output:* $B_{t_1}, B_{t_2}, \ldots, B_{t_n}$
1) $(B_{t_1},\ldots,B_{t_n},B_T) \leftarrow$ Standard_Brownian_Motion($t_1,\ldots,t_n,T$) 2) For i from 1 to n 3) $\quad R_{t_i} \leftarrow x + B_{t_i} - (t_i/T)(x + B_T - y)$

Generating from Brownian motion and Brownian bridges is easy, but generating from general SDE's will require more tools, in particular, an exponential Bernoulli factory.

10.2 An exponential Bernoulli factory

The algorithm for generating from more general SDE requires the use of acceptance/rejection. In Section 2.2, it was shown how to use AR to sample from one density f given a draw from another density g. To use the algorithm, it is necessary to have c such that $c \geq f(x)/g(x)$ for all x in the state space Ω, and the ability to draw from $\mathsf{Bern}(f(x)/[cg(x)]$ for all $x \in \Omega$. The method for doing so will utilize what is known as a *Bernoulli factory* .

To understand what a Bernoulli factory does, it helps to have some examples. Suppose that $X_1, X_2, \ldots$ are iid $\mathsf{Bern}(p)$, and so give an infinite stream of independent coin flips with probability p of heads. Suppose the goal is to build a new coin out of these X_i that has probability $2p - p^2$ of heads. That turns out to be easy. Just let

$$Y = \mathbb{1}(X_1 = 1 \text{ or } X_2 = 1).$$

This Bernoulli factory uses either 1 or 2 coin flips. If $X_1 = 1$, then there is no need to look at X_2. The chance that the second coin is necessary is $1 - p$. Therefore if T represents the number of coins that need to be examined, $\mathbb{E}[T] = 1 + (1 - p) = 2 - p$.

An early Bernoulli factory that uses a random number of coin flips is due to Von Neumann [127]. In this factory, the goal is to generate $Y \sim \mathsf{Bern}(1/2)$, and this technique was used to go from possibly unfair coin flips to fair coin flips. This Bernoulli factory is

$$T = \inf\{t : (X_{2t+1} = 1, X_{2t+2} = 0) \text{ or } (X_{2t+1} = 0, X_{2t+2} = 1)\}, \quad Y = X_{2T+1}.$$

This is essentially AR: keep flipping coins in pairs until either $(1,0)$ or $(0,1)$ is reached. Conditioned on one of these two events, each is equally likely to occur, so $Y \sim \mathsf{Bern}(1/2)$.

Following [50], a Bernoulli factory can be defined as follows.

Definition 10.3. *Given $p^* \in (0,1]$ and a function $f : [0,p^*] \to [0,1]$, a* Bernoulli factory *is a computable function $\mathscr{A}$ that takes as input a number $u \in [0,1]$ together with a sequence of values in $\{0,1\}$, and returns an output in $\{0,1\}$ where the following holds. For any $p \in [0,p^*]$, $X_1, X_2, \ldots$ iid $\mathsf{Bern}(p)$, and $U \sim \mathsf{Unif}([0,1])$, let T be the infimum of times t such that the value of $\mathscr{A}(U, X_1, X_2, \ldots)$ only depends on the values of $X_1, \ldots, X_t$. Then*

1. *T is a stopping time with respect to the natural filtration and $\mathbb{P}(T < \infty) = 1$.*
2. *$\mathscr{A}(U, X_1, X_2, \ldots) \sim \mathsf{Bern}(f(p))$.*

Call T the running time *of the Bernoulli factory.*

Densities for SDE's employ Girsanov's Theorem [42], which gives the density of the form $\exp(-p)$ for a specific p. Therefore, what is needed for these types of problems is an exponential Bernoulli factory where $Y \sim \mathsf{Bern}(\exp(-p))$. Beskos et al. [11] gave the following elegant way of constructing such a Bernoulli factory.

Recall how thinning works (discussed in Sections 2.2 and 7.2.) Take P, a Poisson point process of rate 1 on $[0,1]$, and for each point in P flip a coin of probability p. Then let P' be the points of P that received heads on their coin. P' will be a Poisson point process of rate p. So the number of points in P' is distributed as Poisson with mean p, which makes $\mathbb{P}(\#P' = 0) = \exp(-p)$.

More generally, suppose that $A \subset B$ are sets in $\mathbb{R}^n$ of finite Lebesgue measure. Let P be a Poisson point process of rate 1 on B. Then $P \cap A$ is a Poisson point process of rate 1 on A, and the chance that there are no points in A is $\exp(-m(A))$, where $m(A)$ is the Lebesgue measure of A.

With this in mind, we are now ready to perfectly simulate some SDE's.

10.3 Retrospective exact simulation

Retrospective exact simulation is a form of acceptance/rejection introduced by Beskos et. al [11] to handle SDE's. Start with Equation (10.1).

The SDE has a unique solution, that is, there exists a unique distribution of $\{X_t\}$ for $t \in [0,T]$, given that the functions a and b satisfy some regularity conditions. These conditions ensure that the SDE does not explode to infinity in finite time. See Chapter 4 of Kloeden and Platen [83] for details.

Our original SDE had the form $dV_t = a(v)\,dt + b(v)\,dB_t$. The first step to note is that for many a and b it is possible to restrict our attention to SDE's where $b(v)$ is identically 1.

To see this, consider the differential form of Itō's Lemma (see for instance [110]).

Lemma 10.2 (Itō Lemma (1st form)). *Let f be a twice differentiable function and $dV_t = a(v)\,dt + b(v)\,dB_t$. Then*

$$df(V_t) = [a(v)f'(v) + b(v)^2 f''(v)/2]\,dt + b(v)f'(v)\,dB_t. \tag{10.4}$$

Now let $\eta(v) = \int_z^v b(u)^{-1}\,du$ for any z in the state space Ω. Then $\eta'(v) = b(v)^{-1}$ and $\eta''(v) = -b'(v)/b(v)^2$. Hence for $X_t = \eta(V_t)$,

$$dX_t = [a(v)/b(v) - b'(v)/2]\,dt + dB_t. \tag{10.5}$$

If more convenient, the transformation $\eta_2(v) = -\int_z^v b(u)^{-1}\,du$ can be used. This leads to $dX_t = -[a(v)/b(v) - b'(v)/2]\,dt - dB_t$, but since $B_t \sim -B_t$, this also gives an SDE with unit coefficient in front of the dB_t:

$$dX_t = -[a(v)/b(v) - b'(v)/2]\,dt + dB_t. \tag{10.6}$$

Let $\alpha(x) = b'(\eta^{-1}(x))/2 - a(\eta^{-1}(x))/b(\eta^{-1}(x))$. When α is identically 0, then $dX_t = dB_t$, and the solution is just $X_t = B_t$, standard Brownian motion. If the function α is close to 0, then X_t will be close to, but not exactly, Brownian motion. If it is close to Brownian motion, then the chance of rejecting a pure Brownian motion as a draw from X_t will be fairly small.

Consider the density of the solution of the diffusion with respect to standard Brownian motion. For example, suppose that X_t has the same distribution as standard Brownian motion, but conditioned on $X_1 \geq 1$ and $X_2 \leq 2$. Then the density of $\{X_t\}_{t\in[0,2]}$ with respect to standard Brownian motion is $f_X(x) = \mathbb{1}(X_1 \geq 1, X_2 \leq 2)$. (This is an unnormalized density.) Since the density lies between 0 and 1 (inclusive), it is straightforward to generate from X_t over $0 < t_1 < \cdots < t_k < 2$. Simply generate $x = (x(0), x(t_1), x(t_2), \ldots, x(t_k), x(2))$ as a standard Brownian motion, and accept the draw as a draw from the law of X_t with probability $f_X(x)$.

In fact, solutions to the SDE (10.6) have a density with respect to Brownian motion started at the same location as the solution to the SDE. This important fact is known as Girsanov's theorem [42].

To be precise, let $C = C([0,T],\mathbb{R})$ denote the set of continuous mappings from $[0,T]$ to $\mathbb{R}$, and $\omega \in C$. Then one way to view Brownian motion starting at $B_0 = x$ is that it is a measure $\mathbb{W}^x$ over C such that a random draw $\{B_t\}_{t\in[0,T]}$ from the measure satisfies three properties: $B_0 = x$, for any $0 \leq t_1 < t_2 \leq t_3 < t_4 \leq T$, $B_{t_4} - B_{t_3}$ and $B_{t_2} - B_{t_1}$ are independent and $B_{t_2} - B_{t_1} \sim \mathsf{N}(0, t_2 - t_1)$.

Let $\mathbb{Q}$ be the probability measure of solutions to the diffusion (10.6). Then Girsanov's theorem states that a draw from $\mathbb{Q}$ has a density with respect to $\mathbb{W}$. In measure theory, this density is known as the Radon-Nikodym derivative between the two measures, and is

$$\frac{d\mathbb{Q}}{d\mathbb{W}^x}(\{\omega_t\}) = \exp\left(\int_0^T \alpha(\omega_t)\,d\omega_t - \frac{1}{2}\int_0^T \alpha^2(\omega_t)\,dt\right). \tag{10.7}$$

Again note that if $\alpha(x) = 0$ for all $x \in \mathbb{R}$, then the density is just 1, indicating that the solution to the SDE is just Brownian motion in this case.

If this density is bounded above by c, then to use AR, just draw a standard Brownian motion $\{B_t\}$, and then accept with probability equal to $(1/c)[d\mathbb{Q}/d\mathbb{W}](\{B_t\})$.

When B_t is substituted for ω_t, the first integral being exponentiated is $\int_0^T \alpha(B_t)\,dB_t$. This refers to the *Itō integral* of $\alpha(B_t)$ with respect to the Brownian Motion B_t. (See [110] for a formal definition.) For the Itō integral, when the function α has a continuous first derivative, Itō's Lemma can be employed to rewrite the integral. The simpler form of the lemma is needed here.

Theorem 10.1 (Itō's Lemma (2nd form) [69]). *For a function $f(x)$ with two contin-*

uous derivatives,

$$f(B_T) - f(B_0) = \int_0^T f'(B_t)\, dB_t + \frac{1}{2}\int_0^T f''(B_t)\, dt. \tag{10.8}$$

Apply this to the function $A(x) = \int_0^x \alpha(s)\, ds$. Then $A'(x) = \alpha(x)$, $A''(x) = \alpha'(x)$, and Itō's Lemma gives

$$\int_0^T \alpha(B_t)\, dB_t = A(B_T) - A(B_0) - \frac{1}{2}\int_0^T \alpha'(B_t)\, dt, \tag{10.9}$$

which means (using that $B_0 = x$ under $\mathbb{W}^x$)

$$\frac{d\mathbb{Q}}{d\mathbb{W}^x}(B_t) = \exp\left(A(B_T) - A(x) - \frac{1}{2}\int_0^T [\alpha'(B_t) + \alpha^2(B_t)]\, dt\right). \tag{10.10}$$

Under $\mathbb{W}^x$, the distribution of B_T is normal with mean x and variance T. Say for simplicity that $x = 0$, and suppose that a user wanted to generate a Brownian Motion A with $A_0 = x$ but with A_T uniform over $[-1,1]$.

Intuitively, acceptance/rejection could be used to accomplish this as follows. Generate a standard Brownian Motion B_t over $[0,T]$ from $\mathbb{W}^0$, then accept as a draw from the target distribution with probability

$$a(B_T) = \frac{\exp(-1/(2T))\mathbb{1}(B_T \in [-1,1])}{\exp(-B_T^2/(2T))}. \tag{10.11}$$

Then A_T has the correct density, and given that this is an acceptance/rejection method whose chance of accepting is $f_{A_T}(B_T)/[c\phi(B_T)]$, for a constant c, that means $f_{A_T}/\phi(B_T)$ is also the density of the measure of $\{A_t\}$ with respect to the measure of $\{B_t\}$.

More generally, the following is true.

Lemma 10.3 (Proposition 1 of [11]). *Let $M = \{M_t\}_{t\in[0,T]}$ and $N = \{N_t\}_{t\in[0,T]}$ be two stochastic processes on C with corresponding probability measures $\mathbb{M}$ and $\mathbb{N}$. Let f_M and f_N be the densities of M_T and N_T respectively, and both densities are supported on $\mathbb{R}$. When it holds that $[M|M_T = \rho] \sim [N|N_T = \rho]$ for all $\rho \in \mathbb{R}$, then*

$$\frac{d\mathbb{M}}{d\mathbb{N}}(\{N_t\}) = \left(\frac{f_M}{f_N}\right)(N_T). \tag{10.12}$$

The rigorous proof is in Appendix 1 of [11].

Recall the pinned Brownian motion $\{R_t\}$ from earlier. Then to draw R_t so that R_T has a desired distribution, first draw R_T from the distribution, then fill in the rest of the points from $R_0 = x$ to R_T using Lemma 10.1.

This lemma can be used to simplify (10.10) as follows. Let H have unnormalized density $f_H(u) \propto \exp(A(u) - (u-x)^2/(2T))$. Assume here that $A(u) = \int_x^u \alpha(u)\, du$ is a function that grows slowly enough that $f_H(u)$ can be normalized. Then for $\{R_t\}$ a draw from Brownian Motion conditioned to have $R_T \sim H$,

$$\frac{d\mathscr{L}\{R_t\}}{d\mathbb{W}^x}(\{B_t\}) \propto \frac{\exp(A(B_T) - (B_T - x)^2/(2T))}{\exp(-(B_T - x)^2/(2T))} = \exp(A(B_T)). \tag{10.13}$$

Now Radon-Nikodym derivatives behave much like regular derivatives, including obeying the chain and inverse rules. So

$$\begin{aligned}\frac{d\mathbb{Q}}{d\mathscr{L}\{R_t\}}(\{R_t\}) &= \left[\frac{d\mathbb{Q}}{d\mathbb{W}^x}(\{R_t\})\right] / \left[\frac{d\mathscr{L}\{R_t\}}{d\mathbb{W}^x}(\{R_t\})\right] \\ &\propto \frac{\exp(A(R_T) - (1/2)\int_0^T [\alpha'(R_t) + \alpha^2(R_t)]\,dt)}{\exp(A(R_T))} \\ &= \exp\left(-\int_0^T (\alpha'(R_t) + \alpha^2(R_t))/2\,dt\right).\end{aligned}$$

At this point, to make acceptance/rejection work, assume that the function $(\alpha' + \alpha^2)/2$ is bounded below by k_α. Then setting

$$\ell(u) = \frac{\alpha'(u) + \alpha^2(u)}{2} - k_\alpha \tag{10.14}$$

makes $\ell(u) \geq 0$, and since $\exp(-k_\alpha T)$ is a constant,

$$\frac{d\mathbb{Q}}{d\mathscr{L}\{R_t\}}(\{R_t\}) \propto \exp\left(-\int_0^T \ell(R_t)\,dt\right) \leq 1, \text{ with probability 1 under } \mathscr{L}\{R_t\}. \tag{10.15}$$

It is necessary here that $A(u)$ be a function such that $\exp(A(u) - (u-x)^2/(2T))$ is a normalizable density. This is equivalent to saying that there is a constant a_1 such that $A(u) = a_1 u^2$ for sufficiently large or small u. When a_1 is large, T might need to be very small for the density to be normalizable.

When the density is normalizable, if $\alpha^2 + \alpha'$ is bounded below, it is possible in theory to use an AR approach to generate draws from $\mathbb{Q}$ using draws of $\{R_t\}$.

Beskos et al. [11] introduced a Poisson point process procedure for accomplishing this. Let $A = \{t, y : t \in [0,T], y \in [0, \ell(B_t)]\}$. Recall that if P is a Poisson point process over B where $A \subseteq B$, then $\mathbb{P}(\#(P \cap A) = 0) = \exp(-\nu(A))$. Using $\nu(A) = \int_0^T \ell(R_t)\,dt$ then gives the desired result.

So how to determine if $\#(P \cap A) = 0$? First consider the case where the function ℓ is bounded above by M. Then $A \subseteq [0,T] \times [0,M]$, and it is possible to generate a PPP P of constant intensity 1 over $[0,T] \times [0,M]$. Consider a point $(t_i, y_i) \in P$. When does this point fall into A?

Since B_t and ℓ are continuous maps, this point falls into A if and only if $y_i \leq \ell(R_{t_i})$. So to determine if any point of P falls into A, it is only necessary to evaluate R_t at the finite set of time values $\{t_1, t_2, \ldots, t_{\#P}\}$.

Alternately, given the time value t_i of the points of P, the chance that $y_i > R_i$ is $\max\{0, (M - R_{t_i})/M\}$. So it is not necessary to draw y_i explicitly.

When the PPP on $[0,T] \times [0,M]$ is projected onto the interval $[0,T]$ by removing the second coordinate, the result is a one-dimensional PPP of rate T. It is possible to generate such a PPP by adding exponential random variables of rate M to 0 until the result is greater than T. An exponential random variable of rate M has the same distribution as $(1/M)\ln(1/U)$, where U is uniform over $[0,1]$.

The authors referred to this method as *retrospective sampling*, since once the

Poisson process is simulated, the user needs to go back retrospectively and simulate the proper number of points from the conditional Brownian motion to determine if acceptance or rejection occurs.

The following pseudocode generates $X \sim X_T$ for $dX_t = \alpha(X_t)\,dt + dB_t$ where $\ell(u) = (\alpha'(u) + \alpha(u)^2)/2 - \inf_w(\alpha'(w) + \alpha(w)^2)/2$ and $\ell(u) \leq M$ for all u.

`Retrospective_sampling_for_SDEs`
Input: T, x_0 *Output:* $X \sim X_T$

---

1) $X \leftarrow x_0$
2) Repeat
3) Draw R from density $f_R(u) \propto \exp(A(u) - (u - X)^2/2)$, $a \leftarrow 1$
4) $i \leftarrow 1, U \leftarrow \mathsf{Unif}([0,1]), t_1 \leftarrow -(1/M)\ln(U)$
5) While $t_i \leq T$
6) $U \leftarrow \mathsf{Unif}([0,1]), i \leftarrow i+1, t_i \leftarrow t_{i-1} + (1/M)\ln(1/U)$
7) Draw $(B_{t_1}, \ldots, B_{t_{i-1}}, B_T) \leftarrow$ `Standard_Brownian_Motion`$(t_1, \ldots, t_{i-1}, T)$
8) For j from 1 to $i-1$ do
9) $V_j \leftarrow \mathsf{Unif}([0,1])$
10) $R_{t_j} \leftarrow x_0 + B_{t_i} + (t_j/T)(R - x_0 - B_T)$
10) $a \leftarrow a \cdot \mathbb{1}(V_j \geq \ell(R_{t_j})/M)$
11) Until $a = 1$
12) $X \leftarrow R$

Lemma 10.4. *Suppose* `Retrospective_sampling_for_SDEs` *uses N random variables in a run. Then*

$$2(1+MT) \leq \mathbb{E}[N] \leq 2(1+MT)e^{MT}. \tag{10.16}$$

Proof. One random variable R is chosen at line 3. In line 4, one random variable is used to determine $\#P$, then if $\#P > 0$, $2MT$ more are needed on average to determine the PPP P. Note that for a random variable X that is nonnegative, $\mathbb{P}(X > 0)\mathbb{E}[X|X > 0] = \mathbb{E}[X]$, so $\mathbb{E}[N] \geq 1 + 1 + 2MT = 2(1+MT)$.

Suppose that rejection occurs. This can only happen when $\#P > 0$, but in this case, the process starts over and $\mathbb{E}[N]$ variables are needed. The chance that $\#P > 0$ is $1 - \exp(-MT)$. Hence

$$\mathbb{E}[N] \leq 2(1+MT) + (1 - \exp(-MT))\mathbb{E}[N], \tag{10.17}$$

and solving for $\mathbb{E}[N]$ gives $\mathbb{E}[N] \leq 2(1+MT)e^{MT}$. □

This bound is far too wide to be useful when MT is large. Therefore, a better approach is to break the interval $[0,T]$ into n different intervals of width T/n. The diffusion $\{X_t\}$ has the Markov property, which means that one can simulate $X_{T/n}|X_0$, then $X_{2T/n}|X_{T/n}$ and so on until reaching X_T.

```
Divided_retrospective_sampling_for_SDEs
Input: T, x0, n    Output: X ~ X_T
1)  X ← x0
2)  For i from 1 to n
3)      X ← Retrospective_sampling_for_SDEs(T/n, X)
```

Lemma 10.5. *Suppose* `Divided_retrospective_sampling_for_SDEs` *uses* N *random variables. Then*

$$2n(1+MT/n) \leq \mathbb{E}[N] \leq 2n(1+MT/n)\exp(MT/n). \tag{10.18}$$

In particular, if $n = MT\phi$, *where* $\phi = (1+\sqrt{5})/2$ *is the golden mean, then* $3.61MT \leq \mathbb{E}[N] \leq 9.72MT$.

Proof. There are n intervals each of width T/n, so the previous Lemma immediately gives (10.18).

So if $x = MT/n$, then $2(1+\phi)MT \leq \mathbb{E}[N] \leq MT(2/x)(1+x)\exp(x)$. The right-hand side is minimized at $x = 1/\phi$, and the result follows. □

One more comment: usually acceptance/rejection algorithms are interruptible, since the number of times a proposal is made does not affect the distribution of the final outcome. But here there is an interesting situation where the time to determine if acceptance or rejection occurs depends on the number of points in the PPP over $[0,T]$, and that does affect the chance of acceptance. So if time is measured by the number of random variates generated, this algorithm and the rest of the algorithms presented in this chapter) are noninterruptible.

10.3.1 Somewhat bounded ℓ

Now suppose that ℓ is unbounded, but $\sup_{x\in[M,\infty)} \ell$ exists for all M. Then the techniques of the previous section still can be used, but the execution is a bit trickier.

Again, the presentation of this algorithm follows [11] closely. The first thing that is needed is a way of finding the minimum of a Brownian bridge.

To accomplish this, first consider the distribution of the first time that Brownian motion reaches a fixed level. In this literature this is known as the inverse Gaussian distribution. The following lemma connects this distribution to first passage times.

Lemma 10.6. *Let* B_t *be standard Brownian motion, and* $X_t = k_1 t + k_2 B_t$ *for parameters* k_1 *and* k_2 *that satisfy* $k_1 > 0$, $k_2 \neq 0$. *Let* $T_\alpha = \inf\{t : X_t = \alpha\}$. *Then*

$$T_\alpha \sim \mathsf{InvGau}(\alpha/k_1, \alpha^2/k_2^2). \tag{10.19}$$

An inverse Gaussian can be simulated easily with the ability to generate normally distributed random variables using an algorithm of Michael, Schucany, and Haas [92].

`Inverse_Gaussian` *Input:* μ, λ, *Output:* $X \sim \mathsf{InvGau}(\mu, \lambda)$
1) Draw $W \leftarrow \mathsf{N}(0,1)$, let $Y \leftarrow W^2$
2) $X_1 \leftarrow \mu + \mu^2 Y/(2\lambda) - \mu\sqrt{4\mu\lambda Y + \mu^2 Y^2}/(2\lambda)$
3) Draw $C \leftarrow \mathsf{Bern}(\mu/(\mu + X_1))$
4) $X \leftarrow X_1 C + (1-C)\mu^2/X_1$

Given the story behind an inverse Gaussian, it is unsurprising that this distribution will enter into the simulation of the minimum of the Brownian bridge.

`Minimum_Brownian_Bridge` *Input:* T, a *Output:* $Z_1 \sim \min_{t\in[0,T]} R_t^{0,a}$, $Z_2 \sim \arg\min_{t\in[0,T]} R_t^{0,a}$
1) Draw $A \leftarrow \mathsf{Exp}(1)$
2) $Z_1 \leftarrow (a - \sqrt{2\cdot A\cdot T + a^2})/2$
3) $c_1 \leftarrow (a - Z_1)^2/(2T)$, $c_2 \leftarrow Z_1^2/(2T)$
4) Draw $U \leftarrow \mathsf{Unif}([0,1])$
5) Draw $I_1 \leftarrow$ `Inverse_Gaussian`$(\sqrt{c_1/c_2}, 2c_1)$
6) Draw $I_2 \leftarrow 1/$`Inverse_Gaussian`$(\sqrt{c_2/c_1}, 2c_2)$
7) $V \leftarrow I_1 \cdot \mathbb{1}(U < (1+\sqrt{c_1/c_2})^{-1}) + I_2 \cdot \mathbb{1}(U \geq (1+\sqrt{c_1/c_2})^{-1})$
8) $Z_2 \leftarrow T/(1+V)$

Lemma 10.7 (Proposition 2 of [11]). *The output of* `Minimum_Brownian_Bridge` (Z_1, Z_2) *is that of the value and location of the minimum of Brownian motion conditioned to have* $B_0 = x$ *and* $B_T = y$.

See [11] for the proof.

After running this procedure the minimum value b (and the time t at which the minimum occurs) for the Brownian bridge is known. To get a Brownian bridge between x and y, find a Brownian bridge between 0 and $y - x$, and add x to each point. Let $M = \max_{u\in[b+x,\infty)} \ell(u)$ be the largest that $\ell(B_t)$ can be for times in $[0,T]$. Then generate the PPP over $[0,T]$ of rate M as before, and compare to the Brownian motion simulated over $[0,T]$ at those times to see if acceptance occurs.

The slightly tricky part at this stage is simulating the Brownian bridge conditioned on the minimum value b occurring at time t. The Brownian bridge over $[0,t]$ is conditioned to have $B_0 = 0$ and $B_t = b$, and $B_s \geq b$ for all $s \in [0,T]$.

Such a path can be built by considering the time interval $[0,t]$ and the interval $[t,T]$ separately. To build these paths, it is necessary to be able to simulate a Brownian motion conditioned to stay nonnegative. Such a path is called a *Brownian meander*. Because it is also an example of a Bessel Process conditioned on the endpoints, it is also known as a *Bessel bridge*. Exact simulation of such entities follows from a result of Bertoin and Pitman [7].

Lemma 10.8. *Suppose* $\{B_t^1\}$, $\{B_t^2\}$, *and* $\{B_t^3\}$ *are Brownian bridges over* $[0,1]$ *that begin and end at* 0. *Then*

$$B_s^{meander,y} = \sqrt{(ys + B_s^1)^2 + (B_s^2)^2 + (B_s^3)^3} \qquad (10.20)$$

is Brownian motion over $[0,1]$ *that is conditioned to stay nonnegative, and* $B^y(1) = y$.

This Brownian meander over $[0,1]$ can be flipped and rescaled to provide a way of connecting the point $(0,0)$ to (t,b). Then a second Brownian meander can be used to connect the point (t,b) to (T,a).

Lemma 10.9. *Let $\{B_t^{meander,a}\}$ be a Brownian meander over time interval $[0,1]$ with value 0 at time 0 and value a at time 1. Let B_t be Brownian motion conditioned to pass through the points $(0,0)$, (t,b), and (T,a) with $B_t(t) \geq b$ for all $t \in [0,T]$. Set*

$$B'_s = \sqrt{t} \cdot B_{(t-s)/t}^{meander,-b/\sqrt{t}} + b \qquad \text{for } s \in [0,t]$$
$$B'_s = \sqrt{T-t} \cdot B_{(s-t)/(T-t)}^{meander,(a-b)/\sqrt{T-t}} + b \qquad \text{for } s \in [t,T].$$

Then $\{B_s | B_0 = 0, B_t = b, B_T = a, (\forall r \in [0,T])(B_r \geq b)\}_{s \in [0,T]} \sim \{B'_s\}_{s \in [0,T]}$.

This lemma is Theorem 2 of [11], which in turn was Proposition 2 of Asumussen et al. [4] rescaled to be over the interval $[0,T]$. Putting this lemma in algorithmic form gives us a way to generate the Brownian motion through $(0,0)$ (t,b), and (T,a) at an arbitrary number of values.

`BM_given_min` *Input:* $b, t, a, (t_1, \ldots, t_n), T$ *Output:* $(B_{t_1}, \ldots, B_{t_n})$

$k \leftarrow \max\{j \leq n : t_j < t\}$
If $k > 0$
  $t' \leftarrow ((t - t_1)/t, \ldots, (t - t_k)/t)$
  For j from 1 to 3, $(X_1^j, \ldots, X_k^j) \leftarrow$ `Standard_Brownian_Bridge`(t'')
  For j from 1 to k
    $B_{t_j} \leftarrow b + \sqrt{t} \cdot \sqrt{([-b/\sqrt{t}]t'_j + X_j^1)^2 + (X_j^2)^2 + (X_j^3)^2}$
If $k < n$
  $t'' \leftarrow ((t_{k+1} - t)/(T - t), \ldots, (t_n - t)/(T - t))$
  For j from 1 to 3, $(X_1^j, \ldots, X_{n-k}^j) \leftarrow$ `Standard_Brownian_Bridge`(t'')
  For j from $k+1$ to n
    $B_{t_j} \leftarrow b + \sqrt{T-t} \cdot \sqrt{([(a-b)/\sqrt{T-t}]t''_{j-k} + X_{j-k}^1)^2 + (X_{j-k}^2)^2 + (X_{j-k}^3)^2}$

So the outline of the procedure is

- Find the minimum of B_t over $[0,T]$.
- Use the Brownian meanders to piece together a Brownian motion over $[0,T]$ that passes through the minimum.
- Use the knowledge of the smallest the Brownian motion can be to obtain a bound $\ell(B_t) \leq M$.
- Now use retrospective sampling as before.

Putting all this together gives the following algorithm.

`Retrospective_Sampling_for_SDEs_2` *Input:* T, x_0 *Output:* $X \sim X_T$

$X \leftarrow x_0$
Repeat
  Draw R from density $f_R(u) \propto \exp(A(u) - (u - X)^2/2)$, $a \leftarrow 1$
  $(b, t) \leftarrow$ `Minimum_Brownian_Bridge`$(T, y - x)$
  $M \leftarrow \max_{u \in [b+x, \infty)} \ell(u)$
  $i \leftarrow 1, U \leftarrow \mathsf{Unif}([0,1]), t_1 \leftarrow -(1/M)\ln(U)$
  While $t_i \leq T$
    $U \leftarrow \mathsf{Unif}([0,1]), i \leftarrow i+1, t_i \leftarrow t_{i-1} + (1/M)\ln(1/U)$
  If $i > 1$
    $(X_1, \ldots, X_{i-1}) \leftarrow$ `BM_given_min`$((z_1, z_2), y - x, (t_1, \ldots, t_{i-1}))$
    $(V_1, \ldots, V_{i-1}) \leftarrow \mathsf{Unif}([0,1]^{i-1})$
    $a \leftarrow \prod_{i=1}^{i} \mathbb{1}(V_i \leq X_j/M)$
Until $a = 1$
$X \leftarrow R$

10.3.2 *Example: stochastic logistic growth model*

As an example of the algorithm `Retrospective_Sampling_for_SDEs_2`, consider the stochastic logistic growth model:

$$dV_t = rV_t(1 - V_t/K)\,dt + \beta V_t\,dB_t. \tag{10.21}$$

Here r, K, and β are known positive parameters, and say $V_0 = v_0 \in (0, K)$ is also a given parameter.

This model is used for population size when resources are limited. The parameter K controls how large V_t can get, and is known as the carrying capacity of the model. As noted in [11], it can be shown that with probability 1, for $v_0 \in (0, K)$, $V_t \in (0, K)$ for all t with probability 1 (see [76, Section 15.7].) The parameter β controls the size of the random variation in population.

Following [11], use a slightly different transformation than given previously, namely, $X_t = \eta(V_t)$, where $\eta = -\int_z^x b(v)^{-1}\,du$. Here $b(v) = \beta v$, so $\eta(v) = -\beta^{-1}[\ln(v) - \ln(z)]$. Set $z = 1$ to make $x = \eta(v) = -\beta^{-1}\ln(v)$, or equivalently, $v = \exp(-\beta x)$.

So $\alpha(x) = [a(v)/b(v) - 1/2] = \beta r(1 - V_t/K) - 1/2 = \beta r(1 - \exp(-\beta X_t)/K) - 1/2$ and the SDE of interest is

$$dX_t = \left(\frac{\beta}{2} - \frac{r}{\beta} + \frac{r\exp(-\beta X_t)}{\beta K}\right)\,dt + dB_t. \tag{10.22}$$

The function in front of dt is $\alpha(X_t)$. The next step is bounding $(1/2)(\alpha^2 + \alpha')$:

$$(1/2)(\alpha^2 + \alpha')(u) = \frac{1}{2}\left[\left(\frac{\beta}{2} - \frac{r}{\beta}\right)^2 - 2\frac{r^2}{\beta^2 K}\exp(-\beta u) + \frac{r^2}{\beta^2 K^2}\exp(-2\beta u)\right]. \tag{10.23}$$

The right-hand side is bounded below by $(1/2)(\beta^2/8 - r/2$ for all $u \in \mathbb{R}$. Hence

$$\ell(u) = \frac{1}{2}\left[\frac{r^2}{\beta^2} - 2\frac{r^2}{\beta^2 K}\exp(-\beta u) + \frac{r^2}{\beta^2 K^2}\exp(-2\beta u)\right]. \quad (10.24)$$

The function $\ell(u)$ was designed so that $\ell(u) > 0$. When $u > 0$ $\ell(u) \leq (1/2)(r^2/\beta^2)$, but that leaves the situation when $u < 0$ to worry about.

So use the ideas of the last section: find the minimum value of B_t over $[0,T]$, then that gives a maximum value of $\ell(B_t)$ over $[0,T]$. With that maximum value in place, it once again becomes possible to generate a Bernoulli with mean $\exp(-I)$ where $I = \int_0^T \ell(B_t)\,dt$.

Therefore the techniques of the last section can be applied as long as $A(u)$ does not grow too quickly. So consider $A(u) = \int_0^x \alpha(s)\,ds$. Here

$$A(u) = \left(\frac{\beta}{2} - \frac{r}{\beta}\right)u - \frac{r}{\beta^2 K}\exp(-\beta u). \quad (10.25)$$

Since $A(u)$ grows more slowly than quadratically, it will be possible to normalize the density of B_T.

To be specific, for $x_0 = -\ln(v_0)/\beta$, it is necessary to be able to sample B_T from density

$$\exp(A(u) - (u - x_0)^2/(2T)) \propto \exp(-(u - g_1)^2/(2T) - g_2\exp(-\beta u)), \quad (10.26)$$

where $g_1 = x_0 + T(\beta/2 - r/\beta)$ and $g_2 = r/(\beta^2 K)$ are constants. This can be accomplished quickly by using AR from a normal with mean $g_1 + T\beta g_2 \exp(-\beta g_1)$ and variance T.

10.3.3 *Unbounded* ℓ

To deal with the case where ℓ is an unbounded function, it is necessary to find a value $M > 0$ such that the Brownian motion stays between M and $-M$ from $[0,T]$. In [10], Beskos et al. found such a method by generating a Brownian bridge whose extreme values were bounded.

An alternate method comes from [22]. Recall from Section 2.6.1 that for a Brownian motion $\{B_t\}$ with $T_1 = \inf\{t : |B_t| = 1\}$, it is possible to simulate from T_1 in finite expected time using AR.

Suppose that the goal is to find the solution to the SDE at time T. If $T_1 > T$, then $|B_t| \leq 1$ for all $t \in [0,T]$, and the value of ℓ can be bounded appropriately using this fact.

On the other hand, if $T_1 < T$, let $T_2 = \inf\{t > T_1 : |B_t - B_{T_1}| = 1\}$. Then for all $t \in [T_1, T_2]$, $B_{T_1} - 1 \leq B_t \leq B_{T_1} + 1$. In general, letting $T_{i+1} = \inf\{t > T_i : |B_t - B_{T_i}| = 1\}$ gives a sequence of passage times where the Brownian motion has moved at most one away from its previous value. By the symmetry of Brownian motion, $B_{T_{i+1}}$ is equally likely to be either $B_{T_i} + 1$ or $B_{T_i} - 1$. See Figure 10.1 for an illustration.

Recall that to generate the Brownian motion $\{X_t\}$ that is a candidate for a solution

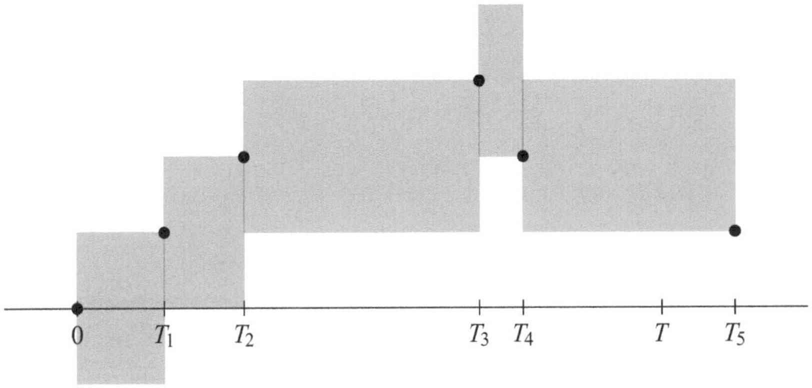

Figure 10.1 *Generating the passage times gives an upper and lower bound on the Brownian motion. The Brownian motion must pass through the circles and stay in the gray shaded area.*

to the ODE, first X_T is drawn, then a Brownian bridge that runs from x_0 at time 0 to X_T at time t can be constructed as

$$X_t = x_0 + (t/T)(X_T - x_0 + B_T - B_t). \tag{10.27}$$

Since $|B_t| \leq N = \sup\{i : T_i < T\}$, $|X_t| \leq x_0 + |X_T - x_0 + N|$, which gives a corresponding bound $M = \max_{u:|u| \leq x_0 + |X_t - x_0 + N|} \ell(u)$ on the ℓ function. What is needed at this point is a way to generate B_t for a finite set of times in $[T_{i-1}, T_i]$ given $B_{T_{i-1}}$ and B_{T_i}. From the Strong Markov Property of Brownian motion, this is equivalent to generating B_t for a finite set of times in $[0, T_1]$, and then just adding $T_{t_{i-1}}$.

Note that if B_t is Brownian motion conditioned to start at 0 and end at 1 at T_1, then $\tilde{B}_t = 1 - B_{T_1 - t}$ is a Brownian meander conditioned to start at 0 and end at 1 at T_1. The original B_t satisfies $|B_t| \leq 1$ when $\tilde{B}_t \leq 2$.

So to draw from the conditioned B_t, it suffices to draw from $\tilde{B}_t$, then accept only when $\tilde{B}_t \leq 2$. The following theorem gives a means for deciding when to accept or not, given $\tilde{B}_t$ at a finite number of time points in $[0, T_1]$.

Lemma 10.10 (Theorem 4.2 of [22]). *Suppose for times $t_1, \ldots, t_n$ in $[0, T_1]$, $\tilde{B}_{t_i} = y_i$. The chance that $(y_1, \ldots, y_n)$ should be accepted as a draw from $[\tilde{B}_t \,|\, \max \tilde{B}_t \leq 2]$ is*

$$\mathbb{P}(\max_{t \in [0, T_1]} \tilde{B}_t \leq 2)^{-1} \prod_{i=1}^{n} p(t_i, y_i; t_{i+1}, y_{i+1}) \cdot q(t_1, y_1) \cdot \mathbb{1}(y_i \in (0, 2)), \tag{10.28}$$

where

$$p(s,x;t,y) = \frac{1 - \sum_{j=1}^{\infty} (\theta_j(s,x;t,y) - \vartheta(s,x;t,y))}{1 - \exp(-2xy/(t-s))},$$

$$\theta_j(s,x;t,y) = \exp\left(-\frac{2(2j-x)(2j-y)}{t-s}\right) + \exp\left(-\frac{2(2(j-1)+x)(2(j-1)+y)}{t-s}\right),$$

$$
\begin{aligned}
\vartheta_j(s,x;t,y) &= \exp\left(-\frac{2j(4j+2(x-y))}{t-s}\right) + \exp\left(-\frac{2j(4j-2(x-y))}{t-s}\right), \\
q(s,x) &= 1 - \frac{1}{x}\sum_{j=1}^{\infty}(\rho_j(s,x) - \tilde{\rho}_j(s,x)) \\
\rho_j(s,x) &= (4j-x)\exp\left(-\frac{4j(2j-x)}{s}\right) \\
\tilde{\rho}_j(s,x) &= (4j+x)\exp\left(-\frac{4j(2j+x)}{s}\right).
\end{aligned}
$$

Once again the probability of accepting has been bounded by an alternating series, and so the AR method of Section 2.6 can be used to obtain draws.

Chapter 11

Applications and Limitations of Perfect Simulation

There are more things in heaven and earth, Horatio, than are dreamt of in your philosophy.

William Shakespeare

In the previous chapters, obtaining the perfect samples themselves was the goal. In this chapter, we explore applications of perfect simulation, situations where finding a perfect draw acts as a subroutine in a larger method or algorithm. In addition, it is shown how to run perfect simulation algorithms where the running time is concentrated around the expected running time, and the limitations of perfect simulation are discussed.

11.1 Doubly intractable distributions

Again consider the basic Bayesian framework for inference. The ingredients are a prior density f_Θ on a random variable Θ that represents the parameters of the model. Next suppose there is a density $f_{Y|\Theta}$ for the data Y given the value of the parameters. Then the posterior density on Θ is

$$f_{\Theta|Y}(\theta|Y=y) \propto f_\Theta(\theta)f(y|\Theta=\theta). \tag{11.1}$$

Note that notation is often collapsed in the Bayesian literature, and θ is used for both the dummy variable and the random variable. Here our notation distinguishes between the two in order to make notation precise. The capital letter theta, Θ, refers to the random variable, while the lower case theta, θ, refers to the dummy variable in the density.

In the statistics literature, the problem of calculating the normalizing constant for $f_{\Theta|Y}$ is often referred to as *intractable* when no practically fast algorithm exists to find it. The meaning of fast will depend on the context of the problem and the statistical model. This is in contrast to the use of intractable in computational complexity, where it usually means that finding the normalizing constant is known to be a Number P complete problem.

As noted back in Chapter 1, even though the normalizing constant is unknown, given a random walk that uses density $q_b(\cdot)$ to propose the next state starting from

b, a Metropolis-Hastings chain with stationary density f accepts the move to state θ' with probability $\min\{1, [f(\theta')q_b(\theta)]/[f(\theta)q_b(\theta')]\}$. When the target density is the posterior density, this makes the acceptance probability

$$\min\left\{1, \frac{f_\Theta(\theta')f(y|\Theta=\theta')}{f_\Theta(\theta)f(y|\Theta=\theta)}\right\}. \tag{11.2}$$

However, when $f(y|\theta)$ is not known, but itself is given by an unnormalized density, it is not even possible to evaluate this ratio, and which point not even a Metropolis-Hastings approximate sampler can be run!

For example, suppose that the data is the quality of soil samples in a rectangular field broken down into square plots. For simplicity, the quality of each plot is measured as either being "good" or "bad." Then a natural model is the Ising model, where 1 represents a good plot and -1 is a bad plot. The Ising model captures the propensity of good plots to be close to other good plots spatially. The parameter β for the Ising model measures the strength of the spatial attraction.

However, there is no closed-form solution for the normalizing constant of the Ising model for general graphs, and even for planar graphs, the $O(n^3)$ time to calculate the normalizing constant could be too slow.

In general, if $f(y|\Theta=\theta) = w(y|\Theta=\theta)/Z_\theta$ for a normalizing constant Z_θ, then the ratio to be computed is

$$r = \min\left\{1, \frac{q_{\theta'}(\theta)f_\Theta(\theta')w(y|\Theta=\theta')/Z_{\theta'}}{q_\theta(\theta')f_\Theta(\theta)w(y|\Theta=\theta)/Z_\theta}\right\}. \tag{11.3}$$

This situation where the normalizing constant of the posterior as well as that of the statistical model are both unknown is referred to as a *doubly intractable* problem in the literature [108]. Møller et al. [101] first came up with a way around this problem, and used it for Gibbs point processes in [6].

Add an auxiliary variable X to the chain, where $[X|\theta]$ follows the same distribution as $[Y|\theta]$. The state of the chain is now (θ, x), where θ is a parameter value, and x is a configuration.

Given current parameter θ and configuration x, propose a new parameter θ', and from the statistical model propose a new configuration x' given parameter θ'. So the density for the move from (θ,x) to (θ',x') is $q_\theta(\theta')w_{Y|\Theta}(x'|\Theta=\theta')/Z_{\theta'}$. That makes the acceptance ratio for moving from (θ,y) to (θ',y') the minimum of 1 and

$$\frac{q_{\theta'}(\theta)[w_{Y|\Theta}(x'|\Theta=\theta)/Z_\theta]f_\Theta(\theta')[w_{Y|\Theta}(y|\Theta=\theta')/Z_{\theta'}]}{q_\theta(\theta')[w_{Y|\Theta}(x|\Theta=\theta')/Z_{\theta'}]f_\Theta(\theta)[w_{Y|\Theta}(y|\Theta=\theta)/Z_\theta]}. \tag{11.4}$$

The $Z_{\theta'}$ and Z_θ factors cancel out of this expression, and so this ratio can be computed without needing to know the (difficult to compute) normalizing constant.

Because this adds a single auxiliary random variable draw from the statistical model, this method was later referred to as the Single Auxiliary Variable Method (SAVM) [108].

11.2 Approximating integrals and sums

One of the main uses of generating samples from unweighted distributions is to approximate the partition function for the distribution. In Chapter 9 the partition function for a particular set of weights on permutations is the permanent of a matrix, a known #P complete problem. The partition function of the Ising model is also known to be #P complete [70].

11.2.1 TPA

One of the primary uses of perfect samplers is in constructing unbiased estimators of integrals. In computer science, this is often accomplished in polynomial time through the use of *self-reducible* problems; see [72] for details.

A more advanced version of the self-reducible idea is the Tootsie Pop algorithm (TPA). This is a simple means for estimating an integral given the ability to sample from a particular family of distributions. The name is a reference to a popular commercial for a Tootsie Pop, which is a candy that consists of a soft chocolate center surrounded by a hard candy shell. The commercial asks the question "How many licks does it take to get to the center of a Tootsie Pop?"

Any integral can be written as the measure ν of a set B. This set B is the shell. The center will be a set A that is a subset of B where $\nu(A)$ is known. If one can perfectly sample from B, then the percentage of times the sample falls in A forms an estimate of $\nu(A)/\nu(B)$, which gives a means for approximating $\nu(B)$.

However, as has been seen time and again, this probability is typically exponentially small in the problem size for many applications. TPA offers a way around this problem. Following [57], TPA can be broken down into four ingredients.

1. A measure space $(\Omega, \mathscr{F}, \mu)$.
2. Two finite measurable sets $A \subseteq B$ with $\mu(A) > 0$. The set A is the *center* and B is the *shell*.
3. A family of sets B_β such that $\beta' < \beta$ implies that $B_{\beta'} \subseteq B_\beta$. These sets should not jump too much in the sense that $\nu(B_\beta)$ should be a continuous function of β. Also, $\lim_{\beta \to -\infty} \nu(B_\beta) = 0$.
4. Two special values of the parameter β_B and β_A such that $B_{\beta_B} = B$ and $B_{\beta_A} = A$.

Given these ingredients, the outline of the TPA protocol is as follows.

TPA *Output:* $N \sim \text{Pois}(\ln(\nu(B)/\nu(A)))$

1) $\beta \leftarrow \beta_B, N \leftarrow -1$
2) Repeat
3) Draw X from ν conditioned to lie in B_β, $N \leftarrow N+1$
4) $\beta \leftarrow \inf\{b : X \in B_b\}$
5) Until $X \in A$

Figure 11.1 illustrates a run of TPA when B_β is the ball of radius β in $\mathbb{R}^2$. The distribution of the output of a run of TPA turns out to have a Poisson distribution with mean $\ln(\nu(B)/\nu(A))$.

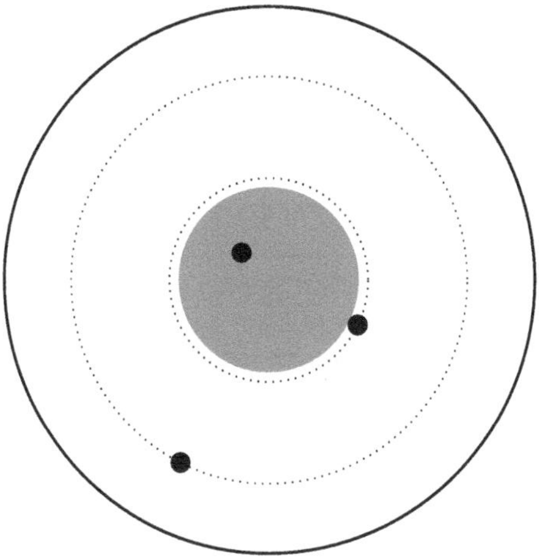

Figure 11.1 *A single run of* TPA *when B is the ball of radius 3, A is the ball of radius 1, and* B_β *is the ball of radius* β. *The number of samples needed before X lands in the center is 3, giving output* $N = 2$.

Note that when X is drawn in TPA, the next value of β cuts off anywhere from 0% to 100% of the region being drawn. A perhaps surprising fact is that this percentage is uniform over these two extremes.

Lemma 11.1. *Suppose* $X \sim B_\beta$ *and* $\beta' = \inf\{b : X \in B_b\}$. *Then* $\nu(B_{\beta'})/\nu(B_\beta) \sim \textsf{Unif}([0,1])$.

Proof. Let $U = \nu(B_{\beta'})/\nu(B_\beta)$. Fix $a \in [0,1)$ and consider $\mathbb{P}(U \leq a)$. Since $\nu(B_\beta)$ is a continuous function of β that goes to 0 as β goes to $-\infty$, there exists a $b \in (-\infty, \beta]$ such that $\nu(B_b)/\nu(B_\beta) = a$. The chance that $X \sim \textsf{Unif}(B_\beta)$ falls in B_b is $\nu(B_b)/\nu(B_\beta) = a$, and when this happens $U_i \leq a$. So $\mathbb{P}(U_i \leq a) \geq a$.

On the other hand, there exists N such that for any $n \geq N$, there is a value b_n such that $\nu(B_{b_n})/\nu(B_\beta) = a + 1/n$. Using the nested property of the family of sets, if $X \notin B_{b_n}$, then $\beta' \geq a + 1/n > a + 1/(n+1)$. Therefore, if $U_i \leq a$, then $X \in B_{b_n}$ for all $n \geq N+1$. But that means that $\mathbb{P}(U_i \leq a) \leq a + 1/n$ for all $n \geq N+1$, which means $\mathbb{P}(U_i \leq a) \leq a$, completing the proof. □

So by starting with measure $\nu(B)$, and after t steps of taking a draw X and generating a new value of β, $\nu(B_{\beta_i}) \sim \nu(B)U_1U_2U_3\cdots U_t$. Or in negative log space,

$$-\ln(\nu(B_{\beta_i}))) \sim (-\ln(\nu(B))) + (-\ln(U_1)) + (-\ln(U_2)) + \cdots + (-\ln(U_t)). \quad (11.5)$$

The reason for working with the negative logarithms of the uniforms is that these have an exponential distribution with rate 1.

Fact 11.1. *For* $U \sim \textsf{Unif}([0,1])$, $-\ln(U) \sim \textsf{Pois}(1)$.

Lemma 11.2. *The output of* TPA *is a Poisson distributed random variable, with mean* $\ln(\nu(B)/\nu(A))$.

Proof. From the last fact, the points $-\ln(U_1), -\ln(U_2) - \ln(U_3), \ldots$ form a Poisson process of rate 1. The number of such points that fall in the interval $[-\ln(\nu(B)), -\ln(\nu(A))]$ will be Poisson distributed with mean $-\ln(\nu(A)) - (-\ln(\nu(B))) = \ln(\nu(B)/\nu(A))$. □

Poisson random variables, like normals, have the nice property that when independent values are added, you simply add the individual parameters. So if TPA is run k times and the results added, the result N_k is a Poisson random variable with mean $k\ln(\nu(B)/\nu(A))$. For any choice of k by the simulator, it is possible to construct an exact confidence interval for the resulting estimate of $\nu(B) \approx \hat{a}$ where $\hat{a} = \nu(A)\exp(N_k/k)$.

So far we have assumed that $\nu(A)$ is easy to compute, but $\nu(B)$ is hard. Of course, TPA just estimates between the two, so it is just as useful when $\nu(B)$ is easy to compute, and the goal is to approximate $\nu(A)$.

11.2.2 TPA for Gibbs distributions

Suppose that a density can be written as $f(x) = \exp(-\beta H(x))/Z_\beta$ for a function $H(x)$. A density of this form is said to represent a *Gibbs distribution.* TPA can be applied to Gibbs distributions when $H(x) \leq 0$. (Actually, any upper or lower bound on $H(x)$ suffices, but usually $H(x) \leq 0$ in practice, so that is the case considered here.) For instance, the Ising model with no external magnetic field has $H(x) = -\sum_{\{i,j\}\in E} \mathbb{1}(x(i) = x(j)) \leq 0$. As usual, let the set of configurations be Ω.

Let $B_\beta = \{x, y : x \in \Omega : y \in [0, \exp(-\beta H(x))]\}$. The restriction that $H(x) \leq 0$ ensures that for $\beta' < \beta$, $B_{\beta'} \subseteq B_\beta$. Now $\nu(B_0) = \#\Omega$, which is usually easy to compute. For example, in the Ising model $\Omega = \{-1,1\}^V$, so $\nu(B_0) = 2^{\#V}$.

To draw (x,y) uniformly from B_β, first draw X from the distribution with parameter β, and then $Y|X$ can be drawn uniformly from $[0, \exp(-\beta H(X))]$. When $Y \leq 1$ then $\beta < 0$ and the center has been reached. When $Y > 1$, the next value for β is then the smallest value of b such that $Y \in [0, \exp(-bH(X))]$. So $b = -\ln(Y)/H(X)$. (When $H(X) = 0$ just set $b = -\infty$.)

`TPA_for_Gibbs` *Input:* β *Output:* $N \sim \text{Pois}(\ln(Z_B/\#\Omega))$
1) $N \leftarrow 0$
2) While $\beta > 0$
3) Draw X from the Gibbs distribution with parameter β
4) Draw Y uniformly from $[0, \exp(-\beta H(X))]$
5) If $Y \leq 1$ or $H(X) = 0$ then $\beta = 0$
6) Else $\beta \leftarrow -\ln(Y)/H(X)$ and $N \leftarrow N + 1$

11.3 Omnithermal approximation

For Gibbs distributions, it is not only possible to use TPA to generate estimates for a single Z_β in this fashion, but also to simultaneously give estimates of Z_β for all β in some interval $[\beta_A, \beta_B]$. Since β is also known as the inverse temperature, getting an estimate of Z_β for many values of the temperature simultaneously is known as an *omnithermal approximation.*

A single run of TPA generates a sequence of points that form a Poisson process of rate 1 in the interval $[-\ln(\nu(B)), -\ln(\nu(A))]$. A nice fact about Poisson processes is that the union of two PPP (called a *superposition*) forms a third PPP whose rate is the sum of the first two rates.

Theorem 11.1. *Let P_1 and P_2 be two Poisson point processes on Ω with rates λ_1 and λ_2 respectively. Then for processes such that for all $x \in \Omega$, $\mathbb{P}(x \in P) = 0$, $P_1 \cup P_2$ is a PPP on Ω of rate $\lambda_1 + \lambda_2$.*

See [105, p. 23] for a proof.

By running TPA k times from β_A to β_B, and taking the union of the resulting output, it is possible to create a set of points P that form a Poisson point process of rate k over $[-\ln(\nu(B)), -\ln(\nu(A))]$. Each of these points corresponds to a specific β value. Call these β values $\beta_1 > \beta_2 > \cdots > \beta_{\#P}$.

For $\beta \in [\beta_A, \beta_B]$, retaining only the points that have $\beta_i \geq \beta$ gives a Poisson point process of rate k over $[-\ln(\nu(B_\beta)), -\ln(\nu(A))]$. (See the discussion of thinning from Section 7.2 to see why.) Let $n(\beta) = \sup\{i : \beta_i \geq \beta\}$ count these points. Then

$$\hat{a}(\beta) = \nu(A)\exp\left(\frac{n(\beta)}{k}\right) \tag{11.6}$$

gives an estimate for $\nu(B_\beta)$ for every value of β.

It should be noted that TPA is not the only way to construct an omnithermal approximation. Bridge sampling [40] and nested sampling [117] also allow for omnithermal sampling. The difference is with TPA it is possible to explicitly bound the probability that the error is large for all values of β simultaneously.

Lemma 11.3. *Let $\varepsilon \in (0, 1/2)$ and $I = [-\ln(\nu(B)), -\ln(\nu(A))]$. Then,*

$$\mathbb{P}\left(\exp(-\varepsilon) \leq \sup_{\beta \in I} \frac{\hat{a}(\beta)}{\nu(B_\beta)} \leq \exp(\varepsilon)\right) \geq 1 - 2\exp\left(-\frac{k\varepsilon^2}{2[\ln(\nu(B)/\nu(A)) + \varepsilon]}\right). \tag{11.7}$$

For a proof see [57].

There are many applications for this type of estimate.

11.3.1 Maximum likelihood estimators for spatial processes

Suppose that the statistical model for data is that X is a draw from the Ising model for some β. To find the maximum likelihood estimator (MLE) for β over some range $[a, b]$, it is necessary to compute

$$\arg\max \frac{\exp(-\beta H(X))}{Z_\beta}. \tag{11.8}$$

One way to obtain a β where $\exp(-\beta(H(X)))/Z_\beta$ is close to the maximum value is to use an omnithermal approximation to obtain Z_β for all $\beta > 0$.

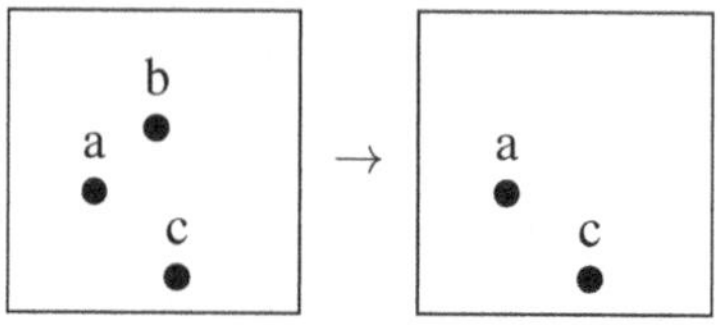

Figure 11.2 *Matern type III process where* $R = 0.35$, $dist(a,b) = 0.3$, $dist(a,c) = 0.4$, $dist(b,c) = 0.5$, $t_a = 0.63$, $t_b = 0.87$, $t_c = 0.22$. *Point c is born first, then a, but b is not because it is too close to a.*

11.3.2 Doubly intractable distributions with one spatial parameter

The doubly intractable distributions from Section 11.1 can also be dealt with using TPA and omnithermal sampling when the spatial component only has one varying parameter β. In this case the posterior becomes

$$f_{\beta|Y=y}(b) \propto f_\beta(b)\exp(-\beta H(y))/Z_\beta. \tag{11.9}$$

With an omnithermal approximation of Z_β in hand, it is possible to use any numerical integration method to find the normalizing constant for $f_{\beta|Y=y}$ approximately, as well as find the posterior mode and construct credible intervals.

11.3.3 Example: Matérn type III processes

In the examples looked at previously, the integral (or sum) was the normalizing constant that appeared in the denominator of the density. In this section an example is given of a model where the normalizing constant is known, and it is the weight in the numerator that has a difficult high-dimensional integral.

Point processes such as the Strauss model operate by giving a density with respect to the standard Poisson point process. Essentially, they are upweighting or downweighting point processes that have certain properties.

An alternative way of modeling point processes is constructive—give an algorithm for generating the point process. This can also be used to generate process with certain properties.

For instance, consider the Matérn type III process [88] illustrated in Figure 11.2 This model (in its simplest form) has two parameters, the intensity $\lambda \in (0,\infty)$ and a radius $R \in [0,\infty)$, together with a reference measure $\nu(\cdot)$ on S.

First, generate a Poisson point process P over the original space S using rate $\lambda \cdot \nu(\cdot)$. Next, generate for each point x_i a birth time $t_i \sim \mathsf{Unif}([0,1])$, where the t_i are all independent.

Using a city metaphor, the birth time represents the time that the city is founded. Once a city is founded, it immediately claims all the area within distance R of the center of the city. If later on a city attempts to be founded within distance R of the first city, it is not allowed and the later city is not added to the process.

This continues until all of the points have been either added or passed over. This

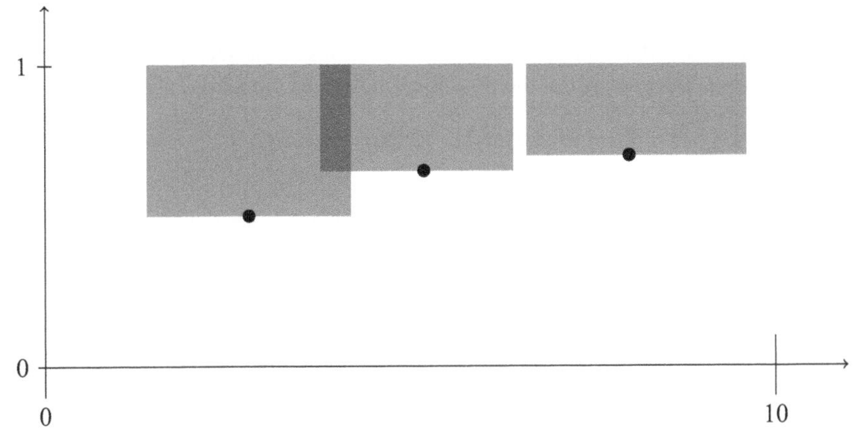

Figure 11.3 *Example of a shadow region for* $S = [0,10]$, $x_1 = 2.8$, $x_2 = 5.2$, $x_3 = 8.0$, $t_1 = 0.5$, $t_2 = 0.65$, $t_3 = 0.7$, $R = 1.4$. *If there are only hidden points in the shadowed region, they will not be seen at time* 1.

is an example of a repulsive point process, since points will be farther apart than they would be otherwise if close points were not being eliminated.

The data that is actually seen is the set of points at time $t = 1$; earlier points that were eliminated are not seen. Note that the set of all points (both seen and unseen) together with their times forms a Poisson point process on $S \times [0,1]$ with intensity equal to the product measure of $\lambda \cdot \nu(\cdot)$ and Lebesgue measure. See Figure 11.3 for an illustration.

Definition 11.1. *Consider a Poisson point process over* $S \times [0,1]$. *Then let*

$$D_R(x,t) = \{(x_i,t_i : (\exists x_j \in x)(dist(x_j,x_i) < R \text{ and } t_j < t_i\} \tag{11.10}$$

be the shadow *of the configuration* (x,t). *If* $(x_i,t_i) \in D_R(x,t)$, *call point* i hidden, *otherwise it is* seen. *Let*

$$A_{R,\lambda}(x,t) = \lambda \cdot (\nu \times m)(D_R(x,t)) \tag{11.11}$$

be the expected number of hidden points given the seen points in (x,t).

In general, the data consists of the seen point locations (without time stamps). The hidden points are completely unknown. It is possible to write down a formula for the density of the Matérn type III process with respect to a standard Poisson point process.

Lemma 11.4 (Theorem 2.1 of [67]). *The density function of the seen points (locations and times) created using intensity* $\lambda \cdot \nu(\cdot) \times m$ *and exclusion radius* R *with respect to a PPP with intensity measure* $\nu(\cdot) \times m$ *over* $S \times [0,1]$ *is*

$$f_{seen}(x_s,t_s) = \mathbb{1}\left(\left[\min_{\{p_1,p_2\} \subset x_s} dist(p_1,p_2)\right] > R\right) \lambda^{\#x_s} \exp(A_{R,\lambda}(x_s,t_s) - \nu(S)(\lambda - 1)). \tag{11.12}$$

See [67] for the proof.

Of course, the data only consists of the locations of the seen points x, and not the time stamps t. Integrating out the possible times gives the marginal density

$$f_{\text{seen}}(x|R,\lambda(\cdot)) = \mathbb{1}(\min_{i \neq j} \text{dist}(x_i,x_j) > R)\exp(-\lambda \cdot \nu(S)) \int_{t\in[0,1]^{\#x}} \exp(A_{R,\lambda}(x,t)). \tag{11.13}$$

This is also (given parameters R and λ) the likelihood of the data.

It is this last factor, the integral with dimension equal to the number of points in the data, that requires the use of Monte Carlo and perfect simulation to solve.

The first step is to build a monotonic update function where the distribution with unnormalized density $f_{\text{target}}(t) = \exp(A_{R,\lambda}(x,t))$ is stationary. Usually, Metropolis-Hastings Markov chains are not monotonic, but in this case they are. It operates as follows. At each step, choose a point $x_i \in x$ uniformly at random. Then draw u uniformly from $[0,1]$, let $t'_i = u$, and $t'_j = t_j$ for all $j \neq i$.

Then if $u \leq t_i$, then accept and change t to t'. Otherwise, accept t' as the new set of time stamps with probability $\exp(A_{R,\lambda}(x,t'))/\exp(A_{R,\lambda}(x,t)) = \exp(-(A_{R,\lambda}(x,t) - A_{R,\lambda}(x,t'))$.

Even in one dimension, calculating $A_{R,\lambda}(x,t) - A_{R,\lambda}(x,t')$ can be difficult, but in two or higher dimensions it is a very difficult task. Quantities such as the area of intersection of three or more circles can show up in the expression. Once again, the Bernoulli factory of Beskos et al. [11] can help.

Generate a Poisson point process in $D_R(x,t) - D_R(x,t')$ by generating a PPP of constant intensity λ over $B_R(x) \times [t',1]$ and thinning any points that appear in $D_R(x,t')$ or which have time stamp greater than t. This only involves finding the distance of each point in the PPP to each point in x, and so it can be done quickly.

The chance that the number of thinned points is 0 is exactly $\exp(-(A_{R,\lambda}(x,t) - A_{R,\lambda}(x,t')))$, and so it is possible to draw a Bernoulli random variable with this mean without actually calculating the mean explicitly.

Now suppose for any other set of times $s \leq t$, the same PPP P is used to determine if $s_i \leftarrow u$. If $u > t_i$, then u is accepted for both. If $s_i < u \leq t_i$, then $s_i = u$ and t_i might accept the move to u or not. Either way, after the step $s_i \leq t_i$.

If $u < s_i$, then if t_i accepts the move down to u, it is because the thinning process over $D_R(x,t) - D_R(x,t')$ had zero points. But $D_R(x,s) \subseteq D_R(x,t)$, so $D_R(x,s) - D_R(x,t')$ will also have zero points, and so s_i will also be moved to u. Hence the update function is monotonic.

In this pseudocode for this update, $i \in \{1,\ldots,\#x\}$, $u \in [0,1]$, and P is a subset of points in $B_R \times [0,1]$.

`Matérn_Type_III_update` *Input:* (x,t), i, u, P *Output:* (x,t)

1) Let $t'_i \leftarrow u$ and for all $j \neq i$, $t_j \leftarrow t_j$
2) Let $N \leftarrow \#\{(v_j,s_j) \in P : s_i \leq t_i \text{ and } (v_j,s_j) \notin D_R(x,t')\}$
3) If $N = 0$ then $t \leftarrow t'$

The maximum state is all 1's, and the minimum is all 0's. When used with `Monotonic_coupling_from_the_past`, this returns samples from the unnormalized density $f_{\text{target}}(t|\ell) = \exp(A_{R,\ell}(x,t))$.

So now let us see how the ability to TPA can be employed to approximate $Z_\ell = \int_t f_{\text{target}}(t)\,dt$. First note that $Z_0 = 1$, so the value of the normalizing constant is known for at least one value of the parameter. Since $A_{R,\ell}(x,t)$ is an increasing function of ℓ, so is Z_ℓ. So to use TPA, just use a single auxiliary variable as with the Gibbs distributions, $y \in [0, \exp(A_{R,\lambda}(x,t))]$. Then Z_λ is just the Lebesgue measure of $B_\ell = \{(t,y) : t \in [0,1]^{\#x}, y \in [0, \exp(A_{R,\ell}(x,t))]\}$.

Note this is one of the cases mentioned earlier when we know the measure of $A = B_0$, but are looking for the measure of $B = B_\lambda$.

As mentioned earlier, omnithermal sampling can be used to approximate $f_{\text{seen}}(x|R,\lambda)$ over a range of λ values. This can then be used to find the MLE for the λ value.

The description of Matérn Type III processes given here was over a set S of finite size. As with other spatial point processes, both the model and the perfect simulation procedure can be extended to a finite window looking onto an infinite space. See [98] for details.

11.4 Running time of TPA

Let $r = \nu(B)/\nu(A)$. Then from Lemma 11.3, to get an estimate $\hat{r}$ of r such that $e^{-\varepsilon} < \hat{r}/r < e^{\varepsilon}$ with probability at least $1 - \delta$ can be accomplished by setting $k = 2\varepsilon^{-2}[\ln(r) + \varepsilon]\ln(2\delta^{-1})$.

At first glance, that might appear strange, since this says that to estimate r to within a factor of e^{ε}, it is necessary to set k to a value that depends on r, the very thing we are trying to estimate!

To solve this dilemma, run two phases of TPA. The first phase is quick, and gives an estimate $\hat{r}_1$ such that $|\ln(\hat{r}_1) - \ln(r)| \leq (1/2)(1 + \sqrt{1 + 4\ln(r)}$ with probability at least $\delta/2$. The second phase then draws the estimate $\hat{r}$ using k derived from $\hat{r}_1$.

Two_phase_TPA *Input:* ε, δ *Output:* $\hat{r}$

---

1) $N \leftarrow 0,\ k_1 \leftarrow \lceil 2\ln(4\delta^{-1})\rceil$
2) Repeat k_1 times
3) $\quad N \leftarrow N + \texttt{TPA}$
4) $\hat{r}_1 \leftarrow N/k_1,\ k_2 \leftarrow \lceil 2\varepsilon^{-2}\ln(4\delta^{-1})(\ln(\hat{r}_1) + \sqrt{\ln(\hat{r}_1)+1} + 1 + \varepsilon)\rceil,\ N \leftarrow 0$
5) Repeat k_2 times
6) $\quad N \leftarrow N + \texttt{TPA}$
7) $\hat{r} \leftarrow N/k_2$

Lemma 11.5. *The output of* Two_phase_TPA *satisfies*

$$\mathbb{P}(e^{-\varepsilon} \leq \hat{r}/r \leq e^{\varepsilon}) > 1 - \delta. \tag{11.14}$$

The expected number of samples drawn is bounded above by

$$(\ln(r)+1)\left[\lceil 2\ln(4\delta)^{-1}\rceil + 2\varepsilon^{-2}\ln(4\delta^{-1})(\ln(r) + \sqrt{\ln(r)+1} + 1 + \varepsilon) + 1\right]. \tag{11.15}$$

Proof. In the first phase, $\varepsilon_1 = (1/2)(1+\sqrt{1+4\ln(r)})$. This was chosen so that $\varepsilon_1^2 = \ln(r)+\varepsilon_1$. So putting that into Lemma 11.3 gives

$$\mathbb{P}(e^{-\varepsilon_1} \leq \hat{r}_1/r \leq e^{\varepsilon_1}) > 1-2\exp\left(-\frac{2\ln(4\delta^{-1})\varepsilon_1^2}{\ln(r)+\varepsilon_1}\right) = 1-\frac{\delta}{2}. \tag{11.16}$$

So that means that with probability at least $1-\delta/2, \hat{r}_1 \geq \ln(r)-(1/2)\sqrt{1+4\ln(r)}$, or equivalently $\ln(r) \leq \ln(\hat{r}_1)+\sqrt{\ln(\hat{r}_1)+1}+1$.

Using Lemma 11.3 once more with k_2 then completes the proof of (11.14).

From Lemma 11.2, each run of TPA uses on average $\ln(r)+1$ samples. The value of k_1 is fixed; for k_2, note that $\mathbb{E}[\hat{r}_1] = r$, and $-[\ln(x)+\sqrt{\ln(x)+1}+1+\varepsilon]$ is a convex function, so $\mathbb{E}[-[\ln(\hat{r})+\sqrt{\ln(\hat{r})+1}+1+\varepsilon]] \leq -[\ln(r)+\sqrt{\ln(r)+1}+1+\varepsilon]$. Using $\lceil x \rceil \leq x+1$ completes the proof. □

11.5 The paired product estimator

In the previous section, the running time of TPA was shown to be quadratic in $\ln(r)$. Since r is usually exponentially large in the problem size, this tends to make TPA require a number of samples that is quadratic in the problem size.

The examples where TPA was used were all Gibbs distributions. In this case, TPA coupled with importance sampling can be a way to get more accurate estimates.

First return to TPA. Suppose it is run k times over $b \in [0,\beta]$, and all the b values from the runs are collected. Then if $r_i = \ln(b_i)$, and $r_{(1)} < r_{(2)} < \cdots r_{(N)}$ are the order statistics of those points, these form a Poisson point process of rate k over $[-Z_{B_\beta}, -Z_0]$.

So if every kth point is considered, $r_{(k)}, r_{(2k)}, \ldots$, these will (on average) be about distance 1 apart. Let $a_i = Z_{b_{ik}}$. Then a_{i+1}/a_i will be near $\exp(1)$. Let $s(i) = e^{r_{(i)}}$ be the parameter value associated with this point.

In TPA, the estimate used is $a_i/a_{i+1} = e^1$. But importance sampling can be used to get a tighter estimate. The idea is as follows. Draw a sample $X_1, \ldots, X_n$ using parameter value $s(i)$. Then let $W_j = \exp(H(X_j)(s(i)-s(i+1)))$.

$$\mathbb{E}[W_j] = \int_x \frac{\exp(H(x)(s(i)-s(i+1)))\exp(-H(x)s(i)}{Z_{s(i)}} = \frac{Z_{s(i+1)}}{Z_{s(i)}}. \tag{11.17}$$

For each level i then draw $W_1, W_2, \ldots, W_\ell$ iid from the distribution using parameter value $s(i)$. Then average them to get

$$\hat{\mu}_i \approx Z_{s(i+1)}/Z_{s(i)}. \tag{11.18}$$

If there are ℓ such parameters, let $s(\ell+1) = \beta$. This gives

$$\prod_{i=1}^{\ell+1} \hat{\mu}_i \approx \frac{Z_{s(1)}}{Z_{s(0)}} \cdots \frac{Z_{s(\ell)}}{Z_{s(\ell-1)}} \cdot \frac{Z_{s(\ell+1)}}{Z_{s(\ell)}} = \frac{Z_\beta}{Z_0}. \tag{11.19}$$

This type of estimate was introduced in [125] and later in [72]. This was called a *product estimator* by Fishman [38]. To understand the variance of such an estimator, it helps to have the notation of relative variance.

Definition 11.2. *A random variable X with finite second moment has* relative variance

$$\mathbb{V}_{\text{rel}}(X) = \frac{\mathbb{V}(X)}{\mathbb{E}[X]^2} = \frac{\mathbb{E}[X^2]}{\mathbb{E}[X]^2} - 1. \tag{11.20}$$

Then the mean and relative variance of a product estimator are as follows.

Lemma 11.6. *For $P = \prod P_i$ where the P_i are independent random variables with finite second moment,*

$$\mathbb{E}[P] = \prod \mathbb{E}[P_i], \quad \mathbb{V}_{\text{rel}}(P) = -1 + \prod (1 + \mathbb{V}_{\text{rel}}(P_i)). \tag{11.21}$$

Proof. Since the P_i are independent, $\mathbb{E}[P] = \prod \mathbb{E}[P_i]$ and $\mathbb{E}[P^2] = \prod \mathbb{E}[P_i]^2$. Hence $\mathbb{E}[P^2]/\mathbb{E}[P]^2 = \prod \mathbb{E}[P_i^2]/\mathbb{E}[P_i]^2$ and the rest just follows from the definition of relative variance. □

For the importance sampler given earlier,

$$\mathbb{E}[W_j^2] = \int_x \frac{\exp(H(x)(s(i) - s(i+1)))^2 \exp(-H(x)s(i))}{Z_{s(i)}} = \frac{Z_{2s(i+1)-s(i)}}{Z_{s(i)}}. \tag{11.22}$$

which gives

$$\mathbb{V}_{\text{rel}}(W_j) = -1 + \frac{Z_{2s(i+1)-s(i)} Z_{s(i)}}{Z_{s(i+1)^2}} \tag{11.23}$$

Unfortunately, this expression is rather unwieldy to work with, since $2s(i+1) - s(i) \notin [s(i), s(i+1)]$, and so involves values of the partition function at parameter values outside of the interval of interest.

This problem can be solved by the paired product estimator. Let $X_1, \ldots, X_n$ be iid draws from the parameter value $s(i)$, and $Y_1, \ldots, Y_n$ be iid draws from the parameter value $s(i+1)$.

To simplify the notation, let $m_i = (s(i+1) + s(i))/2$ be the midpoint of the parameter values, and $h_i = (s(i+1) - s(i))/2$ be the half length of an interval. Then set

$$R_j = \exp(-h_i H(X_j)), \quad S_j = \exp(h_i H(Y_j)). \tag{11.24}$$

A similar calculation to (11.17) gives

$$\mathbb{E}[R_j] = \frac{Z_{m_i}}{Z_{s(i)}}, \quad \mathbb{E}[S_j] = \frac{Z_{m_i}}{Z_{s(i+1)}} \tag{11.25}$$

and

$$\mathbb{V}_{\text{rel}}[R_j] = \mathbb{V}_{\text{rel}}[S_j] = -1 + \frac{Z_{s(i+1)} Z_{s(i)}}{Z_{m_i}^2}. \tag{11.26}$$

For level i, let $\hat{\mu}_i = (R_1 + \cdots + R_n)/n$ and $\hat{\kappa}_i = (S_1 + \cdots + S_n)/n$. Then let $R = \prod_i \hat{\mu}_i$ and $S = \prod_i \hat{\kappa}_i$. Then $\mathbb{E}[R]/\mathbb{E}[S] = Z_\beta / Z_0$, and if n is large both R and S will be close to their means. Therefore, the final estimate is R/S.

The point of this more complicated procedure is that when the schedule is well-balanced, the relative variance of the estimate R and S will be close to 0.

Theorem 11.2. *Let X be a draw from the distribution with parameter value β, and $\eta \leq 4e\varepsilon^{-2}/(2+\ln(2\mathbb{E}[H(X)]))$. Then if for all i, $Z_{s(i+1)}/Z_{s(i)} \leq e^{\eta}$, then*

$$\mathbb{V}_{\text{rel}}(R) = \mathbb{V}_{\text{rel}}(S) \leq \varepsilon. \tag{11.27}$$

The proof is relatively complex, see [58] for details.

For many problems, it is possible to obtain an upper bound on $\mathbb{E}[X]$ based on the problem input, which can then be used to get a usable bound on η. For instance, in the Ising model, $\mathbb{E}[X]$ is upper bounded by the number of nodes of the graph.

11.6 Concentration of the running time

Many of the results shown in this text have been of the form of bounds on the expected running time of the procedure. While it is also usually possible to bound the standard deviation of the running time using similar procedures, such arguments tend to be far more complex than the analysis of the expected value.

Therefore, it is useful to understand how well the running time concentrates around the mean for various forms of perfect simulation. In general, the goal is to obtain an exponential tail on the running time.

Definition 11.3. *Let X be a nonnegative random variable with finite mean where there exist nonnegative constants α_1 and α_2 such that for all sufficiently large k, $\mathbb{P}(X \geq k\mathbb{E}[X]) \leq \alpha_1 k^{-\alpha_2}$. Then say that X has a* polynomial tail. *Suppose there exist positive constants α_1 and α_2 such that for sufficiently large k, $\mathbb{P}(X \geq k\mathbb{E}[X]) \leq \alpha_1 \exp(-\alpha_2 k)$. Then say that X has an* exponential tail.

In general it is not possible to guarantee an exponential tail on the running time of a particular perfect simulation procedure. That being said, for interruptible algorithms, or for CFTP, it is possible to guarantee a tail that goes down faster than any polynomial in k. In order to describe these algorithms, it is helpful to have the notion of quantile.

Definition 11.4. *For a random variable T, an α-*quantile *of T is any number a such that $\mathbb{P}(T \leq a) \geq \alpha$ and $\mathbb{P}(T \geq a) \leq 1-\alpha$. Denote the smallest such α-quantile by $q_\alpha(T)$.*

In this section, the quantile $q_{3/4}(T)$ will be the one considered, so $\mathbb{P}(T \geq q_{3/4}(T)) \leq 1/4$. For random variables that are finite with probability 1, $q_{3/4}(T)$ is finite whether or not T has a finite mean. In the situation that T does have a finite mean $\mathbb{E}[T]$, then by Markov's inequality $q_{3/4}(T) \leq 4\mathbb{E}[T]$.

Theorem 11.3. *Consider an interruptible perfect simulation algorithm $\mathscr{A}$. Let $T_{\mathscr{A}}$ denote the random number of steps taken by $\mathscr{A}$. Then there exists an interruptible algorithm $\mathscr{B}$ that takes $T_{\mathscr{B}}$ steps where $\mathbb{E}[T_{\mathscr{B}}] \leq 6q_{3/4}(T)$, and for $k \geq 1$, $\mathbb{P}(T_{\mathscr{B}} \geq k\mathbb{E}[T_{\mathscr{B}}]) \leq \exp(-k)$.*

To prove the theorem, it is necessary to build an algorithm with the desired concentration properties. Before getting to that algorithm, first a subroutine is needed that takes input t, and then runs algorithm $\mathscr{A}$ forward as most t steps. If algorithm $\mathscr{A}$ returns an answer X, then `Bounded_time_A` returns X as well. Otherwise it returns the symbol $\perp$, indicating that it failed to complete in time.

`Bounded_time_A` *Input: t*

1) Run $\mathscr{A}$ for at most t steps
2) If the algorithm had finished, output the result of the run X
3) Else output $\perp$

Note that because $\mathscr{A}$ is interruptible, the output of `Bounded_time_A` conditioned on $X \neq \perp$ (or $T \leq t$) is the same as the target distribution. The same fact does not hold if the algorithm is noninterruptible!

Now algorithm `Doubled_Concentrated_A` is run as follows.

`Doubled_Concentrated_A` *Input: t* *Output: Y*

1) $t \leftarrow 1$
2) Repeat
3) Let Y be the output of `Fixed_time_A`(t)
4) $t \leftarrow 2t$
5) Until $Y \neq \perp$

This algorithm has a polynomial tail on the running time.

Lemma 11.7. `Doubled_Concentrated_A` *returns output Y from the target distribution. Let $T_{\mathscr{A}}$ and $T_{\mathscr{B}}$ be the running times of $\mathscr{A}$ and* `Doubled_Concentrated_A`. *Then $\mathbb{E}[T_{\mathscr{B}}] \leq 4q_{3/4}(T_{\mathscr{A}})$. Moreover, for $k \geq 2$ it holds that $\mathbb{P}(T_{\mathscr{B}} \geq kq_{3/4}(T_{\mathscr{A}})) \leq 4/k^2$. If $\mathbb{E}[T_{\mathscr{B}}] < \infty$, then $\mathbb{P}(T_{\mathscr{B}} \geq k\mathbb{E}[T_{\mathscr{B}}] \leq (1/2)\exp(-[2\ln(2)]^{-1}\ln(k)^2 + (3/2)\ln(k))$ and $\mathbb{E}[T_{\mathscr{B}}] \leq 6.52\mathbb{E}[T_{\mathscr{A}}]$.*

Proof. Let algorithm $\mathscr{B}$ be `Doubled_Concentrated_A`, and as before, let $T_{\mathscr{A}}$ denote the random number of steps taken by algorithm $\mathscr{A}$. From our AR theory in Chapter 2, the output of $\mathscr{B}$ is a draw from the output of `Bounded_time_A` for some input t, call it Y, which is conditioned on $Y \neq \perp$. No matter what the value of t was, this is a draw from the target distribution conditioned on the number of steps $T_{\mathscr{A}} \leq t$. Since the original algorithm $\mathscr{A}$ was interruptible, this conditional distribution of X is the same as the unconditional distribution of X no matter what the final t used was.

Consider the ith time through the repeat loop. The number of steps of $\mathscr{A}$ taken during this loop is 2^i. Now consider $\mathbb{P}(T_{\mathscr{B}} > s)$. In s steps, there must be a block of size at least $s/3$. The next smaller block must be of size at least $s/4$. After that the block size decreases by a factor of 2, giving $s/8$, $s/16$, and so on. Now suppose $s = kq_{3/4}(T_{\mathscr{A}})$ for integer d.

Then the number of blocks of size at least $q_{3/4}(T_{\mathscr{A}})$ is at least 2 when $k \in [4,8)$, at least 3 when $k \in [8,16)$, and so on.

The chance that n of these blocks of size $q_{3/4}(T_{\mathscr{A}})$ all return $\perp$ is at most

$$(1/4)^n \leq 4^{-(\log_2(k)+1)} = 4/k^2.$$

The value of $\mathbb{E}[T_{\mathscr{B}}]$ can then be upper bounded using the tail sum formula:

$$\begin{aligned}
\mathbb{E}[T_{\mathscr{B}}] &= \int_{s=0}^{\infty} \mathbb{P}(T > s)\, ds \\
&= 2q_{3/4}(T_{\mathscr{A}}) + \int_{s=2q_{3/4}(T_{\mathscr{A}})}^{\infty} \mathbb{P}(T > s)\, ds \\
&= 2q_{3/4}(T_{\mathscr{A}}) + q_{3/4}(T_{\mathscr{A}}) \int_{k=2}^{\infty} \mathbb{P}(T > kq_{3/4}(T_{\mathscr{A}}))\, dk \\
&= 2q_{3/4}(T_{\mathscr{A}}) + q_{3/4}(T_{\mathscr{A}}) \int_{k=2}^{\infty} 4/k^2\, dk \\
&= 4q_{3/4}(T_{\mathscr{A}}).
\end{aligned}$$

When the mean of $T_{\mathscr{A}}$ is known to be finite, a better result is possible using Markov's inequality. Consider once again $\mathbb{P}(T_{\mathscr{B}} > s)$. The largest block in these s steps is of length at least $s/3$, the next largest at least $s/4$, and then it goes down by a factor of 2.

For a block of length $r\mathbb{E}[T_{\mathscr{A}}]$, Markov's inequality says that the chance the output is $\perp$ is at most $1/r$. Suppose $k = 2^d$ (if $k \in (2^d, 2^{d+1})$, rounding k down to the next power of two can only decrease the chance of failure). The chance of failure for the largest blocks when $s = 2^d\mathbb{E}[T_{\mathscr{A}}]$ is at most $3/2^d$, followed by $4/2^d$, $8/2^d$, and so on up to 1. Now

$$\frac{3}{2^d} \prod_{i=2}^{d} \frac{2^i}{2^d} = 2^{-(1/2)d^2+(3/2)d-1} = (3/k)k^{-\log_2(k)/2} \cdot k^{3/2} \cdot (1/2),$$

and simplifying gives the result.

A tail sum analysis then gives $\mathbb{E}[T_{\mathscr{B}}] \leq 6.52\mathbb{E}[T_{\mathscr{A}}]$. □

11.6.1 Subpolynomial tails on the runtime

This accomplishes our goal of a tail that is lighter than any polynomial when $\mathbb{E}[T_{\mathscr{A}}]$ is finite, but only gives a polynomial tail when $\mathbb{E}[T_{\mathscr{A}}]$ is unknown.

To do better, suppose that the time allotted to run is of the form 3^d. Use this time as follows: in the first 3^{d-1} steps, call the algorithm recursively. In the the second 3^{d-1} steps, call the algorithm recursively. In the final 3^{d-1} steps, use `Bounded_time_A`. This gives the following algorithm.

`Tripled_Concentrated_A` *Input: d* *Output: Y*
1) If $d = 0$ then
2) $\quad Y \leftarrow$ `Bounded_time_A`(1)
3) Else
4) $\quad Y \leftarrow$ `Tripled_Concentrated_A`$(d-1)$
5) $\quad$ If $Y = \perp$ then
6) $\quad\quad Y \leftarrow$ `Tripled_Concentrated_A`$(d-1)$
7) $\quad\quad$ If $Y = \perp$ then
8) $\quad\quad\quad Y \leftarrow$ `Bounded_time_A`(3^{d-1})

Lemma 11.8. *Let Y denote the output of* `Tripled_Concentrated_A`(d). *When $Y \neq \perp$, Y is a draw from the target distribution. Moreover,*

$$\mathbb{P}(Y = \perp) \leq \left(\frac{1}{4}\right)^{\lceil 2^{d-2}/q_{3/4}(T_{\mathscr{A}})^{\ln(2)/\ln(3)} \rceil - 1}. \tag{11.28}$$

Proof. The fact that Y conditioned on $Y \neq \perp$ comes from the target distribution follows from induction on d. When $d = 0$, Y is the output from `Bounded_time_A`(1), and so $[Y | Y \neq \perp]$ is just $[Y | T_{\mathscr{A}} = 1]$, which from the interruptibility of A gives the target distribution.

For $d > 1$, assume that it holds for $d - 1$. Then the output Y is either one of the draws from `Tripled_Concentrated_A`$(d-1)$ or a draw from `Bounded_time_A`(3^{d-1}). Either way, the result (conditioned on $Y \neq \perp$) comes from the target distribution.

Let $Y(d)$ denote the output of the algorithm with input d. then the recursive structure of the algorithm gives

$$\begin{aligned} \mathbb{P}(Y(d) = \perp) &= \mathbb{P}(Y(d-1) = \perp)^2 \mathbb{P}(T_{\mathscr{A}} > 3^{d-1}) \\ &\leq \mathbb{P}(Y(d-1) = \perp)^2 (1/4)^{\mathbb{1}(q_{3/4}(T_{\mathscr{A}}) \leq 3^{d-1})}. \end{aligned}$$

So to get an upper bound on this probability, define $p(d)$ recursively as

$$p(d) = p(d-1)^2 (1/4)^{\mathbb{1}(q_{3/4}(T_{\mathscr{A}}) \leq 3^{d-1})} \tag{11.29}$$

and $p(0) = 1$.

Then a simple induction gives $\mathbb{P}(Y(d) = \perp) \leq p(d)$ for all d. Note $p(d) = (1/4)^{r(d)}$ for some nonnegative integer $r(d)$. The recursion equation for $p(d)$ then gives $r(d) = 2r(d-1) + \mathbb{1}(q_{3/4}(T_{\mathscr{A}}) \leq 3^{d-1})$, $r(0) = 0$.

Therefore,

$$r(d) = 2^n + 2^{n-1} + \cdots + 1 = 2^{n+1} - 1, \tag{11.30}$$

where n is the number of powers of three between 3^{d-1} and $q_{3/4}(T_{\mathscr{A}})$. This is at most $d - 2 - \log_3(q_{3/4}(T_{\mathscr{A}})) - 1$.

Since $2^{\log_3(x)} = x^{\ln(2)/\ln(3)}$, this gives $r(d) \leq \lceil 2^{d-2}/q_{3/4}(T_{\mathscr{A}})^{\ln(2)/\ln(3)} \rceil - 1$. □

To actually be useful as an algorithm, it is necessary not to have to set d ahead of time, but to run the chain forward the proper number of steps. Fortunately, it is easy to develop the vector of times to give `Bounded_time_A`. In what follows, `length`(a) denotes the number of elements of the vector a, $[a\ b\ c]$ denotes the concatenation of vectors a, b, and c. (For example if $a = (3,4)$, $b = (1)$, and $c = (0,5)$, then $[a\ b\ c] = (3,4,1,0,5)$.)

Then for $d = 0$, the first time to run `Bounded_time_A` is 1, so start with $a = [1]$. When $d = 1$, first the 1 time is used, then 1, then 3^{d-1}. So now $a = [1\ 1\ 1]$. When $d = 2$, first the $[1\ 1\ 1]$ times are used, then $[1\ 1\ 1]$ again, and finally `Bounded_time_A` is run for $3^{d-1} = 3$ steps. So $a = [1\ 1\ 1\ 1\ 1\ 1\ 3]$. At $d = 3$, $a = [1\ 1\ 1\ 1\ 1\ 1\ 3\ 1\ 1\ 1\ 1\ 1\ 1\ 3\ 9]$, and so on. The vector a can be constructed in this fashion to have as many elements as needed.

Forward_Tripled_Concentrated_A *Input: d* *Output: Y*

```
i ← 0, d ← 0, a ← [1], Y ← ⊥
While Y = ⊥
   i ← i + 1
   If length(a) < i
      a ← [a a 3^d]
      d ← d + 1
   Y ← Bounded_time_A(a(i))
```

Lemma 11.9. *The output Y of* Forward_Tripled_Concentrated_A *comes from the target distribution. Let T be the total number of steps taken by the interruptible algorithm. Then for all $k \geq 1$,*

$$\mathbb{P}(T > kq_{3/4}(T_{\mathscr{A}})) \leq 4\exp(-(1/4)\ln(4)k^{\ln(2)/\ln(3)}). \quad (11.31)$$

Proof. The output is always a draw from Bounded_time_A, and so even conditioned on the input, if $Y \neq \perp$, then Y is a draw from the target distribution by the interruptibility of A.

Let d be the largest value of d such that $kq_{3/4}(T_{\mathscr{A}}) \geq 3^d$. Then from the last lemma the running time is at most

$$\begin{aligned}
\mathbb{P}(T > kq_{3/4}(\mathscr{A})) &\leq \mathbb{P}(T > 3^d) \\
&\leq \exp(-\ln(4)([2^{d-2}/q_{3/4}(T_{\mathscr{A}})^{\ln(2)/\ln(3)}] - 1)) \\
&= 4\exp(-(1/4)\ln(4)[2^d/q_{3/4}(T_{\mathscr{A}})^{\ln(2)/\ln(3)}]) \\
&= 4\exp(-(1/4)\ln(4)[(3^d/q_{3/4}(T_{\mathscr{A}}))^{\ln(2)/\ln(3)}]).
\end{aligned}$$

Since $3^d > kq_{3/4}(T_{\mathscr{A}})/3$, this gives

$$\mathbb{P}(T > kq_{3/4}(T_{\mathscr{A}})) < 4\exp(-(1/4)\ln(4)3^{\ln(2)/\ln(3)}k^{\ln(2)/\ln(3)}),$$

which completes the proof.

□

11.6.2 *Concentration for CFTP*

While the algorithms above were written for interruptible algorithms, the same procedure can be applied to block lengths for CFTP. If the block length is generated be taking the vector a of length 3^d and replacing it with $[a\ a\ 3^d]$ to get the next set of block lengths, nearly the same result as Lemma 11.9 holds. The only difference is that CFTP (being read twice) requires twice as many steps as a forward run algorithm. The subpolynomial tail, however, is exactly the same.

11.7 Relationship between sampling and approximation

What methods such as TPA and PPE show is that given the ability to draw samples exactly from a distribution in polynomial time, then it is usually possible to approximate the partition function of the distribution to any desired degree of accuracy in polynomial time.

When it is not possible to perfectly sample, but only approximate sampling is available, then it is still possible to use TPA and PPE, although the fact that samples are not coming exactly from the correct distribution has the be taken into account.

Now consider the opposite problem of sampling given the ability to compute the partition function. For instance, for the Ising model on planar graphs, it is possible to compute Z in $O(n^3)$ time using the Pfaffian approach [78, 79, 80]. Consider a node v. If the partition function when $x(v) = 1$ (call it Z_1) and when $x(v) = -1$ (call it Z_2) are computed, then $\mathbb{P}(X(v) = 1) = Z_1/[Z_1 + Z_2]$. So it is possible to draw $x(v)$ exactly from its marginal distribution.

Conditioned on the draw for $x(v)$, the value of $x(w)$ could then be drawn for some $w \neq v$. Continue until all the labels have been decided. The point is that given a polynomial time method for computing the partition function, there is a polynomial time method for generating perfect samples.

An approximate method for computing the partition function will only yield an approximate sampling method. The result is that it is possible to approximate the partition function in polynomial time for these types of problems if and only if it is possible to approximately sample from the associated distribution.

If it is possible to perfectly sample from the distribution in polynomial time it is possible to approximate the partition function in polynomial time. However, given the ability to approximate the partition function, it is not necessarily possible to perfect simulate from the distribution.

11.8 Limits on perfect simulation

Perfect simulation algorithms have been found for a wide variety of distributions, but not all. Moreover, as more and more algorithms have been discovered, a pattern has emerged. Distributions where the label of a node only depends on immediate neighbors, and where there is a chance of being able to ignore the neighbors, are the most easily handled by perfect simulation protocols. Such high noise models have long been used to get approximate sampling methods using Markov chain Monte Carlo. Today these same models can be sampled from exactly.

Statistical models in particular tend to fall into this category, as they often do not wish to restrict the outcome too severely, instead giving the data a chance to show where the model is incomplete or incorrect.

However, in statistical physics, a different story emerges. Here models (such as the Ising model) exhibit *phase transitions*, where the character of the distribution changes sharply at a particular value of the parameters for the model. It is not surprising, then, that the perfect simulation algorithms also display these types of phase transitions, provably running in polynomial time only for restricted values of the parameters.

Recent work such as that of Sly and Sun [118] has made this comparison with the phase transition explicit. Consider the hard-core gas model from Definition 1.20.

Recall that in this model on a finite graphs, the weight of a configuration x in $\{0,1\}^V$ is $\lambda^{\sum_{v \in V} x(v)}$ if no two nodes labeled 1 are adjacent to each other, and zero otherwise.

In an infinite graph, it is possible to define the Gibbs measure for the hard-core gas model by specifying the distribution at a single node conditioned on the neighbors.

Definition 11.5. *The* hard-core Gibbs measure *has a specification given by*

$$\mathbb{P}(X(v) = 1|X(V \setminus v)) = \frac{\lambda}{1+\lambda} \cdot \prod_{w:\{v,w\} \in E} (1 - x(w))$$

$$\mathbb{P}(X(v) = 0|X(V \setminus v)) = 1 - \mathbb{P}(X(v) = 1|X(V \setminus v)).$$

Definition 11.6. *An infinite graph $G = (V,E)$ is a d-*regular infinite tree *if there is a unique path connecting any two distinct nodes of the tree, and the degree of each node is d.*

For d-regular infinite trees, the phase transition for the hard-core Gibbs measure is known. In this case, the phase transition is a value of the parameter λ_c such that for $\lambda < \lambda_c$, there exists a unique measure that satisfies the local properties of Definition 11.5, and for $\lambda > \lambda_c$, there are multiple measures that do so.

Lemma 11.10. *The uniqueness threshold for the d-regular tree is $\lambda_c(d) = (d-2)^{-1}[(d-1)/(d-2)^{d-1}]$.*

When d is large, $\lambda_c(d) \approx e(d-2)^{-1}$. Compare that to the analysis of the running time for the bounding chain for the swap chain for the hard-core gas model in Lemma 4.2, which showed that it was possible to sample exactly from the model when $\lambda \leq 2(d-2)^{-1}$.

Recall that an (ε, δ)-randomized approximation scheme (Definition 2.13) returns a value with absolute relative error greater than ε with probability at most δ.

Definition 11.7. *A* fully polynomial-time randomized approximation scheme *(fpras) is an (ε, δ)-ras that runs in time polynomial in the input size of the problem, δ^{-1}, and ε^{-1}.*

Using TPA or PPE together with the bounding chain approach gives a fpras for the hard-core gas model when $\lambda \leq 2(d-2)^{-1}$. Sly and Sun [118] were able to show that this is only possible because λ falls below λ_c. To state their result properly, one more complexity class definition is needed.

Definition 11.8. *A decision problem is in the class* randomized polynomial time *(RP) if the algorithm runs in time polynomial in the input; if the correct answer is false, then the algorithm returns false; and if the correct answer is true then it returns true with probability at least* 1/2.

The question of whether NP = RP (like that of whether NP = P) is still open, but the wealth of NP complete problems indicates that it is unlikely that a randomized polynomial time algorithm to solve a problem that is NP complete will be found any time soon.

Theorem 11.4 (Theorem 1 of [118])**.** *For $d \geq 3$, unless NP = RP, there exists no fpras for the partition function of the hard-core model with parameter $\lambda > \lambda_c = (d-1)^{d-1}/(d-2)^d$ on d-regular graphs.*

Now, Weitz gave a fpras for the hard-core model partition function on graphs of maximum degree d whenever $\lambda < \lambda_c(d)$. Therefore, for $\lambda < 2(d-2)^{-1}$ it is known how to perfectly sample in polynomial time, for $\lambda < \lambda_c(d) \approx e(d-2)^{-1}$ it is known how to approximate the partition function with a fpras, and for $\lambda > \lambda_c(d)$ there exist graphs where no fpras can exist unless NP = RP.

For this reason, the value $2(d-2)^{-1}$ is known as an *artificial phase transition* of the perfect simulation algorithm. Currently it is unknown if it is possible to perfectly sample from all graphs with maximum degree d in polynomial time when $\lambda < \lambda_c(d)$, or if this artificial phase transition reflects something deeper about the problem.

In other models, the question of how good the perfect simulation can be is still unknown. For example, the ferromagnetic Ising model with unidirectional magnetic field is known to have a polynomial time algorithm for all values of the parameter [70]. However, despite the apparent success in practice for CFTP on this model, it is unknown if the method provably runs in polynomial time on these problems.

For other classes of problems the situation is different. For linear extensions (as has been seen), the perfect simulation algorithms are known to be the fastest possible for the problem at hand, beating even the approximate sampling algorithms. Another example is in sampling from weighted permutations for the permanent on dense matrices, where the perfect sampling algorithms have known bounds that beat the approximate sampling methods.

11.9 The future of perfect simulation

Perfect simulation has come far from its humble beginnings in the acceptance/rejection protocol. Coupling from the past opened up a multitude of problems for simulation, including the Ising model. While the Ising model does have a polynomial method for approximate simulation [70], this is far from a linear time method. For many decades it had resisted the creation of a near-linear time algorithm, to the point where many did not think that such an algorithm was even possible.

In the twenty years since the creation of CFTP, a wealth of new perfect simulation protocols have been invented. Many, such as partially recursive acceptance/rejection, and retrospective sampling, are variations on the original acceptance/rejection that have opened up broad new problem areas to perfect simulation. Others such as the randomness recycler are a hybrid of Markov chain and acceptance/rejection type methods.

Fill's method showed that looked at from a certain point of view, CFTP could be seen as a variant of AR run in the reverse direction in time. Are all perfect simulation algorithms in the end variants of AR? Is the artificial phase transition phenomenon intrinsic to certain problems, or merely an artifact of the algorithms in use? The set of problems addressable by perfect simulation protocols continues to grow, albeit more slowly than after the introduction of CFTP jumpstarted the field. Many of these algorithms run into problems that are NP or #P complete. The most interesting open question about perfect simulation is this: how far can these ideas be taken?

Bibliography

[1] D. Aldous. Some inequalities for reversible Markov chains. *J. London Math. Soc.*, 25(2):561–576, 1982.

[2] D. Aldous. A random walk construction of uniform spanning trees and uniform labelled trees. *SIAM J. Discrete Math.*, 3(4):450–465, 1990.

[3] D.J. Aldous and P. Diaconis. Strong uniform times and finite random walks. *Adv. in Appl. Math.*, 8:69–97, 1987.

[4] S. Asmussen, P. W. Glynn, and J. Pitman. Discretization error in simulation of one-dimensional reflecting Brownian motion. *Ann. Appl. Probab.*, 5(4):875–896, 1995.

[5] K. K. Berthelsen and J. Møller. Likelihood and non-parametric Bayesian MCMC inference for spatial point processes based on perfect simulation and path sampling. *Scand. J. Stat.*, 30:549–564, 2003.

[6] K. K. Berthelsen and J. Møller. Non-parametric Bayesian inference for inhomogeneous Markov point processes. *Aust. N. Z. J. Stat.*, 50:257–272, 2008.

[7] J. Bertoin and J. Pitman. Path transofrmations connecting Brownian bridge, excursion and meander. *Bull. Sci. Math.*, 118(2):147–166, 1994.

[8] J. Besag. Spatial interaction and the statistical analysis of lattice systems (with discussion). *J. R. Statist. Soc. Ser. B Stat. Methodol.*, 36:192–236, 1974.

[9] J. Besag. On the statistical analysis of dirty pictures. *J. R. Statist. Soc. Ser. B Stat. Methodol.*, 48:259–302, 1986.

[10] A. Beskos, O. Papaspiliopoulos, and G. O. Roberts. A factorisation of diffusion measure and finite sample path constructions. *Methodol. Comput. Appl. Probab.*, 10:85–104, 2008.

[11] A. Beskos, O. Papspiliopoulous, and G. O. Roberts. Retrospective exact simulation of diffusion sample paths with applications. *Bernoulli*, 12(6):1077–1098, 2006.

[12] I. Bezáková, D. Stefankovic, V. V. Vazirani, and E. Vigoda. Accelerating simulated annealing for the permanent and combinatorial counting problems. *SIAM J. Comput.*, 37(5):1429–1454, 2008.

[13] L. M. Bregman. Some properties of nonnegative matrices and their permanents. *Soviet. Math. Dokl.*, 14(4):945–949, 1973.

[14] L.A. Breyer and G. O. Roberts. Catalytic perfect simulation. *Methodology and Computing in Applied Probability*, 3(2):161–177, 2001.

[15] A. Brix and W.S. Kendall. Simulation of cluster point processes without edge effects. *Adv. in Appl. Probab.*, 34:267–280, 2002.

[16] A. Broder. Generating random spanning trees. In *Proc. 30th Sympos. on Foundations of Computer Science*, pages 442–447, 1989.

[17] R. Bubley and M. Dyer. Path coupling: a technique for proving rapid mixing in Markov chains. In *Proc. 38th Sympos. on Foundations of Computer Science*, pages 223–231, 1997.

[18] J. Bucklew. *Introduction to Rare Event Simulation.* Springer, New York, 2004.

[19] Z. A. Burq and O.D. Jones. Simulation of Brownian motion at first-passage times. *Math. Comput. Simulation*, 77:64–81, 2008.

[20] Sergio Caracciolo, Enrico Rinaldi, and Andrea Sportiello. Exact sampling of corrugated surfaces. *J. Stat. Mech. Theory Exp.*, Feb 2009.

[21] D.S. Carter and P.M. Prenter. Exponential spaces and counting processes. *Z. Wahrsheinlichkeitsth*, 21:1–19, 1972.

[22] N. Chen and Z. Huang. Localization and exact simulation of Brownian motion-driven stochastic differential equations. *Math. Oper. Res.*, 38(3):591–616, 2013.

[23] H. Chernoff. A measure of asymptotic efficiency for tests of a hypothesis based on the sum of observations. *Ann. of Math. Stat.*, 23:493–509, 1952.

[24] F. Chung. Spanning treees in subgraphs of lattices. *Contemp. Math.*, 245:201–219, 1999.

[25] H. Cohn, R. Pemantle, and J. Propp. Generating a random sink-free orientation in quadratic time. *Electron. J. Combin.*, 9(1), 2002.

[26] P. Dagum, R. Karp, M. Luby, and S. Ross. An optimal algorithm for Monte Carlo estimation. *Siam. J. Comput.*, 29(5):1484–1496, 2000.

[27] L. Devroye. Simulating perpetuities. *Methodol. Comput. Appl. Probab.*, 3(1):97–115, 2001.

[28] P. L. Dobruschin. The description of a random field by means of conditional probabilities and conditions of its regularity. *Theor. Prob. Appl.*, 13(2):197–224, 1968.

[29] W. Doeblin. Exposé de la théorie des chains simples constantes de Markov à un nombre fini d'états. *Rev. Math. de l'Union Interbalkanique*, 2:77–105, 1933.

[30] P. Donnelly and D. Welsh. The antivoter problem: random 2-colourings of graphs. *Graph Theory and Combinatorics*, pages 133–144, 1984.

[31] R. Durrett. *Probability: Theory and Examples, 4th edition.* Cambridge University Press, 2010.

[32] M. Dyer and C. Greenhill. On Markov chains for independent sets. *J. Algorithms*, 35(1):17–49, 2000.

[33] G. P. Egorychev. The solution of van der Waerden's problem for permanents.

Adv. in Math., 42:299–305, 1981.

[34] D. I. Falikman. Proof of the van der Waerden's conjecture on the permanent of a doubly stochastic matrix. *Mat. Zametki*, 29(6):931–938, 1981.

[35] J. A. Fill. An interruptible algorithm for perfect sampling via Markov chains. *Ann. Appl. Probab.*, 8:131–162, 1998.

[36] J. A. Fill and M. L. Huber. Perfect simulation of Vervaat perpetuities. *Electron. J. Probab.*, 15:96–109, 2010.

[37] J. A. Fill, M. Machida, D. J. Murdoch, and J. S. Rosenthal. Extension of Fill's perfect rejection sampling algorithm to general chains. *Random Structures Algorithms*, 17:290–316, 2000.

[38] G. S. Fishman. *Monte Carlo: concepts, algorithms, and applications.* Springer-Verlag, New York, 1996.

[39] A.E. Gelfand and A.F.M. Smith. Sampling based approaches to calculating marginal densities. *J. Amer. Statist. Assoc.*, 85:398–409, 1990.

[40] A. Gelman and X. Meng. Simulating normalizing constants: From importance sampling to bridge sampling to path sampling. *Stat. Sci.*, 13(2):163–185, 1998.

[41] A. Gibbs. Convergence in the Wasserstein metric for Markov chain Monte Carlo algorithms with applications to image restoration. *Stochastic Models*, 20:473–492, 2004.

[42] I.V. Girsanov. On transforming a certain class of stochastic processes by absolutely continuous substitution of measures. *Theory Probab. Appl.*, 5(3):285–301, 1960.

[43] S.W. Guo and E. A. Thompson. Performing the exact test of Hardy-Weinberg proportion for multiple alleles. *Biometrics*, 48:361–372, June 1992.

[44] O. Häggström and K. Nelander. On exact simulation from Markov random fields using coupling from the past. *Scand. J. Statist.*, 26(3):395–411, 1999.

[45] O. Häggström and J. E. Steif. Propp-Wilson algorithms and finitary codings for high noise Markov random fields. *Combin. Probab. Computing*, 9:425–439, 2000.

[46] O. Häggström, M.N.M. van Leishout, and J. Møller. Characterisation results and Markov chain Monte Carlo algorithms including exact simulation for some spatial point processes. *Bernoulli*, 5:641–658, 1999.

[47] W. K. Hastings. Monte Carlo sampling methods using Markov chains and their applications. *Biometrika*, 57:97–109, 1970.

[48] J. R. Hindley and J. P. Seldin. *Lambda-Calculus and Combinators: An Introduction (2nd edition).* Cambridge University Press, Cambridge, 2008.

[49] C.R. Hoare. Find (algorithm 65). *Comm. ACM*, 4:321–322, 1961.

[50] M. Huber. Nearly optimal Bernoulli factories for linear functions. *Combin. Probab. Comput.* arXiv:1308.1562. To appear.

[51] M. Huber. Perfect sampling using bounding chains. *Ann. Appl. Probab.*, 14(2):734–753, 2004.

[52] M. Huber. Exact sampling from perfect matchings of dense regular bipartite graphs. *Algorithmica*, 44:183–193, 2006.

[53] M. Huber. Fast perfect sampling from linear extensions. *Discrete Mathematics*, 306:420–428, 2006.

[54] M. Huber. Perfect simulation for image restoration. *Stochastic Models*, 23(3):475–487, 2007.

[55] M. Huber. Spatial birth-death swap chains. *Bernoulli*, 18(3):1031–1041, 2012.

[56] M. Huber. An unbiased estimate for the mean of a $\{0,1\}$ random variable with relative error distributions independent of the mean. 2013. arXiv:1309.5413. Submitted.

[57] M. Huber. Near-linear time simulation of linear extensions of a height-2 poset with bounded interaction. *Chic. J. Theoret. Comput. Sci.*, 2014, 2014.

[58] M. Huber. Approximation algorithms for the normalizing constant of Gibbs distributions. *Ann. Appl. Probab.*, 51:92–105, 2015. arXiv:1206.2689.

[59] M. Huber, Y. Chen, I. Dinwoodie, A. Dobra, and M. Nicholas. Monte Carlo algorithms for Hardy-Weinberg proportions. *Biometrics*, 62:49–53, Mar 2006.

[60] M. Huber and J. Law. Fast approximation of the permanent for very dense problems. In *Proc. of 19th ACM-SIAM Symp. on Discrete Alg.*, pages 681–689, 2008.

[61] M. Huber and G. Reinert. The stationary distribution in the Antivoter model: exact sampling and approximations. In *Stein's Method: Expository Lectures and Applications*, pages 79–94. IMS Lecture Notes 46, 2004.

[62] M. L. Huber. Exact sampling and approximate counting techniques. In *Proc. 30th Sympos. on the Theory of Computing*, pages 31–40, 1998.

[63] M. L. Huber. Exact sampling using Swendsen-Wang. In *Proc. 10th Sympos. on Discrete Algorithms*, pages 921–922, 1999.

[64] M. L. Huber. *Perfect Sampling with Bounding Chains.* PhD thesis, Cornell University, 1999.

[65] M. L. Huber. A faster method for sampling independent sets. In *Proc. 11th ACM-SIAM Sympos. on Discrete Algorithms*, pages 625–626, 2000.

[66] M. L. Huber. A bounding chain for Swendsen-Wang. *Random Structures Algorithms*, 22(1):43–59, 2003.

[67] M. L. Huber and R. L. Wolpert. Likelihood-based inference for Matérn type-III repulsive point processes. *Adv. Appl. Prob.*, 41(4):958–977, 2009.

[68] A. Iserles. *A first course in the numerical analysis of differential equations, Second Edition.* Cambridge University Press, Cambridge, 2008.

[69] K. Itō. Stochastic integral. *Proc. Imperial Acad. Tokyo*, 20:519–524, 1944.

[70] M. Jerrum and A. Sinclair. Polynomial-time approximation algorithms for the

Ising model. *SIAM J. Comput.*, 22:1087–1116, 1993.

[71] M. Jerrum, A. Sinclair, and E. Vigoda. A polynomial-time approximation algorithm for the permanent of a matrix with non-negative entries. In *Proc. 33rd ACM Sympos. on Theory of Computing*, pages 712–721, 2001.

[72] M. Jerrum, L. Valiant, and V. Vazirani. Random generation of combinatorial structures from a uniform distribution. *Theoret. Comput. Sci.*, 43:169–188, 1986.

[73] M.R. Jerrum, A. Sinclair, and E. Vigoda. A polynomial-time approximation algorithm for the permanent of a matrix with nonnegative entries. *J. of the ACM*, 51(4):671–697, 2004.

[74] V. E. Johnson. Studying convergence of Markov chain Monte Carlo algorithms using coupled sample paths. *J. Amer. Statist. Assoc.*, 91:154–166, 1996.

[75] B. Kalantari and L. Khachiyan. On the complexity of nonnegative-matrix scaling. *Linear Algebra Appl.*, 240:87–103, 1996.

[76] S. Karlin and H.M. Taylor. *A Second Course in Stochastic Processes*. Academic Press, New York, 1981.

[77] A. Karzanov and L. Khachiyan. On the conductance of order Markov chains. *Order*, 8(1):7–15, 1991.

[78] P.W. Kasteleyn. The statistics of dimers on a lattice, I., the number of dimer arrangements on a quadratic lattice. *Physica*, 27:1664–1672, 1961.

[79] P.W. Kasteleyn. Dimer statistics and phase transitions. *J. Math. Phys.*, 4:287, 1963.

[80] P.W. Kasteleyn. Graph theory and crystal Physics. In F. Harray, editor, *Graph Theory and Theoretical Physics*, pages 43–110. Academic Press, London, 1967.

[81] W. S. Kendall. Perfect simulation for the area-interaction point process. In *Probability Towards 2000*, volume 128 of *Lecture notes in Statistics*, pages 218–234. Springer-Verlag, 1998.

[82] W. S. Kendall and J. Møller. Perfect simulation using dominating processes on ordered spaces, with application to locally stable point processes. *Adv. Appl. Prob.*, 32:844–865, 2000.

[83] P. E. Kloeden and E. Platen. *Numerical Solution of Stochastic Differential Equations*. Springer-Verlag, Berlin Heidelberg New York, 1992.

[84] G. F. Lawler. *Introduction to stochastic processes*. Chapman & Hall/CRC, 1995.

[85] T. Lindvall. *Lectures on the Coupling Method*. Wiley, NY, 1992.

[86] N. Linial, A. Samorodnitsky, and A. Wigderson. A deterministic strongly polynomial algorithm for matrix scaling and approximate permanents. *Combinatorica*, 20(4):545–568, 2000.

[87] E. Lubetzky and A. Sly. Information percolation for the Ising model: cutoff in three dimensions up to criticality. Technical report, 2014. Preprint.

[88] B. Matérn. *Spatial Variation, Lecture Notes in Statistics, 2nd edition*, volume 36. Springer-Verlag, New York, 1986.

[89] M. Matsumoto and T. Nishimura. Mersenne twister: a 623-dimensionally equidistributed uniform psuedo-random number generator. *ACM Transactions on Modeling and Computer Simulation*, 8(1):3–30, 1998.

[90] D. Mattson. *On perfect simulation of Markovian Queuing Networks with Blocking*. PhD thesis, Göteborg University, 2002.

[91] N. Metropolis, A.W. Rosenbluth, M.N. Rosenbluth, A.H. Teller, and E. Teller. Equation of state calculation by fast computing machines. *J. Chem. Phys.*, 21:1087–1092, 1953.

[92] J. R. Michael, W.R. Schucany, and R.W. Haas. Generating random variates using transformations with multiple roots. *Amer. Statist.*, 30:88–90, 1976.

[93] H. Minc. Upper bounds for permanents of (0,1)-matrices. *Bull. Amer. Math. Soc.*, 69:789–791, 1963.

[94] A. Mira, J. Møller, and G.O. Roberts. Perfect slice samplers. *J. R. Statist. Soc. Ser. B Stat. Methodol.*, 63:593–606, 2001.

[95] A. Mira and L. Tierney. On the use of auxiliary variables in Markov chain Monte Carlo sampling. *Scand. J. Stat.*, 29(1):1–12, 2002.

[96] J. Møller. Perfect simulation of conditionally specified models. *J. R. Statist. Soc. Ser. B Stat. Methodol.*, 61:251–264, 1999.

[97] J. Møller and K.K. Berthelsen. Transforming spatial point processes into Poisson processes using random superposition. *Advances in Applied Probability*, 44:42–62, 2012.

[98] J. Møller, M. L. Huber, and R. L. Wolpert. The stationary Matérn hard core process of type III. *Stochastic Process. Appl.*, 120:2142–2158, 2010.

[99] J. Møller and K. Mengersen. Ergodic averages for monotone functions using upper and lower dominating processes. In J.M. Bernardo, M.J. Bayarri, J.O. Berger, A.P. Dawid, D. Heckerman, A.F.M. Smith, and M. West, editors, *Bayesian Statistics 8*, pages 643–648. Oxford University Press, Oxford, 2007.

[100] J. Møller and K. Mengersen. Ergodic averages via dominating processes. *Bayesian Analysis*, 2:761–782, 2007.

[101] J. Møller, A.N. Pettitt, R. Reeves, and K. K. Berthelsen. An efficient Markov chain Monte Carlo method for distributions with intractable normalising constants. *Biometrika*, 93(2):451–458, 2006.

[102] J. Møller and J. G. Rasmussen. Perfect simulation of Hawkes processes. *Adv. in Appl. Probab.*, 37:629–646, 2005.

[103] J. Møller and K. Schladitz. Extensions of Fill's algorithm for perfect simulation. *J. R. Stat. Soc. B. Stat. Methodol.*, 61:955–969, 1999.

[104] J. Møller and R. P. Schoenberg. Thinning spatial point processes into Poisson processes. *Advances in Applied Probability*, 42:347–358, 2010.

[105] J. Møller and R. P. Waagepetersen. *Statistical Inference and Simulation for Spatial Point Processes.* Chapman & Hall/CRC, 2004.

[106] J. Morton, L. Pachter, A. Shiu, B. Sturmfels, and O. Wienand. Convex rank tests and semigraphoids. *SIAM J. Discrete Math.*, 23(2):1117–1134, 2009.

[107] D. J. Murdoch and P. J. Green. Exact sampling from a continuous state space. *Scand. J. Statist.*, 25(3):483–502, 1998.

[108] I. Murray, Z. Ghahramani, and D. J. C. MacKay. MCMC for doubly-intractable distributions. In *Proc. of 22nd Conf. on Uncertainty in Artificial Intelligence (UAI)*, 2006.

[109] E. Nummelin. *General irreducible Markov chains and non-negative operators.* Cambridge University Press, Cambridge, 1984.

[110] B.K. Øksendal. *Stochastic Differential Equations: An Introduction with Applications.* Springer-Verlag, Berlin, 1998.

[111] C.J. Preston. Spatial birth-and-death processes. *Bull. Inst. Int. Stat.*, 46(2):371–391, 1977.

[112] J. G. Propp and D. B. Wilson. Exact sampling with coupled Markov chains and applications to statistical mechanics. *Random Structures Algorithms*, 9(1–2):223–252, 1996.

[113] A. Reutter and V. Johnson. General strategies for assessing convergence of mcmc algorithms using coupled sample paths. Technical Report 95–25, Duke University, 1995.

[114] R. Rivest. On self-organizing sequential search heuristics. *Comm. of the ACM*, 19:63–67, 1976.

[115] C. P. Robert and G. Casella. *Monte Carlo Statistical Methods (2nd ed.).* Springer, New York, 2004.

[116] R. Sinkhorn. A relationship between arbitrary positive matrices and double stochastic matrices. *A. Math. Statist.*, 35:876–879, 1964.

[117] John Skilling. Nested Sampling for general Bayesian computation. *Bayesian Anal.*, 1(4):833–860, 2006.

[118] A. Sly and N. Sun. The computational hardness of counting in two-spin models on d-regular graphs. In *Foundations of Computer Science (FOCS)*, pages 361–369, 2012.

[119] D. J. Strauss. A model for clustering. *Biometrika*, 63:467–475, 1975.

[120] Roy L. Streit. *Poisson Point Processes: Imaging, Tracking, and Sensing.* Springer, New York, 2010.

[121] R. Swendsen and J-S. Wang. Replica Monte Carlo simulation of spin glasses. *Phys. Rev. Let.*, 57:2607–2609, 1986.

[122] Michael E. Taveirne, Casey M. Theriot, Jonathan Livny, and Victor J. DiRita. The complete campylobacter jejuni transcriptome during colonization of a natural host determined by RNAseq. *PLoS ONE*, 8:e73586, 2013.

[123] A. M. Turing. Computability and λ-definability. *J. Symbolic Logic*, 2(4):153–163, 1937.

[124] L. G. Valiant. The complexity of computing the permanent. *Theoret. Comput. Sci.*, 8:189–201, 1979.

[125] J. P. Valleau and D. N. Card. Monte Carlo estimation of the free energy by multistage sampling. *J. Chem. Phys.*, 57(12):5457–5462, 1972.

[126] J. van den Berg and C. Maes. Disagreement percolation in the study of Markov fields. *Ann. Probab.*, 22:749–763, 1994.

[127] J. von Neumann. Various techniques used in connection with random digits. In *Monte Carlo Method*, National Bureau of Standards Applied Mathematics Series 12, pages 36–38, Washington, D.C., 1951. U.S. Government Printing Office.

[128] D. B. Wilson. Generating random spanning trees more quickly than the cover time. In *Proc. 28th ACM Sympos. on the Theory of Computing*, pages 296–303, 1996.

[129] D. B. Wilson. How to couple from the past using a read-once source of randomness. *Random Structures Algorithms*, 16(1):85–113, 2000.

[130] D. B. Wilson. Layered multishift coupling for use in perfect sampling algorithms (with a primer on cftp). *Fields Institute Communications*, 26:141–176, 2000.

Index